SOUND REPORTING

Sound Reporting

THE NPR GUIDE TO AUDIO JOURNALISM AND PRODUCTION

JONATHAN KERN

THE UNIVERSITY OF CHICAGO PRESS | CHICAGO AND LONDON

The University of Chicago Press, Chicago 60637
The University of Chicago Press, Ltd., London

Printed in the United States of America

26 25 24 23 22 21 20 19 16 17 18

ISBN-13: 978-0-226-43177-2 (cloth)
ISBN-13: 978-0-226-43178-9 (paper)
ISBN-10: 0-226-43177-0 (cloth)
ISBN-10: 0-226-43178-9 (paper)
ISBN-13: 978-0-226-11175-9 (e-book)

Library of Congress Cataloging-in-Publication Data

Kern, Jonathan, 1953–
Sound reporting : the NPR guide to audio journalism and production / Jonathan Kern.
p. cm.
Includes index.
ISBN-13: 978-0-226-43177-2 (cloth : alk. paper)
ISBN-13: 978-0-226-43178-9 (pbk. : alk. paper)
ISBN-10: 0-226-43177-0 (cloth : alk. paper)
ISBN-10: 0-226-43178-9 (pbk. : alk. paper) 1. Radio journalism.
2. Radio—Production and direction. I. Title.
PN4784.R2K47 2008
070.4'3—dc22

2008003994

♾ This paper meets the requirements of ANSI Z39.48-1992 (Permanence of Paper).

CONTENTS

FOREWORD

Somewhere, in one of the dimly lit corners of my imagination, I can hear my mind's announcer, rich of voice and expression, use the intonation you hear on a late-night infomercial. He says, "Now, whenever you want—and in the privacy of your own home, office, basement, or garage—you can create reports, interviews, or programs that sound just like stuff you hear on public radio stations or Web sites across the country. Just read this book and call the number flashing on your occipital cortex!"

If it were only that easy. To report the "NPR way" requires a lot of curiosity; the urge to uncover the truth; a deep dedication to balance, fairness, and public service; and mastery of the craft. Oh, and the will to do it.

So I also imagine there may be any number of reasons you're reading this book:

- You're a journalism or communications student: The good news is that you're reading the most useful and interesting text of its kind. The author and the staff at NPR have been extremely generous, and the book is packed with best practices, professional secrets, and great examples.
- You woke up one morning and thought: "That's it! I want to work in public radio! I wonder how they do what they do?" If this is you, please read this book and then contact your local public radio station. We need the next generation of curious and passionate reporters, editors, and producers to carry on this remarkable enterprise.
- You already work at NPR, at a member station, for a commercial radio or TV station, or as an independent producer: And you want to brush up on a technique or approach to help you in your work. Welcome to Radio Journalism 401, or maybe it's 601.

- You are a true fan of NPR: Your car, home, and office radios are pre-set to your local public stations; you are a member of those stations and generously support them; you visit NPR.org and station Web sites and explore them faithfully; you download NPR and other public radio podcasts; you've been to the NPR.org shop and you already own mugs, CDs, and logo T-shirts. You're thinking about buying an HD radio to receive even more local and national programs. So now you want the ultimate behind-the-scenes visit.
- You want to make a lot of money: Put this book down. The one you want is titled something like *Sound Investing*.

By the way, there are many good reasons Jonathan Kern's name is on this book; first of all, because he had the will and tenacity to research and write it. As Executive Producer for Training for NPR News, he leads a unit charged with developing the curriculum—and conducting classes—in everything from story development, to the use of sound, to ethics, to newsroom diversity.

Before earning his current title, Jon was teaching all of us from his position as Senior Editor of *All Things Considered*, where, for more than five years, he came up with ideas for stories, developed program lineups, wrote and edited scripts, and advised producers about editing recorded interviews. He was later the show's executive producer—a job he took over on September 11, 2001.

For those of you who are new to NPR, I want to take a moment to put what's in this book in a larger context.

- Member stations: At this writing, there are 276 NPR member stations, which operate more than 735 stations or signals. For example, WNYC in New York broadcasts an AM and an FM signal—so that's one station, two signals. Many public stations also podcast, multicast on HD radio, and produce local Web sites.
- NPR and public broadcasting: NPR is a nonprofit membership organization and controls no airtime on the member stations; there is no "must carry" program schedule. Individual stations produce their own shows and buy the programs that meet local and regional needs—from NPR, from other producers and distributors, such as American Public Media, and from independent producers. There are many different kinds of public radio stations—news, talk, classical, jazz, contemporary music—and there are hundreds of different programs carried by those

stations. The reason shows like *Morning Edition*, *Talk of the Nation*, *All Things Considered*, and NPR newscasts are everywhere is because hundreds of stations have purchased them, and air them at approximately the same time in each time zone. NPR produces programs and also distributes shows produced at its member stations. *Fresh Air* with Terry Gross comes from WHYY in Philadelphia, *The Diane Rehm Show* is from WAMU in Washington, D.C., and *On Point* with Tom Ashbrook is produced at WBUR in Boston. But *Marketplace* and *A Prairie Home Companion*? They're great shows, but not from NPR.

- Some numbers: NPR programs attract nearly twenty-two million listeners every week, and more than thirty million people listen to the stations that carry NPR programming. Our Web site, NPR.org, attracts more than three million unique visitors a month, and every month visitors stream more than fourteen million on-demand audio files. Since the fall of 2005, NPR and member stations have also been providing some of the most popular podcasts available, and for free. (As of 2007, NPR-produced podcasts were being downloaded more than nine million times each month.)
- Who's out there: That audience of millions is made up of an equal number of women and men who have a love of lifelong learning and participate more than many Americans do in civic activities and their children's education. Our listeners are of all ages; the median age is in the early fifties, although that age is younger for online and podcast users (forty-seven and thirty-two, respectively). In independent surveys, the audience self-identifies as 32 percent liberal, 23 percent moderate, and 29 percent conservative. Some 16 percent purchased mystery novels in the past year; 34 percent watch the Weather Channel; 29 percent go to museums. And 28 percent of NPR news listeners agree with this statement: Just as the Bible says, the world literally was created in six days.
- There's no business like the journalism business: CBS newsman Don Hewitt, the creator of TV's most successful newsmagazine, *60 Minutes*, was fond of saying that he had the greatest job in the world because he could be part Ed Murrow and part Ed Sullivan. The same is true at NPR: as good as you are at digging for facts or making sense of world events, you still need to be able to write well and speak well—to tell and sell the story, and when appropriate, to astonish or entertain.
- The driveway moment: When a report or interview really works, you can tell and we can tell. We can, because the story hits the top of the

> most emailed list at NPR.org. You can tell, because the story keeps you pinned in your car, in a parking lot, in your driveway, or at the side of the road—as you wait to hear how the story will end. In letters and emails, listeners named these occurrences "driveway moments," and say they look forward to them, even when it means being late for work or dinner. So that's your goal: make some driveway moments.

Driveway moments, the NPR shows you love, and ones you may never have heard can be found at NPR.org. You can also read the judgments and musings of NPR's Ombudsman, who serves as an independent advocate for the audience and for better journalism.

Here's one more reason to visit NPR.org: What Jon has distilled in this book is the result of NPR's more than three decades of experimentation, of trying to do it differently, and trying to get it right. The techniques and rules have changed over time and will continue to do so, and that's where you come in: when you love something and it works, let us know. When you disagree, ditto.

One thing that hasn't changed is being able to recognize how sound can tell a story.

I started at a public radio station when I was sixteen years old, but I'm sure my interest in sound began in third grade, when our teacher asked us to stop what we were doing, take out a pencil and a piece of lined paper, and get very quiet. Just listen, she said, and write down everything you hear. But we don't hear anything, Patti said from the front row. Just listen, the teacher said.

I heard the classroom clock clicking as slowly as it ever did, the scuffling of Buster Brown shoes under desks, the piano from a classroom down the hall ("Goodness how delicious, eating goober peas..."), one of the custodians raking leaves in the schoolyard, and then, the noon train coming through town. We could hear the low whistle and the rumbling from the tracks. The sound was many blocks away, but we had all heard it close up—when we used the train to flatten pennies—so the sound was magnified in our minds. We didn't know where the train was coming from or where it was heading, but we had all visited the antique station in the middle of town and knew it no longer carried passengers. It was all freight cars. Much of the town emptied out during the day as our parents went to work in the city, but they drove or took the bus. And then we heard other kids spilling from classrooms—a low crush of voices, then laughter, then shouting—and it was time to walk home for lunch.

It wasn't that difficult to list all of the sounds that occurred when we thought nothing was happening.

The assignment was just one of many gifts from a teacher, and if I remember correctly, the same teacher who asked us another time to sit on our hands and then try to describe a spiral staircase.

So, for your first lesson in radio journalism: just listen.

Jay Kernis
Senior Vice President for Programming
NPR

PREFACE

This handbook has drawn on many different sources. At the top of the list: more than one hundred current or former NPR journalists who agreed to talk about their work candidly and at length. (Some of these people may now have different jobs from the ones they held when they were interviewed; in any case, their positions or job titles are much less important than the knowledge they have to share.) I based the chapters on newscasting, story editing, fairness, and production ethics in part on workshops I conducted as Executive Producer for Training at NPR; they in turn draw extensively on comments and insights offered by seminar participants. The chapter on writing for broadcast started as a handout for new reporters but also cribs from a talk Robert Siegel gave at an *All Things Considered* retreat a few years ago. I've excerpted several memos from Senior Vice President for Programming Jay Kernis. Former NPR Vice President for News Bruce Drake wrote the ethics code; he also assigned this project and edited an early draft, as did Jay Kernis. Other sources are cited within the relevant chapters.

Although nearly all the journalists interviewed for this book work, or have worked, at NPR, I have tried to emphasize principles and practices that apply to all radio news broadcasters—and, in many cases, to TV reporters and anchors as well. The book starts with a couple of the basics—journalistic fairness and writing for the ear—and moves progressively through reporting, producing, and editing stories; anchoring newscasts and news magazines; producing and editing news programs; and applying radio skills to the Web. If you run across terms that are unfamiliar to you, there's a glossary at the end. The different sections are designed to make sense even if they are not read in sequence; for that reason there will be places where two or more chapters touch on the same subject. Issues related to live coverage, for instance, surface in the chapters on booking, program producing, hosting, and directing. Some duplication also reflects how any large news organization operates. Producers and reporters book guests; hosts, reporters, and producers conduct interviews; producers, reporters, and hosts all write for broadcast; studio directors produce music pieces; and so on. Our work lives overlap—every radio program is the product of many different collaborations—and to some extent that is reflected in this book.

Throughout this book, I use examples taken from actual NPR stories. In general, when I am drawing attention to something that is praiseworthy, I include the names of the reporters or hosts; when illustrating something that I think could have been done better, I have left the names out. I do this not only to spare people embarrassment, but as an acknowledgment that we all work under relentless deadline pressure, and mistakes by reporters, editors, and producers are inevitable.

It certainly isn't possible to read your way to a career in broadcast journalism; experience will always be the best teacher. But I hope this guide will give you an idea of what you might learn if you could bring together a group of NPR hosts, or studio directors, or reporters, or editors, and pick their brains about what they do each day, and how they strive to meet the exacting standards they set for themselves.

Jonathan Kern

CHAPTER 1

Sound and Stories

Radio has proven to be quite a survivor as a news medium. After all, radio listening was a big fad of the 1920s (as was the Charleston), and historians of broadcasting will tell you that radio's "Golden Age" ended more than fifty years ago. Television could have put radio out of business in the 1950s and '60s, but it didn't, and the proliferation of cable news channels in the 1980s and '90s could have made radio news irrelevant—but that didn't happen either. In the last decade or so, the Internet has emerged as a popular source of news, especially for younger people, accelerating the decline in newspaper subscriptions. But even as newspapers lost readers to the Internet, public radio's audience actually grew—from 14.6 million weekly listeners in 2000 to 23 million in 2006. These days, "radio" has less to do with a specific kind of receiver or a means of sending signals from a transmitter than with a way of communicating news and information through words and sound. A "radio show" may be broadcast, or streamed on a Web site, or downloaded in a podcast; soon it could be delivered to a mobile phone, or to another sort of handheld device that gets its data from a nearby wireless access point. But even as the technology is changing, the process of reporting and producing audio news and information today is much the same as it was when NPR began in the early 1970s; and "radio" continues to be a convenient way to describe all forms of mass communication relying primarily on the spoken word. So it's worth considering what it is about this aging medium that continues to be so attractive to people, especially when there are so many alternative ways to find out what is happening.

RADIO IS PORTABLE. People have been listening to the radio in their cars since the 1930s, and pocket-sized transistor radios have been around for half a century. Today you can buy headphones with built-in radios and MP3 players that also contain FM tuners. Water-resistant sets work in the shower, and satellite receivers make it possible to hear the same strong radio signal as you travel from one state to another. People can and do listen to the radio while they jog, cook, drive, work, or bathe—something that can't be claimed by either print or TV.

RADIO IS INTIMATE. No matter how big the audience, a good radio host thinks of himself as talking to a single person—the one who's tuning in—rather than to listeners as a group. (For that reason, if you're on the air and asking people, say, to call in to your program with their recollections of Martin Luther King, it's always better to ask, "What do *you* remember about Martin Luther King?"—as opposed to "We'd like to invite listeners to tell us what they remember. . .") Program directors and other executives sometimes underestimate how tight the bond is between the person who talks on the radio and the ones who are listening. The departure of a longtime host of a news magazine can prompt thousands of angry letters, phone calls, and even petitions. People feel that they've lost a friend.

RADIO IS NIMBLE. Most of the time, a radio reporter can carry all of his equipment—a recorder, microphone, and a computer—in one bag. You don't need a camera crew or a satellite truck, as TV reporters do; and it certainly doesn't matter what you're wearing or whether you've had time to comb your hair. As a radio reporter, if you can get to the scene of a news event, you can report on it, even if your gear consists of little more than a cell phone. (In fact, on many breaking stories, TV *becomes* radio—networks just display a still photo of their correspondent or a map of the area where the event is taking place, and have their reporter phone in the story.)

FEW THINGS AFFECT US MORE THAN THE HUMAN VOICE. Certainly there are photographs that touch us, and TV often can tell a story with vividness and immediacy, and newspaper stories often have great quotes. But people convey what they feel both through their words and through the sound of their voices. During a radio interview, we often can hear for ourselves that a politician is dismissive, or that a protester is angry, or that a Nobel Prize winner is thrilled and exhausted; we don't need a reporter to characterize them for us. And public radio especially allows people to speak at some length; an interview in a news magazine might run as long as eight minutes. We don't force ourselves to reduce a person's insights and emotions to a single ten-second sound bite. Even in transcription, this exchange exposes the tremendous sadness and loss of a farmer in

Wales as she describes how the Ministry of Agriculture shot all of her 228 dairy cows after some of them contracted foot-and-mouth disease:

HOST: Did you watch?
JONES: Oh, my God, no! Oh, no! I heard it. That was enough. I heard it.

Watched? No, no. I said goodbye to them all. But they just shot them where they stood. Oh, no.

Watched? No way. I watched them burn afterwards. Of course, I needed to be there for them. I had to watch that, and now I'm living with the horror of it all. I think it's the most harrowing experience I could ever, ever, ever imagine going through.

I say, my ten-year-old daughter knew every one of those cows by name. She didn't have to look at their numbers. She knew who they were by their faces. I could have gone in blindfolded and touched everybody's udder and I could have told you exactly which cow it was.

SOUND TELLS A STORY. The art of public radio journalism entails most of the skills practiced by television or newspaper reporters—finding sources, conducting interviews, digging through documents, getting to the scene of the action, observing carefully—plus one that is unique to our medium: *listening*, or "reporting with your ears." The right sound—the whine of an air raid siren in wartime, the echoes in a building abandoned because of a chemical spill, the roar of a trading pit in Chicago—can substitute for dozens or hundreds of words, and can be as descriptive and evocative as a photograph.

Today, NPR distributes news reports in many different ways—through its member stations, via satellite, over the Internet, in podcasts, even to cell phones—and it often provides written versions of them on the Web. But radio's greatest strengths remain the power of sound to tell a story, the expressiveness of the human voice, and the intimacy of the medium.

There are also some big challenges to reporting news on the radio.

Just as newspapers and Web sites are laid out graphically—in space—radio programs are laid out in time; radio producers argue over *when* a story will be heard, not which page it will be on, and we measure story lengths in minutes and seconds, not in column inches or words. On the radio, you need to find ways to communicate information that a newspaper can easily convey with a headline, or a photo, or a graph. Think about how much you can learn just by scanning the front page of a daily paper. A banner headline—especially in a paper that rarely runs them—tells you there's momentous news. If there are two or three items related to the lead story on A1, you know that the big story of the day has eclipsed most other events—at least in the minds of the newspaper's editors. On the other hand, a more diverse selection of front page stories suggests an average news load. And a big picture of a pumpkin patch, or of children keeping cool in the water from a fire hydrant, or of couples lounging in the park on a spring day, sends the message there hasn't been much news at all. There may also be a front page index to tell you about developments in business, sports, and entertainment—and where to find more details about them in the paper.

In addition, a newspaper's space is flexible; the length of a radio program isn't. Although papers do budget the amount of space they devote for news, they can add pages—or even whole sections—when events demand it. When the news is thin, they can also fill up the paper with photographs or with stories from wire services, or just run fewer pages. With some exceptions for the biggest national and international events, radio programs are the same length on busy news days and on dull ones. *Morning Edition* is two hours long, *Day to Day* an hour, *Talk of the Nation* two hours, and so on, both when there is a lot of news to cover and when there isn't.

As a radio journalist, in other words, you are working both in sound—and in time. You have listeners, not readers. So here are a few things to keep in mind.

THERE ARE NO HEADLINES. That means that we don't have a way to catch a potential listener's ear the way a big headline at a newsstand catches the eye; to get *our* news, people have to make the effort to turn on the radio and tune to a specific station. At NPR, we do write "billboards" or "opens" to tell people what's coming up each hour. But each billboard is always fifty-eight seconds long—whether the hour it previews is loaded with hard news or mostly softer features. As a rule, we can't stretch a billboard when we have more we want to say or shorten it when we have less.

IMPORTANT STORIES COME FIRST AND GET MORE AIRTIME. A billboard may list four or five stories on a typical day. But when there's big news—after a presidential election, a devastating storm, or some other important event—we may devote most or all of the billboard to a single story. Similarly, when there's a major story of the day, a twelve-minute segment that usually comprises three or four pieces or interviews may focus instead on several different aspects of the same story.

THERE IS NO "FRONT PAGE"; THE BEGINNING OF ANY PROGRAM IS THE MOMENT SOMEONE TUNES IN. While we generally will put the most important stories of the day at the top of an hour, we know that people listen to the radio when it's convenient for them. So even though we mention at 8:10 a.m. that there's been a plane crash in Kentucky, we may give an update on that story ten minutes later—and again twenty-five minutes after that—for people who have joined the program in progress.

A RADIO NEWS MAGAZINE MAY NOT HAVE READILY IDENTIFIABLE "SECTIONS." Some programs try to offer listeners certain types of news at predictable days or times—sports on Fridays, or business at the end of a particular hour. But the latest developments in sports or business or any other subject can show up almost anywhere in a show. As a result, it is often harder to know "where you are" when you listen to a radio program than when you read a newspaper. If you want to read commentaries in the newspaper, you turn to the op ed page. A commentary in a radio news magazine may come up in any program segment—which makes it especially important to identify it clearly as a commentary, so listeners don't mistake it for a news report.

RADIO LISTENERS, UNLIKE READERS, CAN'T SKIP A STORY OR SEGMENT OF A PROGRAM. If you're not interested in sports or business, you probably don't read those sections of the newspaper; and if you listen to NPR or other radio news media on the Web or through podcasts, you can pick and choose the items you want. But when you get your news from the radio, you have to listen to—or at least sit through—arts or economics or foreign stories to get to the subjects you might care about more. If listeners get bored with an item, they'll mentally tune it out, or select a different station, or turn off the radio altogether. For that reason, we try to write and produce our stories to keep the attention of people who are not *already* interested in the topics we're reporting on.

IN RADIO, EDITORIAL DECISIONS ARE OFTEN INTERTWINED WITH PRODUCTION DECISIONS. A correspondent may be given only four minutes for her report, even if it is the top story of the day—so if she wants to add more detail or include another voice, she will have to cut something else from her story. And if she is somehow able to wangle an extra thirty seconds, some other piece or interview will need to be trimmed by the same amount. Producing a radio program is a zero-sum game.

PEOPLE CAN'T RELISTEN TO A STORY, THE WAY THEY CAN REREAD A NEWSPAPER ARTICLE. When you're reading a newspaper, you may be interrupted, or let your mind wander, or just get confused. You may need to reread a couple of paragraphs just to grasp the crux of the story, or to make sure you understand the latest development. You don't get that opportunity on the radio. *Time marches on*—and with it, any opportunity a listener has to understand why a story is important or new, or to identify speakers or places or sounds. On the radio you get one chance to tell the listeners your story, and then there's no going back. (This is less true when people's "radio" is actually a computer, and they are listening via the Internet.)

LISTENERS CAN'T "SEE" (OR HEAR) WHAT'S AHEAD. When you read a story in a newspaper, your peripheral vision gives you an idea of the stories that surround it. You may be halfway through an account of a train crash, but you know there's a story about the discovery of a new dinosaur elsewhere on the same page. On the radio, someone needs to tell you explicitly what's coming up.

A "HARD" DEADLINE ON THE RADIO IS VERY HARD. If you're ten minutes late filing a story at a newspaper, no one is likely to notice. But if your story is slated to be on the air at six minutes past noon and you don't get it done until six and a half minutes past, you've failed abominably. *Talk of the Nation* always starts at 2 o'clock Eastern Time—not fifteen seconds earlier or later. Two seconds can be a long time on the radio.

To be sure, there are occasions when we can pry open the time window, even on the radio. The September 11, 2001, terrorist attacks, the crash of the space shuttle in 2003, the outbreak of war with Iraq that same year, and other big and continuing stories have all justified NPR's doing away with the usual broadcast clocks, at least for a few days. On those occasions, the highly produced billboards were scrapped or made longer or shorter than fifty-eight seconds—which is possible only when the shows don't incorporate the hourly newscasts—and the programs observed few of the usual breaks between segments. For a while, the shows even lost their individual identities and blended into one another, as NPR provided special round-the-clock programming to its stations.

But these occasions are indeed rare in the radio news business. On most days, whether we are reporters, editors, producers, directors, or hosts, our working lives are ruled by the clock.

CHAPTER 2

Fairness

For most journalists, no charge stings worse than an allegation of bias. Yet NPR and most other news organizations are the subjects of such accusations daily (and, in this age of email and the Internet, almost hourly).

If you could look through the email sent to any NPR news program on a given day, you'd find listeners writing angrily that the network is showing its bias for or against a host of individuals, groups, or issues. They accuse public radio of being a spokesman for the administration (regardless of the party in power), a tool of the Pentagon, a proxy for the Democratic party, an arm of the Republican party, soft on the pharmaceutical industry, out to get the oil companies—the list goes on and on, no matter how the news of the day varies. Whether the issue is abortion, the death penalty, the Middle East, tax cuts, or politics, listeners are sure to cite what they see as clear evidence that reporters and hosts are trying to stack the deck for one side or the other.

It's easy—and often wrong—to brush off the charges. News organizations know that people at the far ends of the political spectrum will be most inclined to write, and that the letter (or email) writers don't represent the audience as a whole. Bias, they say, is often in the eye of the beholder. Also, editors and managers console themselves with the fact that people on *both* the right and the left are complaining, often about the exact same stories; if both sides don't like the way we did our jobs, they reason, we probably did okay. But that's not a principle to live by. The only measure of a story's worth is whether you got it right—not how many people were for or against it.

Broadcast news is driven by deadlines, and under time pressure, you are sure to make mistakes—about names, affiliations, places, and so on. These errors are regrettable, and you should always correct them on the air when they occur. But they are not nearly as serious as failing to be fair and unbiased. That may not only discourage people from listening; it can undermine your station or network's reputation—one of its greatest assets. Even occasional lapses can have serious consequences. The price of good journalism is eternal vigilance.

Fair, Accurate, Complete, and Honest

The "NPR Journalist's Code of Ethics and Practices"[1] states that the network's coverage "must be fair, unbiased, accurate, complete and honest." And it goes on to define the essential attributes of the network's reporting:

- "Fair" means that we present all important views on a subject. This range of views may be encompassed in a single story on a controversial topic, or it may play out over a body of coverage or series of commentaries. But at all times the commitment to presenting all important views must be conscious and affirmative, and it must be timely if it is being accomplished over the course of more than one story. We also assure that every possible effort is made to reach an individual (or a spokesperson for an entity) that is the subject of criticism, unfavorable allegations or other negative assertions in a story in order to allow them to respond to those assertions.
- "Unbiased" means that we separate our personal opinions—such as an individual's religious beliefs or political ideology—from the subjects we are covering. We do not approach any coverage with overt or hidden agendas.
- "Accurate" means that each day we make rigorous efforts at all levels of the newsgathering and programming process to ensure our facts are not only accurate but also presented in the correct context. We make every possible effort to ensure assertions of fact in commentaries, including facts implied as the basis for an opinion, are correct. We attempt to verify what our sources and the officials we interview tell us when the material involved is argumentative or open to different interpretations. We are skeptical of all facts gathered and report them only when we are reasonably satisfied of their accuracy. We guard against errors of omission that cause a story to misinform our listeners

1. For the complete text, go to npr.org/about/ethics.

> by failing to be complete. We make sure that our language accurately describes the facts and does not imply a fact we have not confirmed, and that quotations are accurate and placed properly in context.

It is not obvious that journalists everywhere would accept these guidelines. Many self-styled news sites on the Web, for instance, are unabashedly partisan and activist, as are many old-fashioned paper-and-ink newspapers and magazines. People are attracted to the news business for lots of reasons—one of which may be their dedication to a specific cause or ideology. Just calling yourself a journalist doesn't mean you're interested in getting both sides of the story.

So it's worthwhile to ask *why* public radio values fairness, balance, and accuracy so highly—*why* they are the bedrock of all the work we do.

First, there are what might be described as *philosophical* justifications for fairness. Simply put, your goal should be to find the truth—and being fair and unbiased is the way to get there. Reporting means finding out all you can about a given topic, and delivering that information in a way that helps listeners make up their own minds about the issue at hand. In addition, most journalists have a visceral sense that being evenhanded is the right thing to do—that it's part of their "mission," or their public service.

There are also some *practical*, even self-interested reasons for going out of your way to get both sides of the story. Your reporting is more credible if it presents a spectrum of viewpoints. Since listeners have a wide variety of opinions, they are more likely to trust you if they hear their own attitudes and experiences reflected by the people you interview. In addition to believing more of what they hear, they may be inclined to listen more often or for longer periods of time. In fact, being fair is a good business practice for any news organization whose success depends on the goodwill—and in the case of public radio, on the contributions—of its audience.

The Echo Chamber

Ask radio journalists to identify the reasons why any particular story may not be entirely balanced, and they say that they often simply run out of time. A story may be assigned at 10 a.m. to be on the air six or seven hours later. So, in a rush to get their pieces written and produced, they frequently resort to saying that a source "did not return repeated phone calls"—an explanation that rarely comforts listeners who feel their point of view has been ignored. It might also lead people who disagree with the absent speaker to assume that he or she had something to hide, or that their position was so weak it wouldn't stand up to a reporter's scrutiny. Sometimes the reporters are able to contact people on all sides of an issue, but can only persuade the principals on one side to let their interviews be recorded. That, too, makes it sound like the other side is getting short shrift. There are some potential interviewees who simply won't speak to NPR, or to public radio more generally, and for whom there are no appropriate stand-ins. In the run-up to the 2004 presidential election, for example, *All Things Considered* and *Morning Edition* both interviewed Democratic candidate John Kerry, but the White House did not make President Bush available to either show. Occasionally reporters complain that some points of view had to be deleted, or condensed, because they weren't budgeted enough airtime; it's hard to be thorough in three minutes, they say.

These are not adequate reasons for a story flunking the fairness test, but at least they are easy to identify.

A subtler problem may arise from the newsroom "echo chamber," in which reporters and editors fail to represent some viewpoints mainly because they all see events from the same perspective. In a February 2004 interview on the NPR program *On the Media*, Brooke Gladstone interviewed author and academic Cass Sunstein about the "echo chamber" effect:

SUNSTEIN: Sometimes what's meant by echo chamber isn't something completely new. It's like a rumor mill, and the idea is that if you have an idea or a claim about a fact that's repeated over and over again, then it will start to assume reality. Sometimes—and it may be a little more interesting—the echo chamber is narrower, to some part of the citizenry. Say, liberals will all hear one thing echoing, and conservatives will hear another thing echoing, and they'll be very different things, so people will in a way live in different social realities, just because they hear such different claims about the facts.

GLADSTONE: And there are some media watchers who call this manifestation of the echo chamber "incestuous amplification."

SUNSTEIN: Yeah. The term "incestuous amplification" is often a military term, and it's used there as a warning against military leaders who keep talking to each other as if they're in a family, and then they don't

> watch out for things that can go wrong. And so incestuous amplification sets a recipe for disaster. That military idea has a lot of parallels for completely nonmilitary groups, like companies like Enron where there wasn't much dissent—people were just echoing one another's optimism in a way that led to disaster, and in terms of sometimes government behavior when there's incestuous amplification for a domestic policy that goes sour and might not have if people who lived outside the echo chamber had had a chance to speak to the people who live in it.

Sunstein was talking about echo chambers as they appear on the Internet, but the phenomenon can afflict newsrooms as well, especially if they are not as politically, culturally, and racially heterogeneous as the rest of the country. After all, Americans have sharply different attitudes toward abortion, the death penalty, nuclear power, politics, and the military—to name just a few issues—in part because they have led very different lives. You would not expect to get similar points of view on these subjects from, say, a coal miner, a stockbroker, an autoworker, a physician, a post office worker, and an unemployed teenager. If newsrooms were as diverse as the rest of America, there would be lively debates whenever a story idea was suggested. But broadcast journalists are more like one another than they are like the population as a whole, and news reporting can suffer because of it.

How do we know? In seminars at NPR, at workshops for member station reporters and editors, and at public radio conventions, we have taken informal and admittedly unscientific polls to assess participants' interests, behaviors, and opinions. The idea is to get a sense of how much diversity of experience there is among the people in the group (and by crude extrapolation, in the newsrooms they represent). The questions include:

Are you a military veteran?

Do you own a gun?

Do you have an undergraduate college degree?

Did you read the *New York Times* this morning?

In the abortion debate, would you describe yourself as "pro-choice"?

Do you support legal agreements for gays that provide many of the same benefits of marriage?

Did you attend church or another religious service in the past week?

There is a remarkable consistency in the results. Most of the time, the surveyed group includes few or no veterans or gun owners, and an overwhelming majority of college graduates. The proportion of *New York Times* readers generally varies from a low of about 50 percent to a high of nearly 100 percent, depending on where the poll is taken. The participants are consistently pro-choice and favor gay civil unions—often unanimously—and between 20 and 40 percent say they attended a religious service.

Compare this with the way Americans more generally answer similar questions. About 13 percent of American adults are veterans, according to the 2000 census; a 2001 Harris poll found nearly 40 percent living in gun-owning households; and only about one in four has a college degree, according to a 2003 Census Bureau report. The total circulation of the *New York Times* on weekdays is about 1.1 million, and even assuming that each copy is read by two or three people, *Times* readers make up a much smaller percentage of the adult population in the U.S. than of public radio newsrooms (or, no doubt, of newsrooms across the country). In various surveys, about half of all Americans say they are pro-choice, and about half answer yes to the question about legal rights for gay couples. Finally, around 40 percent of the general adult population reports attending a religious service in the past week.

In other words, journalists may be surrounded by people who share their opinions and experiences, and who read at least one of the same daily papers. Presumably the same thing could be said of accountants, or Army officers, or farm workers, or subway mechanics. It's no sin to associate with people you agree with. But it *can* be a problem if you and your fellow reporters derive your sense of what's news by reading the same newspaper each day, and hear your own views reflected back to you by your editors—when what you actually need to ensure fairness is some hard, skeptical questioning.

Here's how the echo chamber can distort newsgathering and editing. Imagine you're pitching a story on how a new housing development "is ruining the quality of life for longtime residents" of a rural community. Your editor (or producer or news director) echoes that idea by providing the names of people who used to have a view of a mountain, but now see only row upon row of townhouses. He suggests you look into whether the new homes might place a burden on local water and sewer systems, what it will cost the town to build roads to the development, whether there is a negative environmental impact of cutting down all the trees, and so on. Now, these might all be valid questions to raise in the story. But because you and your editor see eye-to-eye on the subject, neither of you considers talking to some of the people moving *into* the housing area, to hear how they feel about getting the chance to own their own homes. Neither of you asks whether the property taxes paid by the new homeowners will help out the community. You don't investigate what all the homebuilding is going to mean for local construction companies, whether it will help the regional economy, etc. The story ends up being one-sided not out of prejudice or malice, but because your newsroom lacks anyone who is inclined to question the *premise* of the original story pitch.

As a news director or editor, you can also characterize an event in a way that limits discussion or narrows the approach a reporter takes. Let's say a politician, fictional Governor Mary Emerson, has come under fire by journalists and members of the opposition party for her performance during a recent crisis. Members of Emerson's own party are convening a hearing where her aides and other key decision makers will be called to testify. If you say, "We'll be watching this 'pep rally' at the state house to see if it makes any news, which isn't likely," you're sending the signal that the governor's opponents have the facts on their side, that a defense of the governor is a PR ploy and not a legitimate news event. If, on the other hand, you say, "We'll have a report from the state house today, where the governor's people are going to set the record straight about what happened," you're doing the opposite—suggesting her opponents and the press have been off-base, and that this is a forum where the real story will be told. Neither characterization is unbiased.

Journalists, like scientists, also have to be careful of "confirmation bias"—the inclination to observe and record facts or opinions that confirm their hypotheses and to ignore those that don't. Imagine you are a reporter who has heard rumors that a nominee to head the Commerce Department—we'll call him Ronald Reese—has a history of being impossible to work with; one source has told you that "people on Reese's staff at the Bureau of Industry and Security complained about him constantly, and a couple even quit rather than work with him." So you set out to see what you can discover. The first few staffers you interview all tell you that Reese was tough but fair, that he didn't suffer fools gladly, and that he successfully eliminated some "dead wood" at the bureau. But the next person you meet with says Reese "alienated some of the office's best workers," and that she "considered quitting half a dozen times." She offers to put you in touch with two or three other people who "really had problems" with Reese—and you follow up those leads with three more interviews. What's the real story here? Is it that Reese was a strict but even-handed bureau chief who had to deal with a handful of malcontents—or that his managerial style turned off some of the government's best workers and forced others to leave? Fairness to the nominee would argue for more interviews—and for following up contacts of people who *liked* Reese, as well as those who didn't. If you narrow your reporting so that it confirms your preconception that this nominee is not the kind of manager who should be a cabinet secretary, you are doing Reese, and your audience, a disservice.

In fact, a valuable exercise if you are truly committed to reportorial balance is to try to frame *any* controversial story in a variety of ways. The point isn't to look for good news rather than bad news, or to switch from slanting a story in one direction to slanting it in another, but to force yourself to view a potential story through more than one lens. The aim is to guarantee that all important viewpoints get equal treatment—especially those viewpoints that don't spontaneously occur to you. This is hard to do when almost everyone in a newsroom is of one mind on political, social, environmental, and ethical issues.

For instance, in 2003 and 2004, many news organizations reported intensely on the fate of a brain-damaged Florida woman, Terri Schiavo, who had been on life support for many years. A court granted her husband the right to remove her feeding tube, although her parents wanted her kept alive. The Florida legislature then passed a bill giving Republican Governor Jeb Bush the power to overrule the court and reconnect the woman's feeding tube. NPR's stories focused on many issues—on the legal wrangling, on the Constitutional issues raised by the case and on the woman's medical condition. There were also host interviews, commentaries, and talk-show call-ins. The coverage was extensive and thorough—but to many listeners, it still sounded one-sided. Perhaps that's because the pieces were almost always framed as stories concerning the right to *die*, as opposed to the right to *live*. Some listeners believed the issue had both religious and legal facets, some of which were being ignored by NPR reporters. Only a few of the NPR pieces—or those on commercial TV networks, for that matter—mentioned religion at all, and then only in the political context of Christian groups that supported the Florida legislature and the governor. The opinions of religious leaders—of religiously devout people in general—were conspicuously absent.

Reporters and news directors will sometimes justify their programming by insisting it is what their audience wants. But this "preaching to the choir" mentality sidesteps the fact that as a responsible journalist, you have an obligation to present the most thorough and accurate picture of events possible—even if that means presenting people or points of view many listeners abhor. And your listeners might be more diverse than you think. Almost as many public radio listeners describe themselves as conservative (29 percent) as call themselves liberal (32 percent).

Whether you already are or aspire to be a radio reporter, producer, editor, or manager, you need to make a special effort to see beyond your own personal experiences, especially if your newsroom is starting to look less and less like the community it is serving. Many researchers have documented the aging of the U.S., the growing Hispanic population of the country, the rising percentages of American children growing up in single-parent families, and many other ways the nation is changing. As NPR's Senior Vice President for Programming Jay Kernis pointed out in a statement of NPR's programming and diversity goals, these changes have important implications for all of public radio:

> NPR programs will be inclusive and authentic in such a way that listeners recognize themselves—what they have done and what they believe—as they listen on air or visit on line. The simple truth: people want to hear themselves on the radio. If they don't, they soon think, "This isn't for me."
>
> To present the truth means to truly explore the nuances of ideas, events, policies and the public debate. It means questioning the assumptions and attitudes that we have so easily made and perpetuated. It means not treating a particular ethnic or religious group as having monolithic beliefs or behaviors. It means getting beyond the clichés and the stereotypes.

Verifying Assertions

It's important not to let the mandate for fairness turn into an *artificial* sort of balance. It's not enough to give the same time to people on one side of an issue as to the people on another, if some of the statements on either side can be proven false—or are clearly in doubt—or if the spokesmen for one side is far less articulate or knowledgeable than the other. Nothing is less illuminating than "he says/she says" stories, where every claim on one side of an issue is offset by a counterclaim on the other, without any indication of which speaker has the facts on his side. A typical report of this type will feature residents of a community saying, for instance, that widespread reports of cancer in children who live near high-tension power lines prove that the wires are hazardous to people's health; that will be balanced by clips of utility company executives insisting the lines are safe. You might strive to achieve fairness by including the same number of actualities of people representing both sides. But the result will be almost a Rorschach test of one's political views. Any listener inclined to distrust big business will end up siding with the concerned residents; any listener who tends to be scornful of environmentalists or is a business owner himself will find himself agreeing with the power company. And everyone else will be left no closer to the truth.[2] If you produce a seemingly "balanced" story that relies on assertions that are untrue, you do the audience a disservice—even if it appeases one faction that feels strongly about the issue.

Most journalists know this, and try to approach all their interviews with a large dose of skepticism; the problems crop up when reporters are unconsciously inclined to believe and accept the assertions of one side more than another—to adopt conventional wisdom uncritically, or to see one group as the good guys and the other as the bad guys.

Let's say you are doing a piece on a Detroit community whose residents are trying to get the city to pay to have their houses stripped of lead paint, and who are suing paint manufacturers for possible future health damage to their children. (This is a fictitious example, by the way.) In the course of doing interviews for the story, you talk to someone from an organization advocating for people with mental retardation. And she says, "In the U.S. today, lead poisoning affects four million children—that's one out of every six kids under the age of six."

This is a powerful actuality! But is it true?

A skeptical journalist would ask: Where does that figure come from? What does "affect" mean in this context? Is the speaker saying that four million children in the United States are at risk of developmental disorders, kidney damage, coma, and premature death? Does the assertion conflict with common sense? In other words, if we gathered 600 children

2. In fact, NPR reported a few years ago that the largest, most definitive study of the subject—carried out by the National Cancer Institute—found no link at all between power lines and childhood cancer.

at random, would 100 of them really be "affected" by lead poisoning?[3] These are all reasonable questions. But they may never occur to you if you think of yourself as an environmentalist, if you feel that businesses frequently avoid dealing with the health consequences of their products, and if you assume that more government regulation is in the public interest. And an editor who shares your views may not think to challenge you about including the statistic.

That's why it's so important for reporters and editors always to consider the source of a claim, since groups have a vested interest in inflating statistics that support their cause. For example, various people associated with the Academy Awards have said that the annual Oscars ceremony is watched by a billion people around the globe. It's in the interest of the promoters of the awards to be able to boast of an immense audience; and although the claim is preposterous on its surface, it has been repeated in various media.[4] Similarly, dozens of interest groups claim that "twenty-five million Americans" suffer from certain diseases or take part in particular activities—that they have lung disease, go to indoor tanning salons, drink wine, or have implanted medical devices. The assertions are almost invariably made without citing a source for the figures. It's easy for reporters to remember the round number, and it always makes a good quote or sound bite—but that doesn't make it true. Sometimes it's hard to tell if a statistic is overblown. If someone asserted that American journalists write twenty-five million stories every year, could you say whether the number was reasonable, or off by a factor of a hundred one way or another?

In other words, it is not enough simply to attribute a claim to a person or special-interest group; you have to do your best to determine that the assertions that people make in your reports are true, or at least *could* be true.

3. News reports in the 1980s on the number of missing and kidnapped children provide a case study of journalists failing to apply the "common sense test" to a claim. Individuals and organizations repeatedly asserted that as many as two million children were abducted each year. This would have been about one child out of every thirty! Even more conservative estimates said that some 50,000 children disappeared annually—still an unbelievably high figure (50,000 Americans died in the entire Vietnam War). In 1985, two reporters for the *Denver Post* won a Pulitzer Prize for debunking the claim, which they traced to advocacy groups—and the news media.

4. This particular claim was debunked in a *New Yorker* article by Daniel Radosh in 2005.

Not all dubious assertions have to do with numbers, of course. Consider this excerpt from a report on the Air Force Academy's effort to restore its reputation following a scandal that found the school had tolerated rapes and sexual harassment for years.

> REPORTER: Academy critics say for change to happen, the Air Force needs to investigate past allegations and prosecute offenders, even if they're now high-ranking officers in the military. Dorothy Mackey heads STAMP, which stands for Survivors Take Action Against Abuse by Military Personnel.
>
> MACKEY: If you start prosecuting the generals that were involved in these multiple rapes, just like the Catholic priests—if you go back 50 years, like many of the states are doing, to hold priests accountable and prosecuting them, you will see a rapid stop of this when generals are going to federal prisons.
>
> REPORTER: So far, no top Academy leaders have been charged with crimes related to the scandal. There have been charges brought against a number of men in the lower ranks of the Air Force.

Here the reporter includes an actuality that seems to be alleging that men who are now top Air Force officers "were involved in . . . multiple rapes"—a very serious accusation, if it's true. His voice track saying that only lower-ranking men have actually been charged can be taken a couple of ways—as an indication that the allegation is unproven (if you tend to trust the military to investigate itself), or as a sign of some sort of cover-up (if you assume the Air Force would not go after high-ranking officers). But in either case, the listener is left uninformed about whether there is any basis for the charge other than the speaker's anger. At the very least, we expect the reporter to have asked, "Do you actually have any evidence that generals or other top officers committed these crimes?" If the answer was no, then the actuality shouldn't have gotten into the piece; and if it was yes, then we should have heard what the evidence was.

In contrast, when NPR reporter Daniel Zwerdling was looking into the mistreatment of immigrant detainees in U.S. jails, he checked the truth of the allegations in various ways. He would sometimes interview the same eyewitnesses about the same events several times. "I would repeat it over several days," he says, "because I wanted to see, if I asked them three different times, if they'd give me three different accounts, which would be bad news for the story! It would mean they were not remembering right—or lying." He also asked two former inmates, then living thousands of miles apart, to draw a map describing the New Jersey jail where they claimed a beating had taken place, to see if they would both draw the same picture and describe the same incident. And he obtained prison medical records, which supported the allegations the detainees made. All of his reporting confirmed the allegations of mistreatment.

Sometimes an assertion is so vague it's hard to know how to nail it down. After the Massachusetts Supreme Judicial Court ruled that gays have a right to marry in that state, opponents of gay marriage mobilized to amend the state constitution. NPR ran a report whose intro said, "Recent gains by the movement supporting same-sex marriages have prompted something of a political and social backlash in parts of the country." The evidence for that claim came in the middle of the piece:

DE PERCIN: You know, the news, as somebody said to me, has become "all gay, all the time."

REPORTER: Denise de Percin is with the Colorado Anti-Violence Project, a gay and lesbian advocacy group.

DE PERCIN: Every day there's something that's related to the gay, lesbian, bisexual, and transgender community. And we know, historically, that visibility equals targeting.

REPORTER: In this case, de Percin says, gays have been targeted by a backlash that is not only anti-gay-marriage, but also anti-gay. She says incidents of violence and harassment against gays had been declining, but in the last six months, she says, they shot up more than 30 percent over the same time last year.

DE PERCIN: It's a stunning, stunning increase. And, you know, it's concerning to me also that the backlash seems to be not sort of by extreme militant groups or white supremacist groups, you know, who you often think about committing hate crimes, you know, but more by sort of ordinary people that somehow have interpreted everything that's going on as license to sort of act out on their prejudice or act out on their fear or act out on their hatred.

The person interviewed—whom the reporter identified as being with a "gay and lesbian advocacy group"—makes a slew of claims here (some paraphrased in the voice tracks): that the backlash is "not only anti-gay marriage, but also antigay"; that in the last six months, "incidents of violence and harassment" have "shot up more than thirty percent over the same time last year"; that the backlash—which the speaker associates with "hate crimes"—is coming from "ordinary people" who are acting out on their "prejudice," "fear," or "hatred." What she says may be true, but we have no way to judge, in part because it is all so nebulous. The speaker doesn't cite a single specific incident; her "statistic" relies on percentages and not a fixed number (so we don't know if the unidentified incidents of violence and harassment have gone up from 3 to 4, or from 300 to 400), and she somehow assumes she understands the motive behind the behavior. If you're an editor faced with a script like this, you have an obligation to ask the reporter to check these amorphous claims—and to seek information and verification from someone who does not have an axe to grind.

Getting Both Sides

When the allegation concerns a specific individual or institution, you have to make an even more exhaustive effort to represent all important sides of a story. When a reporter or editor fails to do that, the reputation of your entire news organization can be undermined. In 2005, NPR had to apologize on the air for a report on the *Paris Review*, a journal that was under new editorship. The intro to the piece said "some former editors are worried about what's between the covers. They say the *Paris Review* may be new, but it's not improved—and it's betraying the vision of its founder, George Plimpton." There were a number of problems with the story, but the main one was that the reporter had interviewed the new editor months before the story ran, and had not told him about the accusations made against him or his magazine. He could not answer his critics, because he did not know he was being criticized, or for what. "If the subject of the story doesn't know what you're going to report, how can we be fair to them?" asks Bill Marimow, a Pulitzer Prize–winning reporter and former NPR Vice President for News. He says a conscientious reporter must always try to understand both sides of the story—to "master" the point of view of the subject as well as of his critics. "How can we possibly master it," he asks, "if they don't have the opportunity to respond?" That's why every reporter should be candid about the focus of his story when he's asking for interviews—and give each subject an opportunity to lay out his or her case. "They need to have time to examine [the allegation], scrutinize it, and respond to it," Marimow says. "And we also need the time to scrutinize and master their response."

The harsher the criticism leveled against a person or organization, the more you may need to work to verify accusations and present the other side.

When one NPR reporter, Snigdha Prakash, was looking into allegations that the drug company Merck had tried to silence people who raised safety concerns about the painkiller Vioxx, she not only phoned the company—she also sent letters to the PR department and the chairman. As she obtained emails and other documents showing Merck's internal communications, she sent another letter laying out what she was uncovering and asking for an interview. After what Prakash calls "long and tedious negotiations," Merck finally agreed to talk to her, and she was explicit about what she'd found. "One Friday, before I went home that day, I sent them titles and Bates numbers[5] of all the documents I knew I was going to quote from in my pieces. So essentially I laid all my cards face up," she says. "It allowed the [Merck representative] to be really prepared for the interview, so he could respond to specific issues. It also demonstrated that we meant business, and that we were going to be fair—that we weren't trying to ambush them."

Daniel Zwerdling tried a similar approach after he had amassed convincing evidence of the abuse of detainees at New Jersey jails. He sent registered letters to all the people he wanted to speak to, and made dozens of phone calls. Like Prakash, he let everyone involved know what he was finding. "I would call the head of the Passaic County jail, where I had all the allegations of abuse using dogs," he says, "and I'd say, 'I'm doing a story about conditions for detainees put in your jail by Homeland Security, and I've been talking to a lot of detainees who allege that guards rough them up, and that guards use dogs. And I know that nobody in prison likes to be there—that people are angry, and resentful, and they have the incentive to make up stories—but I have heard enough stories now that are consistent that I think there might be something here, and I'd like to talk to you about it.'" He finally got an interview with a spokesman for one jail, and obtained a statement from the warden of another.

Even when you have no expectation of getting a response from the subject of your story, you have to give it your best shot. For example, there is a common kind of analytical piece in which a reporter begins

5. All documents retrieved during the litigation "discovery process" are assigned unique numbers. These document identifiers are called "Bates numbers," after the Bates automatic numbering machine.

with several public statements—the President justifying his position on tax cuts, for instance, or the CEO of a drug company defending its procedures for ensuring product safety—and then turns to others to criticize those statements. Even when equal time is allotted to both points of view, this sort of "set up the pins and knock them down" story structure is inherently unfair, since the people who made the original statements, or their proxies, never get the opportunity to answer their critics.

So make the effort to get all sides of a story, even if you assume that effort may be fruitless. Reporter John McChesney says his editors wanted him to get the CIA's version of events when he was reporting on the death of an Iraqi insurgent in CIA custody. "I don't think the CIA ever comments on something like this, but I tried repeatedly just to make sure we'd covered ourselves," McChesney says. "I also called probably ten or twelve former CIA people, but none of them wanted to go on the record. What I did find was a former Inspector General of the CIA, who had been there eight years, and he said, 'Well, if there's any slack coming from the top, you're going to have abuses, because the guys in the field are under tremendous pressure.' And that was as close as I could get to an explanation of how this had happened."

Avoiding Loaded Language

Political activists of all stripes are masters at using language in a way to skew an argument. The extreme cases are obvious—for instance, calling a missile "the Peacekeeper," or describing zoning plans as "smart growth." (There aren't many people who see themselves as supporting "dumb growth.") But you need to recognize the subtler cases where people on different sides of an issue are using language to frame a debate or undermine an argument.

Take "reform," a word embraced by anyone who wants to change the status quo. News organizations frequently report on tax reform, Medicare reform, education reform, electoral reform, CIA reform, campaign finance reform, health care reform, and many other "reform" plans. In every case, the "reformers" are on one side of the issue—and the label alone seems to put them on the "right" side.

For instance, if your report on the McCain-Feingold campaign finance legislation keeps describing it as "a campaign finance reform law," then its opponents end up being portrayed as against reform—that is, against making things better. If you describe a tax cut (or tax increase) proposal as "a federal tax reform plan," then anyone who doesn't support it appears to be against correcting whatever is wrong with the current tax system. In the same way, opposition to "health care reform" almost sounds like opposition to health care. That's why politicians love the word "reform," and why you should eye it with extreme skepticism. Reporters sometimes adopt it because they're quoting politicians or because it's convenient shorthand, especially on the radio; it's a lot easier to say "the campaign reform law" that to explain "the law that restricted so-called 'soft money' donations—contributions not subject to federal limits because they technically go to state political parties instead of actual candidates." But this might be a case where it's worthwhile to add the explanation, even if it sounds a bit formal and bureaucratic. Otherwise you may seem to be endorsing one faction and demonizing the other.

There are many other loaded words—too many to list, and in any case the list changes all the time. In 2004, news organizations fretted about what to call the structure Israel was erecting in the occupied Palestinian territories. Was it a "wall"—a term that evoked the Berlin Wall? Was it a fence, a more benign term (which might remind us of the picket fences between neighbors in suburbia)? The U.S. State Department called it a "security barrier"; but to adopt that diplomatic phrase might have implied acceptance of Israel's justification for building it—namely, that it was needed to protect the nation's security. NPR generally used the word "barrier" alone.

So it's important to think about the effect even a single word may have on the listener—and on yourself, if you're reporting a controversial story. Is that new housing area out in exurbia a sure sign of "sprawl," "growth," or "development"? Is the city council debating the future location of a nuclear waste "dump" or of a "disposal site"? This is not to say that you should strip every powerful word out of your stories; sometimes a murder is a murder. It is simply a reminder that reporters and editors need to be aware of how language itself can shift the emphasis of a story one way or another.[6]

6. Certain adjectives tell us more about the speaker than of the word they supposedly modify. If a reporter starts his story with the line, "The U.S. Congress passed long-awaited legislation today adding a prescription drug benefit to Medicare," he's failing to tell us who was actually waiting for the legislation. It leaves us to infer whatever we want, including that the reporter himself was eager for the bill to be passed.

CHAPTER 3

Writing for Broadcast

If you want an instant lesson in the difference between writing for a newspaper and writing for radio, try reading a newspaper article out loud. Don't *imagine* what it would sound like if you were actually saying the words, and don't mumble the text so quietly that the person in the cubicle next to you isn't disturbed. Read it aloud the way you would if, say, you were trying to get the attention of a class of fidgety tenth graders who had more exciting things on their minds.

For example, here's the start of an article that appeared on the front page of the *Washington Post*, chosen at random. The story is headlined, "Bush Takes Responsibility For Failures of Response":

> President Bush yesterday said he takes responsibility for the federal government's stumbling response to Hurricane Katrina as his White House worked on several fronts to move beyond the improvisation of the first days of the crisis and set a long-term course on a problem that aides now believe will shadow the balance of Bush's second term.
>
> "Katrina exposed serious problems in our response capability at all levels of government," Bush said at a White House news conference with Iraqi President Jalal Talabani.

That first sentence would never make it on the radio, though it works fine in print. It's packed with information; it's got the news right at the beginning; it's grammatical; it's even fairly readable—a practiced radio newscaster would probably have to take breaths after "Hurricane Katrina" and "first days of the crisis," but he or she could get through it. The problem isn't that it's badly written, but that it's not written *for the ear*. It pours out the information faster than most of us could absorb it *as we listened to it*. In one fifty-six-word sentence, the reporter tells us the federal government's response to the hurricane was "stumbling"; that the

President took responsibility for that yesterday; that the White House improvised a response during the first days of the crisis; that they are working on several fronts to do better—to set a long term course on the problem (presumably the post-hurricane relief effort); and that the President's aides believe it (again, presumably the relief effort, but maybe the stumbling response) will "shadow" the balance of the President's second term. Whew!

A reader who becomes confused at any point in that sentence or elsewhere in the story can just go back and reread it—or even jump ahead a few paragraphs to search for more details. But if a *listener* doesn't catch a fact the first time around, it's lost. So when the NPR White House reporter wrote the same story, he dispensed the information in bite-size chunks:

> President Bush says the buck stops with him, and he takes responsibility for whatever mistakes the federal government made in its response to Hurricane Katrina. N-P-R's David Greene has more from the White House.
>
> GREENE: After days of insisting this is no time for a "blame game," the President shifted his tone a bit. Standing alongside Iraqi President Jalal Talabani at the White House, Mr. Bush took a question on Hurricane Katrina, and answered by saying he takes responsibility for the federal government's actions. He said he wants to know how to better cooperate with state and local government in the future. He did qualify his remarks, by saying he takes responsibility only to the extent that his administration did anything wrong. He's been saying for some time that the question of which level of government is at fault for the slow response to the hurricane will come later. The White House announced the President will be in Louisiana Thursday to deliver a prime-time address on the aftermath of Katrina. David Greene, N-P-R News, the White House.[1]

Now if you're reading that report silently, the writing may seem unsophisticated, almost juvenile. Many of the sentences are short and their structure is simple. They almost overlap one another—one says the President "shifted his tone a bit"; the next two sentences tell us more about what he said; the sentence after that paraphrases the President's state-

1. This is an example of a news "spot"—a short report written for a five-minute newscast, as opposed to a longer "piece" intended for one of the two-hour-long news magazines.

ment; and so on. Several sentences begin with the same pronoun: "He said," "He did qualify," "He's been saying." But read the text out loud and you'll hear that this simplicity allows us to take a sort of mental breath between sentences. We can register what the reporter is telling us and get ready for the next statement. If a listener misses the intro—after all, since there are no headlines on the radio, the intro is our first clue to what the story is about—the information is amplified in the body of the news report.

This is how we write for broadcast. You have to give out information at the pace and in a form that allows people to absorb it. Successful radio writing—at a minimum—has to be intelligible to people who are *listening*, not reading the newspaper or watching TV. (TV reports on this same story *showed* the President standing next to the Iraqi President in front of rows of reporters; a certain amount of information is transmitted on television even when the sound is off.) There are many ways to make acceptable radio writing better, but at the very least it has to be comprehensible the first time a person hears it. For most listeners, there isn't a second time.

Write the Way You Speak

Remember: when you are on the air, you are communicating with one person at a time.

That's how people listen to the radio, after all. When they tune in to *Morning Edition* or *Weekend Edition Saturday* or any other program, they are usually making breakfast, or behind the wheel of their car, or at their desk in the office, or in front of their computer. They are listening one at a time—or at most, two or three at a time. They are not sitting in a concert hall, along with thousands of other listeners, much less the millions that listen to many public radio programs.

So we have to learn to write as if we were talking not to thousands or millions of people, but to one person; we should communicate to that archetypal listener much the way we actually talk to our friends or family. "When we speak on the air," longtime *All Things Considered* host Robert Siegel says, "we are aspiring to be heard as someone who is describing the world—the stories we've decided to pay attention to—to someone we feel pretty close to, whose intelligence we respect, whom we like, and whom we're helping to explain things to."

And we have to remember that—as far as that listener is concerned—we are indeed *talking*, not reading. People listen to radio news with the same set of ears they use when they listen to their spouses, children, colleagues, and friends; as much as possible, we should speak to them as we would if they were sitting across the desk or dinner table from us. Presenting stories orally requires what Siegel describes as "a kind of reverse engineering." When you write for radio (or TV, or podcasts for that matter), you need to think about what you sound like when you *talk*, and then *write* your stories with a similar vocabulary, cadence, syntax, and grammar. "The script should be like a theatrical script, a dramatic script," Siegel says. "We know it is there when we go to the theater—we know this has been written down and they're reading it—but when it starts to intrude, when we actually think people are reciting the script, the theater is failing. It's the same with us. When we sound like we're speaking to one listener at a time, it's working. When they hear us following a script, it's not working."

Playwrights know that dialog is not the same thing as transcribed speech. In real life, we stammer, and hem and haw, and forget what we're saying, and repeat ourselves. So playwrights give us dialog that *mimics* the style of the spoken word, without the imperfections of everyday banter. It's speech that's been washed and ironed.

It's the same with writing news. The goal is to write the way you *wish* you could speak—or the way you speak on your best day, when you've had just the right amount of caffeine and sleep. Great hosts and reporters write in their own voices—using words and sentence structure and phrasing that come naturally to them. They write in such a way that they can "perform themselves" authentically and convincingly.

One way to find out how you speak is to record yourself when you're talking to your friends, and then write down what you said. See what your own speech looks like. Or look at almost any transcript of an NPR interview—especially one with a regular person, as opposed to a politician or pundit or other professional speaker. Here's an excerpt of a report in which *All Things Considered* host Michele Norris spoke to Belinda Bruce—a woman from New Orleans who had escaped Hurricane Katrina. She was suddenly homeless, and living in a shelter in Baton Rouge:

BRUCE: I really can't explain it, except just say maybe it was for the best.

NORRIS: For the best.

BRUCE: Yes, because I'm more happier when I can stay homeless.

NORRIS: Help me understand that. You're sitting at a shelter. Your house is most likely underwater. And you're happier here than you were back home.

BRUCE: Because there was so much going on in my neighborhood. Even though I was comfortable in my house—in my house—blocks up where the kids went to school, it was horrible. They always had shootouts. They just didn't respect the kids. They were selling drugs. The good thing is I'm away from that with my kids. And maybe they'll get into a better school, a better neighborhood.

This exchange should remind us that real people don't talk the way newspaper reporters write. They don't pack everything they've got to say into one long, dense sentence. Like Belinda Bruce, most of us use relatively few adjectives when we speak. We use sentence fragments. ("Because there was so much going on in my neighborhood.") We add force to what we're saying by speaking in short, repetitive sentences. ("They always had shootouts. They just didn't respect the kids. They were selling drugs.") We begin sentences with "and" or "but." ("And maybe [my kids will] get into a better school.") We don't do this consciously; it's just how people talk. So it shouldn't surprise us that Norris speaks the same way in her question. ("Help me understand that. You're sitting at a shelter. Your house is most likely underwater. And you're happier here than you were back home.")

If you understand how people talk, you can learn how to write scripts that sound like speech. With enough practice, you can sound as good as Belinda Bruce.

How to Sound Like a Real Person

Good broadcast writing often requires us to unlearn many of the ways we learned to write in creative writing classes or in graduate school—or while working at newspapers or wire services.

FIRST, AND FOREMOST, *SAY* YOUR SENTENCES BEFORE YOU WRITE THEM DOWN; OR AT THE VERY LEAST, SAY THEM OUT LOUD AFTER YOU'VE WRITTEN THEM. "This is one of the most commonly offered pieces of advice that we give in every writing workshop," Robert Siegel says, "and it's one of the most commonly ignored." As you write, ask yourself: Would I ever say this sentence in my regular life, when I am not writing a news story? If the answer is no, change it.

Consider the beginning of this story, whose grammar and syntax are identical to those of an item that ran on a news wire: "A Supreme Court increasingly concerned about how individual states' authority fares when pitted against the federal government's power is studying a property rights dispute from Connecticut that could yield a key states-rights ruling." There are many things wrong with this sentence, but let's start at the top. Would you ever actually *say* something like, "The judge was nominated to a Supreme Court increasingly concerned about how individual states' authority fares when pitted against the federal government's power"? Even Laurence Olivier couldn't have pulled off a line like that! A radio version of the story might begin something like, "The Supreme Court has agreed to consider a case that could eventually redefine states' rights. The case has to do with hotel owners in Connecticut, who claim . . ."

Remember, expressing your thoughts in short, declarative sentences doesn't require you to eliminate any of your ideas—just to ration them out. You aren't sacrificing anything by writing less convoluted prose.

DON'T USE WORDS ON THE RADIO YOU WOULDN'T SAY AT OTHER TIMES. Many words or phrases were added to the news lexicon by headline writers: to save a few letters, they turned an "increase in the unemployment rate" into "rise in the jobless rate," though no one actually says "jobless" outside of the news business. And when was the last time you actually described a robbery as a "heist"? Or, for that matter, referred to snow as "the white stuff," as weather forecasters often do, since the word has no natural synonym? "There are a lot of words that have died in spoken language and simply live on in news copy," Robert Siegel says. "It's toxic to good writing. We should be striving for originality."[2]

And that goes for adjectives, as well as nouns. Good writers avoid the sort of hyphenated adjectives that show up only in news stories—phrases like "rebel-held," "mineral-rich," "storm-weary," and "tech-heavy" (as in the "tech-heavy NASDAQ"). "It's hard to process an adjective like that," says Renee Montagne, the *Morning Edition* host. "Like 'crime-ridden

2. Siegel says he does occasionally like to adopt the clichés of the newsroom in ironic contexts—to talk about "a bushel of asparagus with a street value of $50," for instance, or to say, "CBS laid off the staff gangland-style."

neighborhood'—it's not natural speech." And things get even worse when a writer in search of a synonym combines a hyphenated adjective with a hyphenated noun. One reporter referred to Motorola on second reference as "the suburban Chicago–based high-tech gadget-maker"—a phrase that would never crop up in real life. ("Great news, Dad! I just got a job with a suburban Chicago–based high-tech gadget-maker!")

DON'T USE SYNTAX THAT DOES NOT OCCUR NATURALLY. News stories frequently begin with phrases like "President Bush today told members of Congress" or "Heads of state tomorrow meet in Geneva." But we never say "I today went shopping" or "My wife tomorrow plans to throw me a party." The story above described "a Supreme Court increasingly concerned about . . . " But you wouldn't ever utter a phrase like "I'm planning on painting my garage increasingly showing signs of wear and tear."

Whenever you write a sentence to say on the air, think about how you'd say an *analogous* sentence in your daily life, and adopt that more colloquial word order. That will often mean writing two shorter sentences to replace one long one.

USE PRESENT PARTICIPLES—THE "PRESENT PROGRESSIVE" TENSE—TO DESCRIBE THINGS THAT ARE GOING ON AT THE MOMENT. Many reporters will depict scenes in a rather literary manner: "It's the beginning of a new shift at the observatory. A clock strikes midnight. Two astronomers sip coffee as they begin their shift. Behind them, a technician looks at photographs from the previous night's observations." Often the reporter is consciously trying to avoid the verb "to be" by using more colorful verbs ("strikes," "sips," "looks"), or is attempting to achieve a dramatic effect. This style of writing can be effective in print, but it isn't really the way people speak; we use "to be" all the time as an auxiliary verb when we're describing ongoing events. Imagine you were the reporter on the scene at the observatory talking over a cell phone, and someone asked you, "What does it look and sound like?" You might say the clock "is striking midnight," the astronomers "are sipping" their coffee, the technician "is looking" at the pictures. The lesson here: if you're inclined to use the present tense to describe scenes, try doing so the same way you would to an imaginary person on the phone—using the "present progressive."

DON'T PARAPHRASE ACTUALITIES AS IF YOU WERE READING A QUOTE FROM THE NEWSPAPER. Newspaper reporters generally put the attributions at the end of their quotes, as in this fictional example:

> "The advantage of being part of an international trading community is that your increased exports can make up for a weaker currency," said Charles Gold, who was a domestic policy advisor to President Bush and now lives in Chicago.

Read that passage out loud and you'll see how ill-suited the construction is for radio. In normal conversations, when we relate what someone has said, we put the attribution at the beginning—e.g., "My boss says I have to get this piece done by midnight," as opposed to, "I have to get this piece done by midnight, says my boss."

Similarly, newspapers frequently invert sentence structure—tacking together a string of prepositional phrases—in a way that sounds bizarre in speech:

> "I'm confident we can get the job done," Smith said last week at a ceremony marking the transfer of the fiber optics initiative and the new technology project from the Commerce Department to his control.

Don't do this in radio scripts.

KEEP YOUR SENTENCE STRUCTURE SIMPLE. Remember, time goes in only one direction; listeners can't go back to try to figure out who or what you were talking about at the beginning of your voice track. Read this sentence out loud: "Growing up in southern Massachusetts during the First World War, when prejudice against Germans or people of German extraction was still widespread in the United States, Helmut Kleinfelder had a difficult life." For three quarters of the sentence, the listener doesn't know who we're talking about.[3] Keep it simple; allot a sentence to each idea, and wherever possible put the subject at the beginning of the sentence. In this case, it would be better to say something like, "Helmut Kleinfelder had a difficult life. He grew up in southern Massachusetts during the First World War. At the time, many people in this country let their feelings about the war color their attitude toward German-Americans."

3. The only time this sentence structure might work on the radio is when the subject is obvious from the context. In the news spot above, the reporter says, "Standing alongside Iraqi President Jalal Talabani at the White House, Mr. Bush . . ." But that sentence follows one that ends "the President shifted his tone a bit." So it's obvious to the listener who's standing next to the Iraqi president.

Other Newswriting Tips

There are a few other tenets of radio news writing to keep in mind, whether you're writing forty-five-second reports for a newscast, an intro for a host to read, or your own story.

AVOID GENERALITIES AND PLATITUDES. Listen to the radio long enough, and you'll hear introductions and pieces that start with phrases like "Many people think," "It's widely accepted," "You may not be aware of the fact." All of these reflect assumptions on the part of the writer. They're often unsupportable or untrue. (We would balk at a blanket statement like "Mexicans are a very happy people," but for some reason we seem confident about asserting, "Most Americans have never heard of folksinger Tom Paxton.")

AVOID MEANINGLESS ATTRIBUTIONS. Beware of the overused terms "officials," "analysts," "critics," and "experts." "I have spent a lifetime trying to pull 'officials' out of *All Things Considered*," says Robert Siegel. A writer can often replace "officials" with "executives," "politicians," "military officers," "labor leaders," and so on. Better yet, attribute the idea or statement to a real person. Siegel also suggests "good old-fashioned metonymy, of having places and institutions speak." Again, this is in keeping with making scripts reflect colloquial speech. People really do say things like, "Ford says it's coming out with a new hybrid car"; they don't say, "Ford *officials* say they're coming out with a new hybrid car."

USE TITLES SPARINGLY. If you're at a dinner party, you would not expect someone to greet you by saying, "Have you met United States Under Secretary of State for Arms Control and International Security John D. Holum?" So don't introduce him to listeners that way. Instead you can say, "John Holum is a U-S diplomat [or Under Secretary of State, if you have to use his title] specializing in arms control."

KEEP SENTENCES SHORT. Avoid run-on sentences, whether you're writing for yourself or for someone else—"no longer than one or two lines," says NPR host Liane Hansen. The desire to pack in information often results in sentences larded with clauses that separate the subject and verb. "The biggest complaint I have about leads I have to read from other people is too many dependent clauses," says host Scott Simon. When you write a sentence like "President Bush—fresh from his meeting yesterday with Democratic Congressional leaders, the first such meeting in eighteen months—said today the United States . . . ," it takes so long to get to "said" that it's easy to forget whether we're talking about the President or Democratic Congressional leaders.

WRITE IN THE ACTIVE VOICE. If you're old enough, you may remember that President Ronald Reagan, speaking of the Iran-Contra scandal, used the passive voice when he said, "Mistakes were made." He pointedly avoided saying by *whom*. That may be good politics, but it would have been bad journalism. Using the active voice[4] ensures that the reporter will attribute actions to someone or something. Instead of saying "The administration is being criticized for its stand," he has to explain "Democrats are criticizing the administration"; in place of "The paramilitary groups were ordered to kill," he's obligated to be more specific and say "Serbian soldiers ordered the paramilitary groups to kill"; instead of "Students have been told to report sexual harassment," he has to identify who did the telling—"Teachers and principals have told the students to report sexual harassment." The active voice also helps the listener create a vivid *picture* of the action. When we hear, "The paramilitary groups were ordered to kill the women and children," we can imagine the paramilitary forces and their victims, but we don't see anyone giving the orders. Rewriting the sentence to put it in the active voice brings the picture into focus.

DON'T USE RHETORICAL OR HYPOTHETICAL QUESTIONS. Starting an intro or piece with a rhetorical question is all too common because it's so much easier than coming up with a real beginning. Think how many times you've heard someone on the radio or TV open a story by saying, "Have you ever wondered . . . ?" (Usually the honest answer is no.) Sometimes the question is also vague and a cliché—a hat trick! One NPR intro included the line, "But what about America's love affair with the automobile?" Well, what *about* it?

DON'T BE OVERCAUTIOUS. It's fair to say something "may be the largest" of its kind, or that a person is "one of the most famous," or that a vote "could be a setback." But often news writers go a step further and say it "may be one of the largest" or that the person is "perhaps one of the most famous" or that the vote "could be a potential setback." Too

4. The active voice is syntax in which the subject of a verb performs the action (e.g., "The editor made a grammatical mistake"). In the passive voice, the subject receives the action described by the verb (e.g., "A grammatical mistake *was made*").

many qualifications essentially rob the sentence of any meaning. Robert Siegel recalls one story from a reporter in Africa that ended by referring to "a war that—at least for the moment—appears to have no end in sight"—a quadruply modified statement! "We would have probably considered it editorializing if he'd spoken of a 'war without end,'" Siegel says. "But I don't know why we felt that 'it appears' to have no end 'in sight'—when those are basically the same statement. And that's 'at least for the moment'!"

REMEMBER THAT A RADIO AUDIENCE CONSISTS OF LISTENERS, NOT VIEWERS. When you write for radio, you can easily emphasize the aural nature of the medium: "Coming up, we'll *hear* from the woman who broke the story" or "We'll *talk about* why zoning ordinances have become such a hot topic in Maryland." When a newly elected Virginia governor gave the Democratic response to a Republican President's State of the Union address, Robert Siegel described him as "a new voice" for his party—not "a new face."

WATCH OUT FOR GRAMMATICAL ERRORS. "I see pronoun misreferences, singular-plural problems—basic grammar school stuff," says Stu Seidel, who has been editing at NPR for years. This is another consequence of sentences that are filled with dependent clauses and parenthetical phrases. Writers often get misled by sentences like, "The letter sent by the four Nobel Prize–winning economists—three of whom teach at the University of Chicago—say they are . . ." If you're in doubt about what form a verb should take—whether it's singular or plural—read the sentence without the parenthetical clause: "The letter sent by the four Nobel Prize–winning economists *says* they are . . ." Or rewrite the sentence to juxtapose the subject ("The letter") with the verb ("says").

REWRITE AWKWARD PHRASES, EVEN IF THEY'RE "CORRECT." There is a sort of grammatically fussy writing that results from too much education. But anything that gets in the way of the meaning of a sentence will break the listener's concentration. You want people to hear *what* you're saying, not focus on how you're saying it. So generally, if it sounds wrong on the radio, it *is* wrong—even if the grammar books say otherwise.

AVOID NEEDLESS REPETITION OF WORDS. It not only sounds dull to say something like "A dozen *workers* were at the site this morning, *working* in the freezing cold," it also doesn't tell us very much about what they were doing. Yes, workers work, and voters vote, and drivers drive, but you can almost always find stronger, more focused verbs to eliminate the redundancies: "A dozen workers were at the site this morning, digging post holes for a new fence in the freezing cold." (But keep in mind that some words have no obvious or natural synonyms—"gold," for example, or "snow." And it's probably better to repeat the word "rain" a few times than to say "precipitation"—or worse, "precipitation activity.")

RECOGNIZE CLICHÉS AND LOOK FOR ALTERNATIVES. Most people know what's wrong with clichés, but unfortunately they immediately come to mind when news writers are under deadline pressure. Some broadcast writing—sports reporting, for example, or business news—often consists of nothing *but* clichés: "The Bulls were stampeding on Wall Street today. Blue chip stock prices skyrocketed . . ." Many reporters often use military metaphors to heighten a sense of tension: "Democrats and Republicans are battling over welfare reform." There are almost always more colorful and precise ways to express the same thoughts.[5]

AVOID UNNECESSARY JARGON, ACRONYMS, AND INITIALISMS. A lot of jargon and arcane language creeps into radio scripts because the people we interview use it. Sometimes it's a sign of a reporter's lack of confidence; if he doesn't really understand physics, he may leave a phrase like "adiabatic change" in an actuality out of fear of scrambling the speaker's point by defining the term incorrectly. But often the audio can be edited to avoid words, phrases, or acronyms that would be unintelligible to most listeners; if it can't, the reporter may be able to elegantly define terms we're going to hear in the actuality. Before a sound bite of someone mentioning "the AIR meeting," for instance, the reporter or host

5. George Orwell, in his classic essay "Politics and the English Language," writes something that should be clipped out and tacked to the computers of almost all writers, but especially journalists:

> A scrupulous writer, in every sentence that he writes, will ask himself at least four questions, thus:
>
> 1. What am I trying to say?
> 2. What words will express it?
> 3. What image or idiom will make it clearer?
> 4. Is this image fresh enough to have an effect?
>
> And he will probably ask himself two more:
>
> 1. Could I put it more shortly?
> 2. Have I said anything that is avoidably ugly?

can sneak in the definition of AIR: "The Association of Independents in Radio—or AIR—has been around since 1988. At this year's AIR meeting, producers plan to discuss . . ." The listener should never be forced to try to figure out what a word or expression means. While he's trying to parse the last sentence, he's missing the rest of the story.

DON'T OVERWHELM PEOPLE WITH TOO MANY NAMES AND NUMBERS. As a *Morning Edition* host, Steve Inskeep says he often has had to deal with overburdened intros; he advises, "Specific names and numbers should be the essential ones. Others should be salted into the piece, or dropped altogether." It's easier said than done. When we know a lot about a subject, it's often tempting to try to give the listener an earful of facts. But we can be too smart for our own good. Nigeria may have 130 million people, and the new President may have been elected with 63 percent of the vote, and only 40 percent of the eligible voters may have cast ballots—but all these numbers almost certainly obscure what's really important in your piece. Ask yourself why you're including the numbers, and then see if there's another way to make the same point: "Nigeria is Africa's most populous nation . . ."

LISTEN FOR TONGUE TWISTERS, RHYMES, AND OTHER ODDITIES. Jane Smith may be Mr. Fister's sister, but you probably don't want to identify her that way in your script. (If you don't know why, go back and read that sentence out loud.) Similarly, you don't want to end an intro with, "Wayne Corey has the story" or "Richard Dorr has more." And beware of unintentional alliteration. (Beware of too much alliteration of any sort; no one thinks it's clever.) If you're inclined to start a scene by telling us that "Fred Fishkin is foraging for fungi," you might want to say he's "looking for a few good mushrooms" instead.

STEER CLEAR OF GIMMICKS YOU'VE HEARD ONLY IN OTHER NEWS BROADCASTS. You may have heard TV anchors hyping a story by holding back the subject—teasing the viewer for a few seconds to try to generate curiosity. "It will be one of the biggest events this city has ever seen, and organizers hope it will draw tourists from around the country. And it won't have any sponsors, or celebrities, or even free give-aways. 'It' is the first annual kite-flying festival . . ." Mentally healthy people do not talk this way. ("It was a total surprise when I came to work this morning. It involved the secretary, the producer, and the vice president of news. 'It' was an announcement that we would all be getting pay raises . . .") Whatever effectiveness this device may once have had has surely been worn away by decades of overuse. For the same reason, you should be careful not to overuse the verbless, headline style of lead sentence you hear all the time on the radio and TV: "In Albany today—a sixty-one-car pileup." This can be a forceful way to get into a subject once in a while, but it's hardly colloquial. ("On the subway today—a crowded car and a kid playing music too loud.")

KNOW HOW TO PRONOUNCE FOREIGN WORDS AND UNUSUAL NAMES AND PLACES, AND INCLUDE THE PRONUNCIATIONS IN YOUR SCRIPT. Even if you think you know how to pronounce "Llangollen," for instance, when the moment comes for you to say it on the air it you'll be glad to see the reminder "lahn-GAHKH-lin" in your script (which is probably as close as an American can get to the real name of the Welsh city). And pronouncers are essential whenever you're writing for someone other than yourself, such as a program host.[6]

CHECK FOR TYPOS, MISSING WORDS, AND OTHER CLERICAL ERRORS. Every writer and editor should think of his or her script as being ready for broadcast; in a pinch, it may end up on the air—or on the Web. Don't assume that someone else will correct your mistakes and expand your idiosyncratic abbreviations. For instance, an intro shouldn't use the reporter's initials rather than his name; after all, the host is not going to say, "LG has this report." And don't count on a computer's spelling or grammar checker to do your work for you. As Stu Seidel points out, even small errors suggest a lack of attention to details—not a good trait in a journalist. "If you have 'their' where it should be 'they're' or 'there,'" he says, "it makes me worry about every aspect of the piece."

6. See Appendix II for a short guide to writing pronunciations of foreign words and unusual names.

CHAPTER 4

Reporting

In 1998, National Public Radio undertook a "visibility campaign"—taking out full-page ads in several news magazines to try to increase the public's awareness of NPR. Whatever their effectiveness, the ads' tag line—"NPR Takes You There"—brought home the idea that public radio programs transport listeners to many places where news is being made.

They do that, most of the time, through reporters.

True, a good interview can put listeners on the scene. And shows like *Talk of the Nation* can take calls from people all over the country—from listeners who can describe what things are like in the place they're calling from. But much of the time, reporters are the eyes and ears of public radio's news magazines, just as they are for commercial news broadcasts. On a typical day, one program included reports from a mega-church in Houston, from a terrorist trial in Birmingham, Alabama, from the Afghanistan-Pakistan border, from a prison in Colorado, from a wildlife preserve in Texas—and from many other places inside and outside the United States.

Some reporters are on the road almost every week. Others have beats that tend to keep them fixed in one location. Some do their reports in only a day or two, trying to unearth news wherever they can and to jump on stories as soon as they break. Other reporters rarely tackle hard news, but focus instead on the stories behind the headlines. Still others follow cultural trends, or advances in technology, or developments in the arts, science, or business. But they all have some common goals: getting the story right, whether it's a piece about a controversial weapons deal or a heavy metal band that's found a new outlet for its music; conducting probing interviews to uncover the most telling facts and details; editing those interviews effectively and fairly; representing events and people vividly, through strong storytelling and on-air performance; and using scenes and sound to make the audience an active participant in the news. When it all works, public radio does indeed take you where news is being made—behind the scenes at the U.S. Capitol, to a refugee camp in Southeast Asia, to the Everglades in Florida, and to virtually any other place a reporter can bring a microphone.

The Reporter's Personality and Skills

The job description for "reporter" comprises many duties, at NPR and at most other news organizations. Two reporters with adjacent cubicles may have almost completely different daily routines. But there are a few qualities and skills shared by successful reporters of all stripes.

A GOOD REPORTER IS CURIOUS. Reporters should be asking questions all the time, because pursuing the answers will often lead to interesting stories. How does the World Bank determine whether and how its funds are spent? What is life like for a dictator-in-exile? How is the White House using the press to promote its political initiatives, and how much is it costing? Who was the first person to come up with the idea of DNA fingerprinting? What's it like for a folksinger to perform the same song at concerts for thirty years? A reporter should view the world "with amusement and fascination," says David Folkenflik, who came to NPR in 2004 as a media correspondent. "You have to be open to outrage and joy"—and then let those emotions lead you to unexplored places. That's true whether your beat is media, Capitol Hill, transportation, education, or cinema.

A GOOD REPORTER IS SKEPTICAL. "You need to be an aggressive, questioning person," says Washington-based correspondent Larry Abramson. "You have to be pushy enough not to simply take what someone tells you as the truth." There are certainly plenty of reasons for a reporter to doubt what he or she hears. Some people have political agendas, and distort the facts to advance their personal causes. Companies hoping to increase their profits may hype the advantages of a particular product to get favorable attention in the press. Governments at all levels like to control information. Entertainers embellish their careers. Even when people don't mean to mislead reporters, their recollections may be muddled. People involved in news events may have told their stories so often that they believe their accounts are more accurate than they are. Others lie outright. For all these reasons, good reporters do not accept assertions as facts. "People who are gullible, naïve, or who are used to taking orders without question probably need a brain transplant before they get into this business," Abramson says.

A GOOD REPORTER LOOKS AND LISTENS FOR THE TRUTH. This might seem to go without saying, but that sense of "outrage" that may motivate a reporter to ask questions and uncover a good story should not skew his or her reporting in any one direction. The best reporters are driven by their zeal to find the facts, no matter where they lie; they don't assume that any one person or group has a monopoly on virtue or veracity. "The idea is to master the two poles of the possible truth," says Bill Marimow, a former NPR news vice president, "and to give listeners as much information as they can possibly get, so they can decide independently where the truth lies."

Curiosity, skepticism, a desire for the truth—those are some of the key attributes of first-rate reporters. But just having the right personality doesn't guarantee success as a reporter; you also have to acquire the right mix of skills. Sometimes it's hard to tell where a personality trait leaves off and a skill begins.

A GOOD REPORTER LISTENS WELL. As reporter Laura Sullivan notes, interviewees will open up to reporters only if they think you care about what they're saying. "People get freaked out if they see you're not listening, and they stop talking real quick." Paying attention to what people are saying also encourages an interviewee to put his trust in the reporter. "The best reporters I know," says Bill Marimow, "are the people who know how to elicit personal, colorful, detailed information—who can instinctively, without trying, establish trust, rapport, and mutual respect."

A GOOD REPORTER ABSORBS INFORMATION QUICKLY. This is especially important if you're doing daily stories and news spots, or working as a general assignment reporter where almost every story requires coming up to speed quickly. "I used to be an academic and I spent months and months studying things before I wrote long papers about them," says Larry Abramson. "This is pretty much the polar opposite." To be successful as a reporter, you have to know where to get information, and you need to be able to make sense of that information rapidly.

A GOOD RADIO REPORTER USES THE MEDIUM EFFECTIVELY. It doesn't matter how good you are at digging up stories if you can't communicate them on the air. Successful radio reporters use all sorts of audio well—whether it's a twelve-second sound bite recorded over the phone for a news spot, or the thunder of a herd of elephants recorded in stereo deep in the African jungle. The best radio reporters also write conversationally and remember that they're talking to one listener at a time, not to millions. And they use their voices to present their reports in a way that keeps that listener interested. "Information is transmitted in many ways, and it's not just in content," says Chris Joyce, who has been both a reporter and editor at NPR. "It's passed on in the way you inflect, and in the words you choose, and the emotion in your voice."

A GOOD REPORTER DOES EVERYTHING POSSIBLE TO AVOID MAKING MISTAKES. There is often some tension between the desire to get things right and the need to get a story on the air as soon as possible. "When you go to break stories," says Chris Joyce, "you skate closer to the edge of saying, 'We don't have this confirmed, and we're not going to be able to before the deadline.' Your chances of getting burned are much higher." But rushing a story to air is no excuse for getting things wrong. Indeed, every reporter should have "an absolute reverence for facts," in the words of Bill Marimow. "The very heart of your job is accuracy, thoroughness, and fairness. Those three attributes supersede everything else, including active sound, mood, ambience—everything."

Developing News Sources

There are as many ways to find news as there are kinds of stories; clearly, a reporter whose beat is police and prisons is going to have different sources than one who covers Congress, or religion, or sports. But if you want to find stories no one else has found, you should start simply by talking with lots of people.

That's one reason beat reporters often have the most impressive records of breaking news—they know lots of people with information that has not been made public yet. For instance, a reporter who covers advanced technology may talk regularly with executives at chip-making, computer, telecom, TV, and Internet companies; with computer programmers and engineers; with materials scientists; with sales reps and advertising professionals; and with dozens of other people in fields related to high tech. Any one of them may know and talk about something that could lead to a good news story. If you are reporting, take the time to meet frequently with the key people in your field—even when you're not pursuing a specific story. Laura Sullivan says she listens for the kind of things that will make someone tell his or her spouse, "You know what I heard today at work?" That, she says, is sure to be a story she wants to hear, too.

You can become a beat reporter, in effect if not in title, just by following a subject that interests you. For example, a reporter at a local station may decide on his own to focus on transportation issues. In that case, Larry Abramson says, "you can go down and introduce yourself to the people in the transit department, and make sure you do a story in that area every few weeks." Eventually, he says, you'll get a sense of who the key players are and what issues are most relevant to the public, and you'll be able to recognize when something new is happening long before it shows up in other media. You can even specialize if you already have a beat. Scott Horsley is usually a business reporter, but he says he's been covering a lot of energy issues in recent years. "Just by accident, I also fell into doing a lot of corporate scandal stories. That meant developing some sources with accounting expertise, legal expertise, and so on." The point is not to let yourself get spread so thin—by your own interests or by your editor—that you're starting every assignment with a blank slate.

The best reporters mingle with their sources. "You have to go to lunch with people," says Laura Sullivan. Even if much of your deadline-driven reporting is done over the phone, it's worthwhile spending time with potential sources face-to-face. "It's the only way they'll return your phone calls," she says, "because they'll know you on a personal level." Even in Washington, D.C., with facilities like WAND (Washington Area News Distribution) providing audio for many public events, reporters need to get out of their cubicles. "It's important to go to news conferences," Larry Abramson says, "and go to people's offices, and shake their hands, so that maybe in the future they'll give you something."

Keep track of people who have been helpful to you in the past. In 2004, when Daniel Zwerdling started to look into allegations of abuse in American jails, he drew on sources he had used years earlier for a story on torture in Turkish prisons. Because he had kept the contact information for those earlier interviews, he knew whom to start calling in what he calls "the torture community."

Developing sources takes time. Mary Louise Kelly had her work cut out for her when she began reporting on intelligence agencies—a beat where people most closely involved with events never speak on the record; in addition, since hers was a new beat, she did not inherit a Rolodex from anyone, and had almost no names or numbers for sources. To get started, she identified three or four reporters at newspapers or magazines who were consistently breaking news on her beat, and whose stories had turned out to be accurate. "I started clipping everything they wrote, every name they had who was identified as being on the record." But her sleuthing for sources only *began* with the names. "You'd see references to 'Joe Smith, a former CIA case officer,' and that's all it said! Or 'Jim Jones, who worked in Army intelligence from 1992 to 2004,' with no indication where to reach them today . . . but I'd start asking around." Once she had a few sources, she could work her way towards the people who were most in the know. "You can ask, 'Did you ever work with Jim Jones? If I were to send an email to you, could you forward it to him?'" Kelly says that's how she gets in touch with people who are undercover and don't care to talk or have their phone numbers made public. "I establish a decent relationship with someone who comes to trust me and my reporting, and ask them, 'Could you forward my request?'" People who would never talk to her if she called them directly will often respond if they hear first from a third party whom they trust. "You have to have an introduction."

The route to the right source can be roundabout. For a story on internal politics at the Corporation for Public Broadcasting, David Folkenflik had coffee with potential sources, sent email to others' home addresses, and talked to people at institutions that had connections to CPB. In short, he used each contact to try to find another, who might in turn give him a bit of the story. "You sort of draw concentric circles around a story in your mind," Folkenflik says. "There are people right at the heart of it, and there are people around them—and around them, and around them."

However you find sources, and no matter how often you meet with them, you need to keep them, figuratively, at arm's length. Bill Marimow won two Pulitzer Prizes as a newspaper reporter, and developed many productive sources covering police issues. He emphasizes that there is a critical line you shouldn't cross—that remaining objective about a story means keeping your relationship with your sources professional. That can be hard to do, Marimow says, because in many respects, cultivating a source is very similar to making a friend. You chat together, you go out to lunch, you may visit the person where he or she works. "You can't easily say to the source, 'Listen, I'm *not* your friend—you are a source, and this is a utilitarian relationship,'" but you can send that message tacitly—by not inviting sources to gatherings usually attended only by friends and family members, and by avoiding situations where your family, and your source's, might both be present.

Other Sources of Original Stories

Even if you don't yet have a Rolodex filled with five hundred names and business cards, there are a variety of ways you can come up with good sources for original stories.

Many investigative stories draw on documents that a person, company, or other organization would prefer not to be made public. When Daniel Zwerdling was reporting on the beating and terrorization of detainees in U.S. prisons, he confirmed their allegations by obtaining both official documents from the New Jersey jail where they had been held and confidential medical records. John McChesney's report on the death of an Iraqi prisoner in U.S. custody relied on hundreds of pages of documents—the product of a CIA investigation into the man's death—that included many eyewitness accounts. Snigdha Prakash got her hands on confidential email, letters, spreadsheets, and other documents from drug maker Merck; they showed that the company exerted pressure on individual doctors and several medical schools to suppress discussion of safety problems associated with its bestselling drug Vioxx. Sometimes a person inside a company or government agency will leak a document because he or she thinks it ought to be made public; sometimes lawyers, or other parties who have an interest in a case, will contact a reporter in hopes of advancing their cause. But you can't count on that. If you have reason to believe a document exists that could be the basis of a good story, you will probably need to ferret it out. "One of the things—and it's a real obstacle when you're working with these groups—is they don't necessarily listen to NPR, they don't know what NPR is, and we don't have much of a track record as a source of investigative news," Prakash says. "So people don't come to you. You have to go to them."

The Freedom of Information Act—which applies to all departments and federal agencies in the executive branch—allows you to request copies of records that are not usually distributed to the public, and sets standards for determining which records must be made available.[1] But simply making a request doesn't mean you'll get much back. When McChesney was starting his piece on the Iraqi prisoner's death, he says, he decided "to FOIA the hell out of everybody—CIA, Naval NCIS, the Army CID—Criminal Investigation Division—the Special Warfare Operations of the Navy." Six months later, he hadn't gotten a single response. They all told him that an investigation was in progress, or that the information was classified, or that his request had been referred to another office at the Department of Defense. When FOIA requests go nowhere, you have to "persist in contacting and persuading everyone involved to allow you to see these documents," McChesney says. Zwerdling also found his *federal* FOIA requests unproductive, but had more luck getting material under *state* open records laws. He was able to get the names of the staff at two of the jails he was investigating—and used that information to verify claims by the detainees.

Sometimes valuable documents may *already* be public. "There's often a lot of gold in those hills, if you know where to dig," says Scott Horsley. Nonprofit institutions have to file tax forms called 990s, which show, among other things, how much money they raise through fund-raising, their capital expenses, and the salaries of the organizations' highest-paid executives; finding out how nonprofits spend their money can often lead you to a good story. Profit-making companies have to file annual reports called 10-Ks, which include profit-and-loss statements; the *footnotes* for those statements may discuss current or pending lawsuits or government regulations that have an impact on company operations—information the firm is required to include, but which may not be widely known. Data collected by the Census Bureau can provide a nearly endless stream of stories—on poverty, income disparity, home ownership, and many other aspects of American life—if you are patient enough to sift through it all. Lots of other information is available if you know where to look. For instance, homes sales are reported publicly; by looking closely at sales in a region, you might be able to determine, say, whether sellers do as well, or better, by working with discount brokers—something full-price real estate agents often deny. Individual states keep data on licenses issued for everything from securities agents to bingo parlors; spotting trends in

1. The Department of Justice lists the principal contacts for federal agency FOIA requests on its Web site.

that data could lead to a reportable story. And so on. In fact, the challenge facing enterprising reporters is not dealing with a *lack* of information, but knowing where to find relevant material, and how to dissect and analyze that data. As Larry Abramson says, "For much of the reporting that we do, that's as important as having good sources who are going to help you break news."[2]

Or you can do both—that is, find someone who is already screening reams of data on a regular basis. "There are a lot of people doing reportorial work out there who are not reporters," Abramson says. "There are people regularly filing Freedom of Information Act requests—people like privacy activists, who are constantly digging around in these documents." He says the activists might not think of their findings as news—they may just post them on their Web sites or write about them in Web logs—but their discoveries can tip a reporter off to a story that no one else has. "That's because you're not reading them in the press—you're gleaning them from the mind of someone who's really informed." Informed, and probably biased, so it's essential for the reporter to look at the data on his own, and to seek out the other side.

Trade association conferences, scientific meetings, and other public events are good places to meet key people, and potential sources, face-to-face. Science reporter David Kestenbaum recalls how he stumbled into a story when he attended a meeting of radiation scientists in Washington, D.C. "There was a guy from Homeland Security talking about new clean-up standards in the case of a dirty bomb," he says. (A "dirty bomb" is a device that uses conventional explosives to spread radioactive material.) "That meant, basically, they were going to say it was okay if you were exposed to the equivalent of a thousand dental X-rays in a year." In other words, Kestenbaum says, the government was in effect presenting the view that if a dirty bomb were detonated, and people near the blast had to choose between suffering radiation exposure and fleeing their homes, they should simply stay put. "That was going to cause a big stir. And because I happened to go to the meeting, I heard about that before anybody else did."

Many of the stories radio reporters cover every day don't arise from their personal enterprise, but are dictated by news events. Listeners simply expect reporters to be on the scene when scientists announce

2. There are a number of Web sites that serve as resources for computer-assisted and investigative reporting, among them poynter.org, ire.org, and powerreporting.org.

breakthroughs, when tornadoes strike, when Congress passes laws, and so on. NPR and most other news organizations are also committed to covering controversial issues—the debates over federal funding of stem cell research, the friction between anti-terrorism laws and civil liberties, the potential consequences of tax law changes, and thousands of others. When the need for reporting a story is itself blindingly obvious, the journalistic discussion usually focuses on which angle to take. Here again, the nature of the story frequently determines the mode of coverage. How much time the reporter devotes to each of the time-honored who, what, when, where, why, and how questions will depend on whether the story concerns a plane crash, a murder trial, or the naming of a Supreme Court nominee. Reporters usually should have at least a short consultation with their editors, even on breaking news, to make sure they are both on the same page about how a story is to be reported.

Getting People on Tape

Some of the saddest words you can hear at NPR are "He (or she) won't talk on tape." And some of the most exciting are "I got *great* tape!" (Ironically, people who have never seen a reel of magnetic tape still refer to it. No one says, "I got great audio!")

It is a tenet of public radio reporting that the best pieces rely on strong actualities and sound. Not surprisingly, reporters put a lot of time and effort not only into finding the right people to interview, but also into persuading them to let the interview be recorded, and then encouraging them to speak fluently and colorfully.

There are certain situations where people are unwilling to let themselves be recorded—or even to let a reporter use their names. Sometimes they are speaking *on background*. In other words, they are willing to give you information that you can quote or paraphrase in your report, but only if you do not attribute it to them by name. As an intelligence correspondent, Mary Louise Kelly says she often can get people to speak to her only on background. In that case, she makes sure they understand what "speaking on background" means, and then she negotiates with them how they can be identified in her report. "You try to get as specific as you can," she says, "because the more information you can give the listener, the better they can evaluate how credible the speaker is. So you push: 'Can I say "a CIA officer who left very recently after a long career in the clandestine service?'" And they'll usually say no! But eventually you'll figure out something—'a former senior intelligence official.'" Here's how she used information she got on background for a report on a memo to CIA station chiefs by the then-new intelligence director, John Negroponte:

> Negroponte's memo is classified. Officials who've read it describe it as Negroponte introducing himself to his troops—the message being, "I'm on the scene now, and I'm in charge." The controversial part is that it invites C-I-A station chiefs to communicate directly with Negroponte on some issues. That has raised the hackles of C-I-A insiders, who fear the memo represents a further blow to the authority of C-I-A Director Porter Goss.
>
> One former intelligence official says at the very least, the memo invites confusion. "If you're a C-I-A station chief in the field," this official says, "you're going to be wondering, 'To whom exactly do I report now and on which issues?'" The official continues, "Say you're about to do an operation, you can't be asking, 'Whose permission do I need?' You just don't have time for that."

Kelly's piece later ends with her quoting "two former senior CIA officials." This particular story doesn't have any actualities, since everyone spoke on background—but at least it had several quotes. "Sometimes people will talk to you off the record completely—they don't want to be quoted at all," Kelly says, even if you agree not to use their names. So as a reporter, make sure you and your source agree on the conditions for the interview, and whether the source can be quoted in any fashion. (Even one who won't allow his or her quotes to be used can point you in the right direction to do more reporting.)

On the other hand, some people will let you use their names, but won't allow their voices to be recorded, for personal or legal reasons. Here again it's up to the reporter to accurately represent what they said—as WNPR reporter John Dankosky did in his piece on Connecticut charging someone with manslaughter in connection with an assisted suicide:

> David Shepack, the state's attorney prosecuting the case, declined to talk on tape. He says although such prosecutions are rare, the statute is applied when someone causes or aids another person to commit suicide.

Some people won't let you record their interviews if they fear reprisals from their employers or other adverse consequences. This can be extremely frustrating to reporters who have uncovered a great story; the last thing you want is to have to tell everything without any actualities from the key sources. Bill Marimow suggests one way around this dilemma, based on his experience as a print reporter. Years ago, he discovered that police dogs in Philadelphia were being trained to attack on command—but not to release on command. His sources could describe the brutal "bite and hold" policy, but they didn't want to go on the record. So he had them talk to a *third* party whom he *could* quote—the former head of the black police officers organization in Philadelphia. Marimow encourages radio reporters to try the same gambit—to "find someone in a position of credibility, expertise, and experience, who is not as vulnerable as your sources on the line, and get that person to hear what you've heard." Then, he says, get *that* person to do a recorded interview with you.

There are a few circumstances where sources are willing to be recorded but do not want their names to be used. NPR's policy on granting people anonymity is quite strict,[3] but it does allow for doing so if naming the speaker would put him or her in economic or physical jeopardy. That's how Janet Babin, a reporter for WCPN in Cleveland, was able to get an actuality of a woman who said her live-in husband had tried to choke her to death.

> The black 61-year-old says, with a weary smile, that she never thought she'd be in a shelter at her age. Fearing retribution from her attacker, she spoke to N-P-R on condition that it withhold her name.

Note that the reporter expressly states the condition of not providing the speaker's name and the reason for the condition.

Again, it's essential that you and the source expressly agree on the terms of the interview—especially if the person you're talking to isn't used to dealing with the media. Someone saying "I want this to be off the record" might mean she doesn't want her name to be used, or that she doesn't want the information to be reported in *any* form. It's always best to clarify this—and if possible, to get the interviewee to acknowledge the agreement after you've started recording, in case there are any difficulties later.

3. See "The NPR Journalist's Code of Ethics and Practices" at npr.org/about/ethics for more details on conditions for granting anonymity.

Finally, if you're using documents in your report, you'll have to decide how to bring them to life. You can read them yourself, or have them read by an actor. "I can listen to myself forever," Snigdha Prakash jokes, but when she wrote her story about Vioxx, a producer auditioned various actors to read the internal memos. "She cast people as characters" in the story, Prakash says, and "was at pains to get people to read the documents with energy, but not with dramatic flourish." (Even so, it's important to tell listeners that they're hearing an actor, and not the author of the documents.) In John McChesney's story on the death of the Iraqi prisoner, the documents included direct quotations. "Only when we can infer the tone of voice from the quotation did we 'punch it up,'" he says—and he gives this example, where the report quoted a Navy SEAL:

MCCHESNEY: A C-I-A security agent, riding shotgun in a Suburban that night, provides this account for investigators from the C-I-A's Office of the Inspector General.

ACTOR (reading): The SEALs picked Al-Jamadi up and tossed him into the back of the Humvee like a bag of potatoes. "Did you see that? Did you see that?" the agent recalled saying to the others. Two SEALs then plopped down on him. The agent demonstrated the action, indicating the SEALs physically sat on top of the detainee.

Even a short excerpt from the CIA report became a very powerful bit of sound.

Getting Great Actualities

Even when people are happy to talk to you, they may be nervous, or intimidated by the microphone. Many radio reporters have had the experience of arranging an interview on the phone, only to show up at the guest's office or lab or factory and have the person say in astonishment, "I didn't think you were going to *record* this!" So it's a good idea when you're setting up an interview to let the guest know that you'll be there with your microphone and recorder, just so there are no surprises that get the conversation off on the wrong foot. (You can always say you'll "only" be bringing a small MiniDisc or flash memory recorder, to allay any potential concerns that you'll show up with an engineer or crew.) Even if you usually record with a shotgun microphone, you may want to put it aside for a smaller, omnidirectional mic if an interviewee seems overly conscious of the equipment.[4] And once you start recording, keep the mic below and to the side of the speaker's mouth so you can easily maintain eye contact.

"I never start straight out with an interview—I always make small talk," says Chris Joyce. Many reporters try to put guests at ease by getting them to talk about themselves—how they got into their field, what they're working on at the moment, and so on. The point is to begin with simple, uncontroversial subjects—subjects the interviewees can talk about without fear of making a mistake or saying something they'd hoped not to divulge. Depending on the subject, you might also want to show them that you're familiar with their work. "With scientists," Joyce says, "you have to make them comfortable that you're going to get all the facts right. And if I know a fair amount about their field, I may show off some of that knowledge."

Showing off *too* much knowledge can backfire, though, if the interviewee is inclined to use jargon, acronyms, or technical language, or is going to assume knowledge that the average listener won't have. So it may be necessary to "play dumb" to a certain extent. Some reporters tell guests to imagine they are speaking to a smart, but easily distracted high school student. Religion reporter Barbara Bradley Hagerty sometimes asks people to pretend she doesn't know English well, or she'll ask an interviewee to imagine he's explaining the same subject to his mother. Other reporters just casually hint that they'd prefer if the speaker avoided technical or arcane references. "I did an interview with a paleontologist about how humans evolved into a running animal," Chris Joyce recalls, "and I said at the beginning, 'We don't run as well as four-legged animals—as you said in your paper, "quadrupeds," but let's try to avoid that term if we can.'" If people use acronyms or initialisms in their daily work, they probably won't be able to avoid using them when they're interviewed, so keep your ears open for them. (If you're a business reporter, you probably know right away what the CPSC does, but to the rest of us, it's just a bunch of

4. An omnidirectional microphone, as the adjective suggests, picks up sound more or less equally from all directions. A shotgun mic, in contrast, is much more sensitive to sounds directly in front of it than to sounds on the left and right or at the rear. Some reporters prefer to use a shotgun mic for interviews, since it reduces unwanted background noise. But shotguns can be intimidating to guests because they are much longer than most handheld mics and are sometimes used with "pistol grips" to minimize handling noise.

letters.)[5] So when an interviewee mentions a term that you suspect listeners won't understand, jump in and ask for an informal definition: "What does PROP stand for—and more important, how would you describe it to someone who wasn't following the Russian space program?"[6] Or, "For someone who's never fired—or even seen—an AR-15, can you describe what the gun is like?" When you know a lot about your interviewee's areas of expertise, Larry Abramson warns, "it's tempting to speak 'geek' to them, and stay within the boundaries of your own common understanding. But you usually end up with pretty bad tape when you have one of those insider conversations."

Getting good tape can be also difficult if the interview is monitored by a company or government agency's press officer. Laura Sullivan says her beat frequently takes her to the FBI and the Justice Department, as well as to prisons and jails and police departments around the country—and nearly all of them have "media relations" offices. Media representatives may see their role as "providing accurate information to reporters in a prompt and comprehensive manner," as one such office described it; but from the reporter's point of view, Sullivan says, "their job is to get in your way." Their very presence during an interview can be a problem. Sullivan's solution is to wait them out— "bore them to tears"—by spending the first part of her interview on questions that even the most cautious media officer would find innocuous. "You want to know everything, for example, about the microfiber of the bulletproof vest that's going to be worn by all the officers. An hour later, the press person will be looking at his watch." Once she's alone with her interviewee, she can get to subjects that the media relations person would have considered off-limits. "Then you can ask the interviewee what happened to the money that was supposed to go in the police department's funds that actually ended up in somebody else's back pocket." Another solution is to get the guest out of his or her office by asking for a tour; you may be able to leave the media relations person behind—and even if you can't, your interviewee may be less mindful of the monitor. If a face-to-face interview fails completely because there was a media person hovering nearby, try calling back and doing the interview on the phone; sometimes you can get through to your interviewee directly, bypassing the would-be censor. (Sometimes the

5. The CPSC is the Consumer Product Safety Commission.

6. "PROP" refers to a device on a Soviet craft sent to Mars. It is an acronym for the Russian words meaning "Surface Evaluation Instrument."

media relations person won't volunteer the interviewee's direct number, so you may need to ask for it as a backup number. And the PR contact might try to listen in over the phone, even if you think you've gotten straight through to the source. If you suspect that there may be a second person on the line, ask.)

And don't be intimidated by the status, or title, or reputation of the person you're talking to. Sometimes a reporter is so grateful even to land an interview with, say, the head of a big computer company or the secretary of education that he or she doesn't follow up vague or misleading answers. A few reporters expressly avoid using formal titles with their guests—instead of "Senator Wilson" or "Secretary McFadden" they'll address the guests as "Ms. Wilson" or "Mr. McFadden"—because they don't want to give the impression the speaker is *deigning* to be interviewed. Try to think of the people you interview as your peers, Sullivan suggests. "It's not your job to make people like you," she says. "It's not your job to make them respect you, or to think that you're smart, and then to walk out of the door with nothing in your notebook or on your MiniDisc. Your job is to get them to answer questions, and to share things with the public."

Needless to say, securing an interview is only the first step. You also want to elicit answers that will help you tell your story. Here are a few practical tips:

REMEMBER THAT YES-OR-NO QUESTIONS DON'T USUALLY YIELD GOOD ACTUALITIES. A simple "yes" or "no" may be enough for a print reporter, but it's hard (though not impossible) to use a one-word actuality. So imagine the kind of answer you *hope* to get, and tailor your question accordingly. Instead of asking a NASA official, "Would you say the manned space program still has the support of the American public?" you might ask something like "How do you determine how much public support there is for the manned space program?" or "How would your future missions change if you discovered public support for the manned space program had declined?"

AVOID "EITHER-OR" QUESTIONS. Many reporters assume that there are two obvious sides to an issue, and frame their questions accordingly. A question like, "Would you say the Pentagon is doing enough to provide armor to U.S. troops in Iraq, or should the top officers be pushing Congress for more money?" presumes that the interviewee takes one of those positions. With a reporter sticking a microphone in his face, he may choose to embrace one or the other. "Don't box your guest in," Chris Joyce advises. "Keep your questions short and open-ended." A question such as "What do you think should be done to maximize the safety of U.S. troops in Iraq?" allows the guest to offer a recommendation that you haven't thought of. Joyce likes to ask people, "What do you *make* of that?" Barbara Bradley Hagerty says, "Sometimes I'll make a statement. And they'll say, 'Yes.' And I'll say, 'Tell me about that.'" Always ask for specific examples; they help even a cooperative interviewee get beyond platitudes and boilerplate responses. One of the best open-ended questions is simply, "Why?"

GET PEOPLE TO USE ANALOGIES TO EXPLAIN TECHNICAL SUBJECTS. That may require you to let the interviewee know what you're looking for. "You say the Earth wobbles on its axis. Help me visualize this." Or "If this tax proposal were a machine, what would it look like?" David Kestenbaum recalls spending half an hour in vain trying to get a scientist to talk animatedly about a three-hundred-million-year-old bacterial spore that had been discovered—and was still alive. He couldn't even get the guest not to refer to the spore by its scientific name "6,3,2." Finally, in desperation, Kestenbaum just asked him, "Can we try looking at it from the *bacterium*'s point of view?" And it worked—he got a snappy narrative of three hundred million years of history, as viewed by "Harry the Bacterium." "It was great," Kestenbaum says. "But I had to get him outside the mode of lecturing at the podium."

PUT THE SAME ENERGY INTO YOUR QUESTIONS YOU'D LIKE TO GET IN THE ANSWERS. If you sound like you're just collecting facts to put in your notebook, you'll tend to elicit a recitation of facts. If you seem interested when you pose your questions, you'll often get interested and energetic replies. Show some emotion! Sound engrossed or surprised—even if you already knew what they just told you or the subject is not intrinsically fascinating. If you're a business reporter doing a piece about the euro-dollar exchange rate and the guest says she suspects "the value of the euro will increase by 20 percent vis-à-vis the dollar some time in the next twelve months"—hardly prizewinning audio—you might say, "But wait a minute. Twenty percent? What's that going to do to the U.S. economy?" Some reporters balk at trying to infuse what they feel is artificial enthusiasm, but most feel it's just a fact of life when you work in radio—a way to get people to talk as freely and passionately as they do when they're not being recorded. "You just have to get the tape, and you have to relate to people on some level," says Laura Sullivan. "If that means you need to talk the way they talk, sit the way they sit, make sure your body language is mimicking theirs—that's just what you've got to do."

MAKE SURE YOU'RE ON MIC WHEN YOU CONDUCT THE INTERVIEW. You may want to use your own questions in your piece, but you often won't know that until you hear the answers. Strong, pointed questions can help advance a story, or provide a dramatic moment. And a two- or three-word reply that may not stand on its own can be a powerful actuality when it is paired with a good question:

REPORTER: So you're saying you spent the entire night on the roof of your house waiting to be rescued? And no one ever showed up until morning?

INTERVIEWEE: No one showed up.

QUOTE THE INTERVIEWEE'S CRITICS. Barbara Bradley Hagerty uses the example of the interviews she did about Father Gene Robinson, who became the first openly gay bishop in the Episcopal Church. "I had someone saying 'What they've done is an end run around the process, an end run around the religion, and an end run around our tradition'—a really strong cut. When I interviewed Gene Robinson, or someone on his side, I'd say, 'I've heard people say, "It's an end run . . ."'" By quoting another interviewee verbatim, you can get a person to react to his or her opponents, more than to you, the reporter.

CONSIDER DOING THE INTERVIEW SOMEWHERE OTHER THAN AN OFFICE. It's not only important to think about *what* you ask people—but also *where*. You might want to interview the technician at the lab where she made the discovery, or ask the archaeologist to go outside, where the excavation is taking place. Even lawyers, business executives, and politicians may talk with less prompting if you can get them out from behind their desks. "I find it makes people more animated if they get up and walk around the office to give a tour, or take you back out to a manufacturing floor," says Chris Arnold. "If you ask them questions while they're moving around and on their feet, and their blood pressure is up by a few points, they're just a little more interesting."

Working on Tight Deadlines

One of the advantages broadcasting has over print is that it can bring new stories to people faster than they can be published. Either because of their beats or because of their temperaments, some radio reporters specialize in "one-day-turnaround stories"—that is, stories assigned, reported, written, edited, voiced, and broadcast in less than twenty-four hours, and often *much* less. There's no substitute for years of experience when you need to report and present a story in less than one day. And the process will be very different depending on whether you're reporting on a public hearing, a shake-up at a government agency, the detonation of a nuclear weapon in a developing country, the merger of two cell phone companies, or the breakup of a big rock group. But there are a few things you can keep in mind if you've got only four or five hours to crash a piece on the air.

ANTICIPATE EVENTS WHENEVER POSSIBLE. Many "one-day" stories build slowly before the news breaks. Here again, having a beat really pays off, Scott Horsley says. "When you have a short time to do a story, it's a huge head start if you already know some of the people you're going to talk to—or people who can turn you on to people to talk to." Mary Louise Kelly often has to report on five-hundred-page studies on intelligence issues, prepared in secret and made public with little or no advance warning. In those situations, getting a coherent story on the air quickly not only requires getting a copy of the study, but also finding people who are familiar with its details and conclusions. And you can't do that at the last minute, she says. "You establish relationships with sources in advance. You work them for a few weeks or months before." Then when the news breaks, you'll know where to turn to find someone with firsthand knowledge of events, who can help you get the facts, verify assertions, and so on.

KNOW WHERE TO FIND PEOPLE WHO CAN GIVE THEIR INFORMED OPINIONS—especially on controversial subjects where the lines are clearly drawn. Larry Abramson—who has reported on privacy issues, antiterrorism laws, and civil liberties questions, among other things—says he often knows where different groups and individuals stand on various issues. "I know that their organization is for or against a certain proposal. And I've seen them quoted saying that, or I've talked to them about it, so I know what they're going to say. And it's just a question of getting them to help me explain it."

There will be times when you are assigned to cover a story in a field you know nothing about. Before he came to NPR, Scott Horsley worked at a number of stations, both inside and outside public radio, and says he often had to come develop a list of sources quickly. His advice: "One quick place to start is to take a poll of your colleagues, or of anyone who sits around you. 'Does anyone know an expert on bankruptcy law?' 'Does anyone know who I can talk to about Chinese energy consumption?' That way you not only have the advantage of your Rolodex, but of everyone else's Rolodex." And "everyone else" might include reporters from other media. "I'll do a quick Lexis-Nexis search and try to read just a few articles quickly—and often I'll notice that the same sources are cited in a number of different articles. And I'll think, 'Okay—that's someone I ought to talk to.'" But even if you *talk* to a person who appears in other journalists' stories, you may want to use him or her mainly to get other names; be careful not to always return to the usual suspects who have the usual points of view. There are a number of listservs for journalists—on business, education, the environment, religion, and many other beats—where you can put out a quick query and get rapid and helpful responses.

Horsley also says it's important to take a moment and think about the different *kinds* of actualities you'll need to get the whole story; you don't want to waste time getting the same perspective from the same sorts of people. "Sometimes you need lawmakers, administration officials, or at the local level, city council members—who aren't experts, but are the decision makers and policy makers." And "time permitting," he tries to get "color"—actualities of people affected by the events in the story.

MAKE YOUR CALLS EARLY. The one variable you cannot control is callbacks. "You can have all the best sources in the world, and know who to go to, and if they happen to be out of the country—or just in the shower—you're not going to get your calls returned," Larry Abramson says. He and other reporters who do quick-turnaround stories generally make as many calls as possible as early as they can. As Barbara Bradley Hagerty puts it, "I try to get a lot of lines out there so some fish will bite." If she calls eight people, she expects at most three to call her back. Reporter Ari Shapiro says he thinks of a "hierarchy" of the people he might want in his piece. "Ideally I would get the president of this organization and the plaintiff in this lawsuit, and this person from this advocacy group, and this ambient sound from a year ago when they did this." But then he makes a second set of calls in case the first choices don't pan out. "I'll say to myself, if I don't get those people, then I can do the piece with this person, this person, and this person—and I'll put in those calls." Since they are his backup interviews, he says, "I may give them some indication that they may not make it into the final report, but that I'd really like to talk to them about the subject."

While you're waiting for people to get back to you, do whatever other reporting you can. Find and read relevant documents. Do background research. Get copies of previous stories on the subject, including your own. In short, inform yourself. And as you do, start thinking about what you want to get from each of your interviewees.

KEEP YOUR INTERVIEWS SHORT AND FOCUSED. Many reporters on tight deadlines try to keep their interviews to ten minutes each, as opposed to a half hour or more for longer stories with later deadlines. Some reporters takes notes while the guest is talking, to remind themselves which parts of the interview yielded the best—i.e., most succinct, ear-catching, surprising, or thoughtful—answers. Scott Horsley says that when he's doing interviews for a piece that will air in a few hours, "I'm literally picking the cuts on the fly."

Don't let the tight deadline keep you from getting both (or multiple) sides of a story, as hard as that can be when the clock is ticking. Mary Louise Kelly says she usually prefers to do interviews where she really doesn't know in advance what the guest is going to say; but deadline pieces are an exception, especially if she's hearing only one side of a story during most of her interviews. Then, she says, "you call someone because you know they're going to have the contrarian view to what everyone else you've been talking to is telling you."

SELECT YOUR ACTUALITIES, AND PUT THEM IN A LOGICAL ORDER. "I usually pull a few extra," says Larry Abramson, "and if I don't use them in that piece I may use them in news spots." But otherwise, he tries not to select much more sound than he can possibly fit into his story—at the most, five short cuts for a three-minute piece, or six or seven for a four-minute piece. Ari Shapiro says he doesn't budget any specific number of cuts. "I think more in terms of the number of *voices* I need to tell the story, and I know I'll use at least one cut of tape of each of those voices." Barbara Bradley Hagerty always makes sure her editor knows exactly which actualities she plans to use, since it takes more time to find a different cut than it does to rewrite a track. "I won't write a word until I've checked out my tape with the editor and she's comfortable with the story structure," Hagerty says.

KEEP THE STORY STRUCTURE AS SIMPLE AND STRAIGHTFORWARD AS POSSIBLE. When you have more time to work on the story, you can consider various ways to use flashbacks, or artful scene changes, or clever production devices; when you're working on a deadline, as Hagerty says, "chronology is your friend."

BUDGET YOUR TIME. If the assignment comes down at 11 a.m. and your piece is going to be on the air at 4:40 p.m., you'll have less than six hours for everything. So backtime your work day. Larry Abramson says he has a mental schedule for a one-day piece that airs around 5 p.m.: "Usually I've done my interviews by 1:30 or 2. I know I need to spend one hour writing. So I'll spend twenty minutes pulling cuts of tape. I mostly just listen through the interviews—which again are ten minutes long—at double speed. Then I try to allow myself an hour or even two to write, and twenty minutes at the end to rewrite." Don't forget to leave enough time for the edit. Ari Shapiro recalls one report that he finished two hours before his *All Things Considered* deadline. "The editor said to me, 'This is fine, but, you know, it doesn't sound like you. I don't hear your voice in it. Take an hour and just work on it.'" Because he hadn't worked right up to his deadline, Shapiro had time to rewrite. "Her telling me, 'Take a step back and make it sound nicer' made it a better piece." Also, don't forget it will take time for the piece to be mixed; audio files can get lost or corrupted or deleted. "Stuff always goes wrong, so you have to build in a buffer," says Hagerty.

DON'T RUSH THE RECORDING SESSION. A bad performance can ruin an otherwise strong story. "I don't mind tracking live," Abramson says, "but I don't like tracking with ten minutes to go. Then you're under more pressure, and the producer is yelling at you, and more mistakes happen that way."

Beyond Acts and Tracks

It is hard to be particularly creative when you're staring at the clock, or waiting for that key person to call you back, or frantically searching for a fact you need. But it's not impossible.

For example, one of Nina Totenberg's trademarks is her reconstruction of oral arguments before the Supreme Court. These dialogs are unlike anything else on public radio—or in broadcasting, for that matter. For example, in her report on two cases dealing with government displays of the text of the Ten Commandments on public property, she included this reenactment (with Nina reading all of the parts):

Inside the courtroom, the debate was more legal, but just as passionate. Duke law professor Erwin Chemerinsky told the justices that in Texas, the six-foot-high monument on the state capitol grounds is a profound statement of religion beginning with the words "I am the Lord thy God." The state was not only endorsing religion in general with the monument, but one religion in particular since it has chosen a version of the Commandments, the Protestant version.

JUSTICE O'CONNOR: "What about if it's packaged in a museum-like setting?"

ANSWER: "This isn't a museum. It's the only religious monument on the capitol grounds, and it's located between the state capitol and the state Supreme Court."

JUSTICE O'CONNOR: "Would it be all right if there were monuments honoring several religions?"

ANSWER: "Yes."

JUSTICE O'CONNOR observed, "The court has long said that state legislatures can open their sessions with a prayer, so," she asked, "could the legislature post the Ten Commandments in the hallway?"

ANSWER: "No."

O'CONNOR: "But it's so hard to draw that line."

ANSWER: "Just because the legislative body can say an opening prayer doesn't mean they can put a giant cross on top of the building."

JUSTICE KENNEDY: "There is this obsessive concern with the mention of religion, and I just don't see how this is not hostility to religion. This is a classic avert-your-eyes situation. The atheists can look away or think of something else."

ANSWER: "What's left of the Constitution's bar to state establishment of religion if the government can put up any religious message and the only answer is 'Avert your eyes'?"

JUSTICE SCALIA: "I bet that 90 percent of the American people believe in the Ten Commandments, and 85 percent don't know what they all are. They are a symbol that the government derives its authority from God, and that seems to me entirely appropriate."

JUSTICE BREYER: "We are a religious nation. I don't see much divisiveness with this."

ANSWER: "There is divisiveness. The Alabama chief justice was forced to resign over a monument he put up in the courthouse. There are crowds outside today. I have gotten hate messages. People care because this is a religious document."

Not a single actuality in the entire exchange, and yet you feel like you're in the courtroom.

At the other end of the spectrum is a piece that is essentially without a script at all. An editor called Scott Horsley around 7 a.m. to say new figures had come out saying that the median selling price for a home in the United States was $206,000, and that Horsley was being assigned to do a quick-turnaround piece illustrating what that would buy in high-priced San Diego. Horsley says he knew right away he'd probably talk to a nearby real estate agent, but since that office didn't open until 10 a.m., he went on the Web and did some of his own research. He found a couple of one-bedroom condos for about $210,000— "rundown, shooting gallery kind of places." But he also found a "nice, single-family freestanding house" for about the same price. "I called the guy who had the MLS [Multiple Listing Service] listing on the Web and asked, 'Is this a misprint?' And he says, 'Oh—that doesn't include the land!'" Horsley knew right away that that line—one way or another—would be the *ending* of his piece. Now he had a sense of all the elements he needed, even though he hadn't done any of the interviews. The rest of the day was spent getting to the scenes—to the run-down condo, to the single-family house—and talking to the realtor. "The whole piece was kind of like a running stand-up. It was all recorded in the field." Here are a few excerpts of the report that illustrate both the technique and the tone:

HORSLEY: And I'm Scott Horsley in San Diego, which *Money* magazine has just dubbed "America's scariest housing market." Home prices here have more than doubled in the last five years, and only about one family in nine can afford the typical home.

I'm here at a real estate office. This happens to be next to my supermarket. I often stop by and look at some of the nice pictures you have in the window here, and I'm here with Xavier McDonald.

Tell me about some of these homes y'all have listed. This is a nice one here, City Heights, two-bedroom, one bath.

MCDONALD: And the price on this home is 385. It would be considered kind of a starter property in that price range. . . .

HORSLEY: . . . After talking with Xavier, I visited a one-bedroom condominium in the central part of San Diego. And I'm here with Cynthia Peterson, the sales manager.

PETERSON: Hi.

HORSLEY: This is being offered in what you call a "value range."

PETERSON: Correct, and that's—this unit is 199 to 210.

HORSLEY: And this is a—What is it?—560 square feet?

PETERSON: Yes. It's definitely cozy, but for a lot of, like, single professionals, they feel happy and comfortable here.

HORSLEY: One thing about a smaller place, you don't wind up spending a lot of money on furniture. . . .

HORSLEY: . . . suppose you're determined to have a single-family house. Here on the outskirts of San Diego, about 20 miles from town, I did find one—three bedrooms, two baths—for the remarkable price of just under 208-thousand-dollars. But there's a catch. That price does not include the land. So in addition to your mortgage, you also have a monthly lease payment of $620. Still, a man's home is his castle—even if someone else owns the ground underneath it.

Because of the time difference between Washington, where the program would originate, and San Diego, where he was doing the reporting, Horsley was facing a 1 p.m. deadline. So he used those first hours to conduct research and to conceive of a way to execute his story—namely, as a series of on-the-scene "stand-up" reports.

Sometimes a reporter can enliven an otherwise conventional story by using sound in unusual or unexpected ways. Here's how White House reporter David Greene took advantage of audio that a lot of people might have discarded:

GREENE: And one sign of a free society, Mr. Bush has often said, is a free press. In Afghanistan, there are dozens of news organizations that have cropped up since the Taliban government was overthrown. But in the East Room today, a seating section reserved for Afghan journalists was nearly empty. White House officials rushed to fill the chairs with American reporters, and the two presidents seemed surprised when they turned to see if any foreign journalists had questions.

PRESIDENT BUSH: Somebody from the Afghan press?

PRESIDENT KARZAI: Anybody from the Afghan press? Do we have an Afghan press? Oh, here he is.

[*Sound of light laughter*]

GREENE: Only one newspaper reporter had traveled to Washington with Karzai. Nine other reporters were supposed to make the trip, according to a spokesman for the Afghan president, but the Karzai government decided in the end to prohibit them from traveling.

Greene uses the awkward moment to make a point—that despite the development of a free (or at least freer) press in Afghanistan, the government allowed only one reporter to come to Washington with the president.

Likewise, when Tom Goldman reported on a Congressional hearing on steroids and baseball, he observed that there was something extraordinary about the sight of ballplayers testifying on Capitol Hill—so he made that the lead of his story:

GOLDMAN: The House Government Reform Committee hearing officially began at 10 in the morning, but it really started several hours later when five star witnesses walked into the room.

[*Sound of photographers*]

GOLDMAN: Photographers clicked away as the hushed crowd gawked at Curt Schilling, Rafael Palmeiro, Mark McGwire, Sammy Sosa, and Jose Canseco. The players were subpoenaed, and for a week, there was talk they'd fight the order to show up, but there they sat in business suits talking about steroids and, for the most part, frustrating lawmakers with their answers. At one end of the table was Schilling. In 2002, he was quoted as saying "muscle-building drugs like steroids had transformed baseball into something of a freak show." Yesterday—a different tune.

SCHILLING: While I agree it's a problem, I think the issue was grossly overstated by some people, including myself.

GOLDMAN: At the other end of the table—and the steroid spectrum—Jose Canseco.

CANSECO: From what I'm hearing, more or less, I was the only individual in major-league baseball to use steroids. So that's hard to believe.

The trick to making these sorts of public events into memorable radio is to see them as something *other* than routine—and then to take just a few moments to represent that insight in your writing.

A Storytelling Sampler

Because these were all stories produced on tight deadlines, the reporters did their best with what they found on site. And to a certain extent they were fortunate. But as Louis Pasteur famously said in another context, "Fortune favors the prepared mind." "If your story is airing tomorrow or later," David Kestenbaum says, "you've got no excuse for not doing something really good." Pedestrian writing, dull actualities, a boring structure, or a lackluster delivery can undermine outstanding reporting—especially if the news you're presenting is not so powerful that it demands people's attention.

While editors can help enliven a dull story, they can't help you pose the right questions in your interviews or collect the right sound when you're on the scene. So think about ways to bring your story to life while you're doing your reporting.

Kestenbaum remembers trying to come up with an imaginative way to report on scientists' success in determining the genetic sequence for a plant called Arabidopsis. "I wondered, 'What is there to say about it?'" But his editors felt it was important—and clearly the scientists did too. "They were having a big, earnest press conference the next day, and the press release had '10 Reasons Why We Sequenced Arabidopsis.'" In hopes of getting something that didn't sound like scientists droning on at a lectern, he arranged to interview one of the researchers the night before the press conference; but when he got to the hotel room, the guest had the "10 Reasons" PowerPoint display up on his computer—a sign that the interview might not be any more interesting than the news conference. "The first reason was something like, 'Arabidopsis grows everywhere,'" Kestenbaum recalls. "So I said, 'Okay—let's go find it.'"

That moment of inspiration—the idea to get the scientist to show him some examples of this plant that grows "everywhere"—ended up shaping the whole top of the story:

KESTENBAUM: People always ask Joe Ecker, "Why Arabidopsis?" Why not the rose? Why not iceberg lettuce? Ecker is a biologist at the Salk Institute. He was in D-C for yesterday's announcement. On a laptop in his hotel room, he's pulled up a top-10 list of reasons why Arabidopsis. Mostly, it was easy, he says. The plant has a small, compact genome, but it is related to broccoli and its genes could be used to design crops that would grow almost anywhere.

ECKER: Arabidopsis is found from the equator to the Arctic. And there are—a number of varieties of Arabidopsis will grow under a whole host of different conditions. So, cold, hot, increased salt, etc.

KESTENBAUM: Ecker says you could probably find one growing in a sidewalk crack. So we go on a little field biology expedition. Ecker checks the soil of plants in the hotel lobby—no luck. Outside in the cold night air, the sidewalk cracks are bare. So Ecker picks through a bed of decorative cabbages on 11th Street.

[*Sound of traffic*]

ECKER: It's kind of dark here. No, it looks like the gardeners have done too good of a job of weeding. You know, one would pull it out if you saw it. It's a garden variety weed. It's a good weed. It grows fast, it sets seeds, and it gets the heck out of there.

KESTENBAUM: That's another reason plant geneticists are fond of Arabidopsis. You can cram a lot of them in a small space and they reproduce quickly...

Even though they couldn't find Arabidopsis growing nearby, Kestenbaum managed to get the scientist out of the hotel. In the process, he created a scene of the researcher examining plants in the lobby and crouching down to peer at sidewalks in the winter darkness.

When Noah Adams wrote a piece on the eightieth anniversary of the Scopes trial, he began it in Dayton, Tennessee, where the trial took place. That's not in itself unusual. But Adams let a local farmer start his story:

ADAMS: If you drive into Dayton and go to the Courthouse Square, the first person you might talk with is O. W. Wooden. The tailgate of his truck is open, and he's selling pickled okra and jams and jellies. He has a farm in the hills outside of town. Mr. Wooden, when he was young, got to travel a bit in the service.

WOODEN: I was in Panama, see, and this Panamanian lived there, you know, and he asked me what part of the States I was from. And I told him I was from Tennessee—Dayton, Tennessee. "Oh," he said, "that's Monkey Town," he said. Yeah. And I reckon that's known all over the world everywhere I've been.

ADAMS: And the second question then, everywhere he's been, is: "Well, what do you think about the theory of evolution?"

WOODEN: They's trying to, you know, tell you that people come from monkeys, you know, and all that stuff, and it couldn't be right. Monkeys, to me, is like a chicken, and you know what a man is. It's just one of them things, and people's people.

Adams doesn't have to say something like, "Eighty years later, the debate over evolution continues"—because he's *shown* us the same thing by introducing us to this man who fervently believes "people's people."

All of the reports we've been considering are constructed around specific locations—from the Supreme Court to downtown Dayton. The most powerful radio stories are often built around scenes, but those scenes can be evoked in different ways.

Reporter Robert Smith is a master of using sound to put us in a place quickly, and to set the tone of a story. Here's how he got us into the subject of a Seattle high school that was teaching science by the "inquiry method"—letting students ask and answer their own questions:

SMITH: Here's today assignment in intro biology: test different parts of the school for the presence of bacteria. Take a guess which room every teenager wanted to check.

[*Sound of toilet flushing*]

SMITH: That's right—the boys' bathroom. Freshman Joseph McMurry is a brave scientist facing the great unknown.

MCMURRY: We're about to put this tape onto the bathroom toilet seat and take it back off, collecting the bacteria.

SMITH: He leans over the rust-stained porcelain.

MCMURRY: Now I don't know whether to touch the rim of the toilet or the actual seat.

SMITH: And he realizes too late why most real scientists wear gloves.

MCMURRY (sighs): I'm going to wash my hands as soon as I'm done with this.

This is a serious education story that looks at the value and purpose of an unconventional approach to teaching science. Stated that way, however, it would itself have come across as a lesson. Smith knew the sound of a toilet flushing would be like nothing else on the air that day—and would be sure to get a laugh. "What we do is both informative and entertaining—part of our mission is to keep people listening, and keep people engaged with the radio. And so you use sound, or a technique or a trick, to make people pay attention."

If you're going to be there to record scenes like this, you have to do plenty of homework; otherwise you're sure to arrive at the school the day *after* the students have done their experiments. So plan ahead. "I went up to West Point to do a story there," Smith says, "and I asked everything like, 'There's some sort of reveille, right? What time does that happen? *Where* does it happen? What do they do there? They fire off a cannon? Where is that? What time is that? They raise the flag? When is that going to happen? Is there some sort of bell system between classes? Do they eat lunch all together?' You start to track all of these moments when you want to be there."

Smith says it's especially important to know about "transition" moments—when a school event begins, when a factory is shut down for the day, when the shifts at the diner change, and so on. For example, he says if you're doing a story on commuting, make sure to think about where and when you'll do your interviews. "These are a lot of things that may seem tangential to the actual content of the story, but when you start to think about how you're going to put the story together, you realize, 'I really want to talk to this guy about his commute and I guess I want to do it in his car.'" Ask yourself how you'll get into the scene. Should you start recording as he enters the car, turns on the engine, and pulls out of the driveway? Do you want to have the sound of him pulling into the parking place at his job? "Any story worth its salt has an action," Smith says, "and the action has to take place somewhere, and you just think about what is that action and how can I be there?"

Tom Goldman asked that question—and included the answer in his profile of fitness guru Jack La Lanne:

GOLDMAN: Be there between 7 and 7:30 in the morning, I was told. That's when Jack La Lanne finishes his daily two-hour workout. Hitting the weights at 5:00 every morning is not bad for an 89-year-old. Still, driving to his home in the rolling hills near the central California coast, I wondered how much old age had changed him from the days of his T-V show when a young, vital Jack La Lanne was frozen in time.

[*Sound from the Jack La Lanne show*]

LA LANNE: I'm not talking about the hangover, the kind that you get from overindulgence. I'm talking about the kind you get from a lack of exercise and eating too much of the wrong foods. You know, you're hanging here and hanging here and everything's hanging. That's the kind of hangover I'm talking about, and I'm going to show you how to firm up, how to...

GOLDMAN: Trim and muscular in his trademark one-piece jumpsuit, La Lanne was the perfect balance of chirpy optimism, "you can do it," and drill sergeant discipline, counting, forever counting as he led the exercises.

[*Sound from the Jack La Lanne show*]

LA LANNE: One, two, three, four.

[*Sound of water splashing*]

GOLDMAN: When I finally meet Jack La Lanne, he is—what else?—counting his way through a morning workout. Wearing a bright red bathing cap and blue swim trunks, he's doing water resistance exercises in an outdoor Jacuzzi. The rippled weightlifter's physique is gone and there's 89-year-old sag here and there, but there is muscle definition in the arms and shoulders. When he turns away, his back is broad. And La Lanne is working as much as he ever did.

Goldman gives us two scenes here—one of La Lanne as he appeared (though of course we can only hear him) many years ago, and another as he looks today.

He also puts himself into the piece—something most reporters are reluctant to do. But there are situations and occasions where that first-person approach is an essential part of the story. When Rob Gifford took listeners along with him as he traveled across China on Route 312—the Chinese equivalent of our Route 66—he interacted with the people he met. And his interactions told NPR listeners a lot about what modern China is like:

GIFFORD: This is the former Communist heartland where the Communist Party first expanded in the 1930s. Now, though, the area is dotted with churches. In the whole of China, according to some estimates, there are as many Christians as there are Communist Party members. That's roughly 60-million people, and it's not just the old.

LOU LIUXIN: (Chinese spoken)

GIFFORD: (Chinese spoken)

GIFFORD: Thirty-year-old Lou Liuxin says she became a Christian when she was 20. "I felt a complete emptiness in my life," she says. "I was attracted to Christianity as soon as I first read the words of the Bible."

Many people in these poor, dry, hopeless villages put all their hope in Route 312 and the journey east to find work, but Lou Liuxin has found the hope that will sustain her right here, and she's not going anywhere. After a long wait, it becomes clear that the pastor is not going to show up today. It's then that the situation for the visiting foreigner takes a surprise turn.

The service was supposed to start at 10:30 and the pastor is not here.

UNIDENTIFIED WOMAN: (Chinese spoken)

GIFFORD: OK. So the people in the church are sitting, waiting for the service, and they've asked me if I will preach the sermon. This is not something I was expecting when I arrived.

UNIDENTIFIED WOMAN: (Chinese spoken)

GIFFORD: OK. Here's an 83-year-old lady who is asking me to preach the sermon.

UNIDENTIFIED WOMAN: (Chinese spoken)

GIFFORD: (Chinese spoken)

UNIDENTIFIED WOMAN: (Chinese spoken)
GIFFORD: I'm trying to explain that this isn't really something I'm used to doing...

In the end, Gifford *does* preach the sermon, just as in other pieces from this series he engages Chinese factory employees, cross-country bicycle riders, migrant workers, and many others. Still, he isn't the focus of these stories; *they* are. He is a guide. His presence helps us to understand and visualize the Chinese people he meets, none of whom speaks English.

Many other reporters put themselves in their stories—and the technique can be effective, as long as it doesn't come across as self-indulgent, or affect the truth or outcome of the story. Consider this opening of a piece by science reporter John Nielsen. The intro described it "the first of three stories on the links between transportation decisions and environmental changes" (which one editor described as "more of a threat than an intro"). But the moment Nielsen starts to speak, you realize he's in a plane, talking over the roar of the engine:

NIELSEN: At dawn, looking down from a metro traffic news plane, the Washington, D-C, freeway system looks like an urban planner's dream. Tiny little cars, evenly spaced, zoom towards the city along three main highways, and then around the city on an eight-lane beltway. Traffic seems to be light. Everything seems to be moving smoothly.
Then people get up and go to work.
TRAFFIC REPORTER: We'll hit about ten miles per hour southbound on the 270 spur. We still have a slow drive. Oooh, eastbound 66, it's an accident, gang, near Nutley Street. If you're heading to the Wilson Bridge out of Virginia, delays begin at Van Dorn Street...

Once again, the reporter has gotten us involved quickly by putting himself in the first scene. And he gave us the first-person point of view without even writing in the first person.

There are many other production devices that radio reporters can use that would not be as effective—or even possible—in other media. Here's how Don Gonyea began a piece on the Bush administration's reaction to a human rights report:

GONYEA: Top administration officials denounced a recent report from the group Amnesty International about alleged human rights abuses at the U-S military detention center at Guantanamo Bay. Listen to the list of adjectives. President Bush.

BUSH: It's absurd.
GONYEA: Defense Secretary Donald Rumsfeld.
RUMSFELD: Reprehensible.
GONYEA: General Richard Myers, chairman of the Joint Chiefs.
MYERS: Irresponsible.
GONYEA: Press secretary Scott McClellan.
McCLELLAN: Ridiculous.
GONYEA: And this from Vice President Cheney.
CHENEY: Frankly, I was offended by it.

With this series of one-second actualities, Gonyea immediately gives us a sense of how the Administration members felt. He could have said—using the language of diplomacy and politics—that they were "sharply critical." Instead, he just let us hear it.

There's no way to illustrate all the ways successful reporters can and do use sound and editing to enhance their reporting, but a few examples may suggest the range of possibilities. When Scott Horsley was doing a historical piece on Franklin D. Roosevelt's wartime acceptance of the presidential nomination in 1940, the historian he was interviewing put him on hold; Horsley worked the historian's "hold music"—the wartime hit "The Boogie Woogie Bugle Boy"—into his story. When Robert Smith was reporting on the launch of the liberal radio network Air America, we heard its star anchor, Al Franken, banging on the door as he was accidentally locked out of the building. When the future of a brain-damaged woman in Florida became a hot political issue in Washington, Andrea Seabrook and Don Gonyea interlaced their reports—handing the story back and forth—to depict what was happening simultaneously in Congress and at the White House. In a piece on Boston using ice-breakers to keep its harbor open, WBUR reporter Fred Thys graphically represented how cold it was by including the moment when his MiniDisc recorder stopped working!

It's easy to scoff at these sorts of creative approaches as mere gimmicks—as distractions from the real story—but they were all very effective. The best production devices are always at the service of the reporting. Christopher Joyce has done sound-rich pieces about coral reefs in the Philippines, about penguins in Argentina, and about a mountain wilderness in China. But he's a reporter first and foremost. "I love ambient sound," he says. "I think it's great for painting a picture. But it's not the main event. You can have fabulous sounds of lions roaring. But after four seconds, I know what lions sound like. People want story first of all."

Even a straightforward radio story can be more powerful than most newspaper articles. Though Bill Marimow spent most of his career as a print reporter and editor, when he came to NPR he quickly came to recognize what radio can do. "Hearing the human voice in all sorts of situations—whether they're emotional, poignant, humorous, somber—has an impact that far outweighs the one-dimensionality of the printed word," he says. As a case in point, consider this excerpt from Linda Wertheimer's piece on women soldiers wounded in the Iraq war.

WERTHEIMER: The explosion blew Connie Spinks out of the top of the Humvee. As she came down, her flak vest caught on the open door. She was hanging from the door of a burning vehicle, struggling and thinking of her mother.

CONNIE SPINKS: I'm not dying over here. You know, my mom is not going to have to bury her daughter from Iraq. It is not going to happen. So I was screaming. I was trying to do everything I can. That's why, like, when I was caught on that door, I wasn't going to just sit there and dangle, you know, I was trying to move, and, like, these two fingers are broken, you know? Like, I could see them, they were split open—like, I could see my bone, I could see the blood. I didn't care that, like, I was on fire. I didn't care that I could smell burning flesh. And I was trying to live.

WERTHEIMER: She got her vest un-Velcroed and fell to the ground screaming, she says, because she could not move her legs. Members of her unit pulled her to safety, and she began the journey to Brooke Army Medical Center. Her parents, who had never flown before, flew to San Antonio to see her. Her mother has stayed since October.

ANNETTE SPINKS: My husband couldn't deal with it. He had walked in the room, he turned around and walked back out. And he went to the room and he just cried. He cried and he cried. "But my daughter is hurt."

The actualities of the soldier and the details added by the reporter combine to tell a story that is as vivid as a war movie. "Someone recounting a story is just as good as being there, and often better," says science reporter David Kestenbaum. So always try to get *stories*—not just sound bites—from of the people you interview. "The odds that you're going to be there at the moment of some exciting action are really pretty small. But the odds that something interesting *did* happen to them—and it's a story they can *tell* you—the odds of that are very high."

CHAPTER 5

Field Producing

The job description of a producer covers a lot of ground in public radio. Producers do technical jobs, like mix pieces for broadcast and book studio time; they suggest story ideas and conduct research; they may arrange interviews, and even write a script or suggest questions for a host to ask. But no producing task combines as many skills as field producing—that is, going on a remote assignment with a program host or reporter. "It's an undercredited role," says *All Things Considered* host Robert Siegel. "I don't know why we don't credit field producers right after stories, and give them more due." The line between reporting and field producing is very fine indeed; a good reporter thinks like a field producer, and vice versa.

Jane Greenhalgh, who started field producing at NPR in 1989, gives a thumbnail description of what her job involves: "The field producer does research, books interviews, goes out and gathers the sound, and helps put the story together. And consequently, the field producer is involved with every single step." Except, of course, that the host or reporter will write the piece, and voice it; like other producers, Greenhalgh almost always works behind the scenes. (Producers at television networks may also write scripts for the correspondents. And even in public radio, people who call themselves "independent producers" frequently write and voice their own stories.)

Remaining in an off-air role never bothered Margaret Low Smith in the years she was working in the field for both *Morning Edition* and *All Things Considered*. "I didn't go on air because there were people I knew who were so much better at it. I felt that I could spot talent, I could work with talent, I could help make people better at what they did," says Smith. "I think one of the great pleasures of field producing was that there was enormous pride of authorship *without* having to be on the air."

Field producing also requires exceptional organizational skills. "It's like being a parent," says producer Sarah Beyer Kelly. "You have to have an answer for everything—practically including 'where's the bathroom?' It's not only the intellectual and creative process of reporting a piece, but it's all the practical stuff—of navigating an unknown city, and being as efficient as possible in a place you don't know."

NPR hosts almost always travel with producers when they're on reporting trips. And producers are often assigned to work with reporters, especially on big stories. For that reason, many producers' first experience working outside of their home base will be when there is breaking news—a war, a hurricane, a terrorism incident, or some other disaster. They may find themselves putting in twenty-hour days with the reporter or host, thinking or talking about their stories during most of their waking moments. Most of the time, reporting trips won't be so intense, but under the best of circumstances the job still demands working closely with another journalist—often for days at a time, and sometimes for weeks. Looking back on her years as a field producer, Margaret Low Smith says, "It was the most collaborative thing I had ever done or will ever do in my life—next to marriage."

Getting Started

One of the first steps many field producers will take is to help focus the story—and to make sure that the reporter's or host's impressions or insights are supported by the facts. That usually means some intense library research—and a lot of telephone calls.

At this point, you're probably not calling to book people for interviews, but to start collecting information, and registering people's points of view. Do anger management classes work? Is taking grizzlies off the endangered species list a good idea? Did the government prepare adequately for the natural disaster that struck Texas? When drug companies sponsor medical research, do they affect how the studies are conducted and publicized? Many of the people you call at this stage of the project may not turn out to be reliable sources—but that shouldn't discourage you. "One rule of thumb that I have," says Jane Greenhalgh, "is if you find *one* person to talk to about the story that you're doing, before you hang up the phone, ask them for two or three more names of people that they think you should call. And if you do that every time, you end up with a long list of contacts."

As the story begins to take shape, you should start lining up people for the reporter or host to talk to. Pre-interview them, and listen for people who have interesting things to say and novel ways to say them. When she was making calls about taking grizzlies off the endangered species list, for instance, Anne Hawke logged all of her conversations on the computer. "I always create a file where I have name, phone number, location, what their availability is. And I also track how many times I've called them: 'Called Friday; no answer. Called again Monday; left voice mail.' And I'll put a star next to things if I feel they said something really great," she says. "I talked to this one woman—a homeowner—and she said, 'I call it bear-anoia.'"

As you make calls and gather information, meet frequently with the host or reporter to explain what you're discovering. He or she will probably suggest more people for you to call, or ask you to dig up various documents, statistics, or reports. You should also be thinking of how to make your trip more than just a succession of interviews. What scenes will bring your story to life? As Matt Martinez was booking a story for *Weekend Edition Saturday* host Scott Simon—a piece on illegal immigrants who have lived for decades in the U.S., but now face deportation—he was already trying to imagine how he could work some scenes into the story. "We're going Labor Day week; there might be a barbecue," Martinez says. "There might be some family gathering. We might sit at the dinner table with them. But I won't know that until I talk to them."

Now comes the time to put together a schedule for the reporter or host you're working with, listing the interviews and events you want to record. "A lot of the planning is backtiming," Anne Hawke says, "making sure that the distances are travelable, and if one interview runs late you won't be screwed up." Hawke's entry for just one interview for her grizzly story gives a sense of how much planning is involved. (We've altered the names, phone numbers, and email addresses to protect people's privacy.)

Mike Giardina
Wyoming Game & Fish
101-101-4567 W
101-101-5678 cell
101-101-6789 H (if calling after Saturday)
mike.giardina@wgf.state.wy.xyz

grizzly/human conflict specialist—works to reduce conflicts between humans and grizzlies in the Shoshone River drainage

TUESDAY 9:00 am—can talk about managing grizzlies locally—ground level

He recommends calling Jay Richards, the statewide person in Cheyenne

Asst Chief Wildlife (on vacation this week)—101-101-7777
in charge of orchestrating Wyoming Game & Fish's position in the delisting process

2820 State Hwy 120 S of airport
Turn L out of airport, R on next road—state highway to Mafitzi (?)—down half a mile on L—Cody Regional Office

Hawke says when there isn't much time to research and book a big story, "you go out and you just cross your fingers that you'll get good talkers." But your goal when you're in the field should always be to leave as little to chance as possible. "Don't expect that you'll get there and just find somebody—because you might not," warns Jane Greenhalgh. "You have to have someone lined up that you think is pretty good and pretty compelling. But you might find someone better."

Remember, too, that you're booking every aspect of the trip, not just an interview extravaganza. Know where you're going to stay at each stop. Have maps to help you get from one interview to the next, and try to get directions from local contacts. (They'll know if the interstate exit has been closed or if the drug store you're relying on as a landmark has recently been replaced by a coffee shop.) If you're traveling with an engineer, make sure he or she knows what the recording conditions will be like. You might even want to check on convenient places to eat. In one of her old notebooks, Margaret Low Smith still has the notes she made while planning a trip to Alabama with host Robert Siegel for what became a prizewinning report on parole. "Here I have just the names," she says as she leafs through the notebook. "'Todd Clear'—he was the person who had done the study about crime and redemption and referred me to this parole board.... I've got 'Mr. Trotter'—I think he ran the parole board. I've got 'YMCA's Barracudas'—that was finding where I could swim when I was in Alabama! 'Brent Smith, University of Alabama, Birmingham; Nancy Kahn McCreary, Attorney General's office.' Restaurants—I've got lists of restaurants," she says. "I've got lists of the people who we were going to hear. 'Andrew McDaniel—kidnapped woman at gunpoint. Forced sex. 6th parole hearing.' I've got, 'Barry Davis—robbery, three times convicted 1972 in St. Clair. Danny Odum, murder, assault—convicted. Louis Downs'—who ended up being the star of our story—'Convicted. Murdered neighbor who had a judgment against him.' In any case, what I have here is the list of everybody I spoke to before I got there—all of their phone numbers."

Once you've got a plan for the trip clearly laid out, meet at least one more time with the person doing the reporting to make sure you haven't forgotten anything. Julia Buckley says she puts everything together in a small binder when she's working with a program host on a story. "They don't need to be overloaded with huge amounts of information, but you put together a briefing book saying where you're going, who you're talking to, why you're talking to them—interesting points about that, some clips that might be helpful. But not too many." And if you're traveling with an engineer, Matt Martinez says, give him or her the same materials you provide the host. "When one person falls out of the loop, there's a kind of momentum that's lost in the field. When everybody's on the same page and knows exactly what we need for the story, then they're actually able to get those things."

In the Field

Art Silverman, who has been producing stories in the field with hosts and reporters for more than twenty-five years, says he always has to play least three roles—"travel agent, bookkeeper, and creative person." But he says he's learned from experience that one of the most important challenges he'll face is time management. "Have I left time for the host to feel—assuming location's important—that they're actually *there*? Have they been able to look around? What you get on tape is the real prize," he says. "But hosts are reporters when they're in the field. They have to see the place, and have time to write notes, and make observations. You can't just helicopter in, and rush them, or pile up too many interviews." Matt Martinez agrees that the biggest mistake new field producers make is underestimating how much work goes into each trip. They know hosts often do five or six interviews a day when they're at home, and assume they can do as many when they're out on assignment. That's a mistake. "Fieldwork is exhausting," Martinez says. "More than likely, you'll have just flown in and you're off and running. You can't put five interviews in a day, and include lunch and dinner and breakfast."

Here's Martinez's schedule for the first day he and Scott Simon spent on a political campaign story in New York. (Again, we've changed a few things to protect people's privacy.)

TUESDAY—September 28 —

SS Delta Airlines Flight# 1946 Class:Y
From: Washington Natl DC, U 28Sep04 09:30am Tuesday
To: New York La Guardia N 28Sep04 10:30am Tuesday

CAR: Pick Up City: New York City Area NY, USA
Hertz Rent A Car Type: Inter Car Auto A/c
Confirmation#: C6893117190 * Rate: 74.40USD
Drop Off: 29Sep Wednesday

Rate Info: USD74.40Day-Ulmtd MI Xtra Hr 47.00
RF-ER374821
Pickup: NYCC15
Arrival Time: 11:00am
Dropoff Time: 06:00pm

222 EAST 40TH STR NEW YORK PHONE 101-101-5060

HOTEL: Hotel Hudson Phone: 101-101-6000
356 WEST 58 STREET, NEW YORK NY 10019, US Fax: 101-101-6001
Number of Rooms: 1 Room Guaranteed
SS Confirmation#: 3088757846
MM Confirmation#: 3098162260 *
Check Out: 29Sep Wednesday (1 Night)

Morning/Afternoon (time not confirmed)—Rick Ulaby, Democratic Strategist—101-101-4567—if not late breakfast, early lunch?

4:15p—John Edwards Campaign Rally in Newark, Robert Treet Hotel
50 Park Place, Newark, NJ 07102

AJ Dyson—NJ Dem Party, call day of if interested in interviews with state dems—Ginny Herrick, state chair of Dem. Party and Sen. Jon Corzine are available. 101-101-5678.

*****RAIN IS FORECAST!*****

6–9:30p—Rutt's Hut, 417 River Road, Clifton, New Jersey—Classic Car Night—Lots of folks come out from around the area to this classic hot dog stand. There'll be classic cars on display Tuesday night and Gary, one of the owners, says it's fine with him if we come out and talk with voters.

Look at all the things that are included here. In addition to the same sorts of notes you'd make if you were going on vacation (flight number, hotel location, rental car company), Martinez has listed the people Simon is scheduled to talk to, some alternative interviewees, a possible scene where they can meet some voters—even a weather forecast.

Politicians are accustomed to being interviewed, but it will be a new experience for many of the people you'll meet when you're on a remote assignment. And you'll be talking to them in their own homes, or factories, or schools—places where they are not used to dealing with journalists. To put people at ease, many field producers make a point of letting the interviewees know what the experience will be like. Martinez says, "People feel ambushed when you just show up—three people—and the sound engineer has all this stuff hanging all over him and sticks a microphone right in your face. It's a little threatening." Art Silverman also tries to leave time for the guests to become comfortable with the idea of being recorded. "Most people are not going to say their best stuff if you run in, rushing, and set it up," he says. "You want the process to leave time for the subject to forget about your entrance, forget about your exit, maybe if you're lucky forget about the microphone, maybe forget about the producer." Build that extra time into your schedule.

And then, after the interview starts, "stand back," Julia Buckley advises. Once the host asks the first question, she says, "I am there for support, and to facilitate, but I am not seen. I just put the host and guest together, and they have their conversation." Buckley even sits *behind* the guest, so she will not be a distraction. If someone else is holding the mic, you should take notes as the interview proceeds to make sure the reporter or host asks all the key questions, and that the guest touches on the same subjects or anecdotes you elicited in your pre-interview; even if you can't write these things down, you should remember them. "I'm listening to make sure that we have usable tape," producer Alice Winkler says. "Sometimes the host, who's in the middle of a conversation, can't hear that. So I may say, 'Let's go back and try this again,' because I've talked to the person on the phone and know they've got a great story, but they didn't really tell it in a way we can use it."

You have to give these sorts of directions subtly when you're in the field. Winkler sometimes just hands the host a note with a few words on it—"Try the first question again." Buckley says she'll just make a gesture or change her expression to encourage the host to follow up on something the guest says. Veteran producer Barry Gordemer says he uses a tap on the shoulder to get the reporter's attention.

Finding Sound and Scenes

The whole point of going *out* on a story is to capture a sense of place. So think in advance about ways to make the interviews you've booked *sound* different from the ones we usually hear. You might want to ask the guest to give the reporter a tour of the lab, or to describe some photographs in a family album, or to point out where he made the arrest. "It's most interesting talking to people while they're *doing* something," Barry Gordemer says. "There might be something about his activity that clues you in to his location." Gordemer says one of the best actualities you can get in the field is someone saying, in effect, "Now look at that. Did you see *that*?" He managed to pull that off when he was producing a piece with former *Morning Edition* host Bob Edwards. "I did a story about a guy who was experimenting with different herbs, and tree barks, for medicines. And he took a leaf and he broke it off and said, 'Now smell this!' And Bob just went, 'Whew!'"

That demonstration was planned, but Gordemer says you often have to improvise to turn a straightforward interview into one with a sense of place. "When I'm in the field, I'm listening to the interview with one ear, but I'm also looking around the room, or the field—wherever I am," Gordemer says. That paid off when he was with Bob Edwards for an interview with country singer Glen Campbell at his home in Phoenix. "I noticed a picture of Glen Campbell and Johnny Cash above the wall," he says. "And I tapped Bob, and alerted him to that. And during the interview Bob asked, 'Now there's a photograph of you and Johnny Cash, two sons of sharecroppers'—I didn't even know he knew that!—'who found music as a way of getting out of poverty.' It told people where we were."

In booking your trip, you will have planned for various scenes—the immigrant family's barbecue in Phoenix, the tourist outing to spot grizzlies in Yellowstone, a hearing of the Alabama patrol board, Classic Car Night in Clifton, New Jersey—but you'll also want to be looking for great scenes you *hadn't* planned on. Anne Hawke was producing a political story on the "swing state" of Ohio, when she and correspondent Don Gonyea—en route to an interview—saw a sign along the side of the road saying, "AFL-CIO—Take Back America." "We looked at each other and did a U-turn," she recalls. "Don said, 'Call our three o'clock interview.' And I did, and said, 'Sorry—we're not coming.' And instead we went to this AFL-CIO convention, where we walked in, and they were selling these political buttons, and there were all these union members, and we got just great talkers and great sound and a good scene. It ended up being a big portion of a piece that we hadn't planned for."

So keep your eyes—and your ears—open. "While I'm paying attention to what's going on in the conversation," Alice Winkler says, "I'm also able to keep a broader view of what's going on to make sure that we have sound that might be meaningful to the piece." And record *everything*. Chances are you won't be able to go back to the same places a second time, so gather as much sound as possible. "It doesn't mean you have to use it," Winkler says. "If there's a great gong sound, but later it seems like a cliché, you don't use it. Sometimes I'll record the car door opening, shutting, the car starting—you never know when you need to get from one scene to another, and maybe that little bit of sound will help you."

Also, since you don't know how your sound will be woven into the final piece, remember to record at different distances, and in various settings. "I like to think of getting a close-up shot of the sound, and a pullback shot of the sound," Jane Greenhalgh says. "So, if you're in a village and people are washing pots, you have close-up shots of dishes clanking—and then you pull back and you have a far-away shot, so you can use that in production." Margaret Low Smith said she always imagined she and the host were "creating a movie on the radio."

Even if you are working with an audio engineer, keep in mind that you are responsible for making sure the right sound is recorded at the right time. "There are engineers who are as much producers as they are engineers," Matt Martinez says. "But with some you have to say, 'Put the mic *there*. We need this, and this and this.'" And directing the production may mean more than just telling the engineer to record the children playing stickball or the sound of the sluice gate as it opens. "You are the on-site manager," Julia Buckley says. "You need to be assertive, and say things like, 'I need you to wait for that airplane to go by.' Or 'I need to stop down [i.e., stop recording] in the middle of that interview. I was having trouble hearing the question because of the airplane, or because a refrigerator was running.' Or 'I need one minute of silence.'"

What you end up doing on a remote assignment will depend on the nature of the story and on the personality and background of the reporter you're paired with. "With a reporter who wants to be in charge of all the technical things," Anne Hawke says, "your job mostly involves generating ideas—'Why don't we talk to this person in this setting?' or 'Why don't we get sound here?'" Jane Greenhalgh has worked for years with science reporters, who have their own areas of expertise. Then, she says, "your role becomes less worrying about the facts of the story and much more, 'How can we tell this story in a compelling way? How can we take these scientific facts and make them make sense to our listeners and make our listeners interested?'"

On the other hand, there will be some occasions when you're called on to record news conferences, conduct interviews, or do any of the other tasks usually left to reporters and correspondents. So be prepared—by having the right equipment and the right attitude. When Anne Hawke went to Sri Lanka to produce stories about a tsunami recovery effort, she arrived days before either of the NPR reporters she was supposed to work with. She ended up hiring a translator, attending an emergency briefing at the Presidential Palace, and going to the airport where the Red Cross was distributing goods. "I could have thought, 'Well, I'm a producer, I'll wait till I have something to produce.' But I figured, 'There's stuff happening. We need to go get it, and it needs to go on the air.'"

Experienced producers offer a raft of other tips for reporting in the field:

FIND A FRIEND WHO CAN HELP YOU WHEN YOU'RE ON SITE. As a field producer, Sara Sarasohn always wanted a local "fixer." "Maybe there's a local newspaper editor who is connected to a bunch of people," she says. "Find a friend you can call up and ask things like, 'Who's the person who knows about X?' or 'Where do you think we should stay?' Find someone who can help you out with all these little things."

BE PREPARED TO RECORD FROM THE MINUTE YOU GET UP IN THE MORNING UNTIL YOU GO TO BED AT NIGHT. Put batteries in your equipment, lay out your mic and cables and notebooks, load your MiniDisc or flash recorder, and so on. "Reporters do not take kindly to your not being prepared," Anne Hawke says.

HAVE BACKUP PHONE NUMBERS FOR EVERYONE YOU THINK YOU MIGHT WANT TO INTERVIEW—a work number, a cell phone number, a home number, and an email address. "Especially a home phone number," Jane Greenhalgh says, "because you might get there and you need to change something around. You need to make sure that you can find your people when you are in the field."

ALWAYS RECORD EXTRA AMBIENCE, at every scene, after every interview, wherever you may need it. You probably won't be coming back. You'll never regret that extra minute or two it takes.

MAKE SURE YOU GET PHYSICAL DESCRIPTIONS OF PEOPLE AND PLACES. Hawke says, "I tell reporters, 'Stop where you are, put a track mark in, and say, "Here's a description"—and just talk.'" Or use a digital camera to record what things look like.

"SLATE" ANY SOUND THAT MAY BE DIFFICULT TO IDENTIFY LATER ON. If you're recording various pieces of equipment in an industrial lab, just say, "This is the sound of the magnetic stirrer" or "This is the analog centrifuge." You may also want to add the day and time, or any other information that could be useful later: "It's 5:20 p.m. on Thursday, March 15th, and we're outside the Macon County Courthouse. The crowd sound is being made by about three dozen people—half proponents and half opponents—who are behind police barricades. Ten officers are present."

MAKE SURE YOUR RECORDED AUDIO IS AIRWORTHY BEFORE YOU LEAVE A LOCATION. "If you're curious about the quality of your tape," Julia Buckley suggests, "get in your car, start the engine and play it there. If you can't understand it in the car, it's not tape that should be played on the air."

LEAVE ROOM IN THE SCHEDULE FOR SERENDIPITY, and for following up leads you have discovered during your reporting. Sara Sarasohn says she never books up her last day in the field. "I may have some idea what I want to do that day, but I don't have a lot of 'must do's,'" she says.

Finally: "Don't wear noisy shoes, because when you're walking around it'll make noise," Sarasohn warns. "Really."

Putting the Story Together

If mixing a radio piece can be likened to following a recipe, producing one in the field is like putting together a gourmet meal—and you're doing it entirely from scratch. You're the one who has to create the structure of the story out of the raw interviews and audio you've collected. The level of responsibility is many times greater than most work in the studio. But so are the rewards.

Many field assignments will result in your recording hours and hours of audio. So you want to think about ways to identify the actualities and sound you want to keep. A lot of that can be done either as the interviews are taking place, or immediately afterward. If you're running the equipment yourself, for instance, make track marks on the MiniDisc or flash memory card to remind yourself which parts of the interview were most interesting. "The fewer track marks the better," Anne Hawke suggests, especially when you're on deadline. And as soon as the interview is over, take a moment to talk to the reporter or host about the parts you thought were strongest. "I will always try to have a conversation away from the interviewee," Sara Sarasohn says, "and ask the host, 'What were the three best things here?' And I'll just jot them down. When you get back to your desk and you've got all this tape and don't really know what to do with it, it really helps you to focus." Hawke says she actually records her conversation with the reporter along with the interview. "It's always hard to remember otherwise. It's like the little hard drive in your head that holds the interview somehow gets erased when you do the next interview!"

While you're out recording interviews and gathering sound, you and the reporter or host may have discussed scenes that you think should be kept in or left out. So you may already have the skeleton of the story's structure in mind. Years ago, when she was working with Daniel Zwerdling, producer Tracy Wahl discovered he had the remarkable ability to "see" the piece based just on her pre-interviews. "On the way there, I'd debrief him about what I'd found out. I get really excited when I talk, and he could take those moments of excitement and channel them into a very logical structure. He'd just say, 'Okay, here's what I think we should do. Here's the first scene, here's the second scene, here's the third scene . . .'" More often, there will be some give and take between the producer and the reporter about how to use all the material they've accumulated. Jane Greenhalgh says she tries to make sure they agree on a basic structure before they head home. "I sketch out the story, how I think we're going to tell the story—working with a piece of paper—and I'll lay out the whole piece in scenes to see if we've got any gaps. Because if you've got a gap, and you're still there, you can fill it in—but it's much more difficult, obviously, if you get back and you have a gap."

Depending on your situation—whether you're going to be filing your piece from the field or assembling it back in your office—you may want to start isolating the best actualities and sound clips at the end of each day. "I usually listen to the tape in the hotel room," Matt Martinez says. "I know what I want, and I just go get it after we're done."

Some reporters and hosts will expect to have transcripts for all of the actualities you select. Sarah Beyer Kelly says she finds transcribing the cuts helps her later, when she's trying to arrange them in a logical order. "I'm kind of a visual person, so I put them down and move them around—I actually do this with the transcription." Veteran producers say you should resist the temptation to "log," or transcribe audio you're not likely to use—even if the journalist you're working with encourages it. "I remember coming back from stories with ten hours of tape, fifteen hours of tape," Margaret Low Smith recalls. "If I had spent the time logging the tape, my head would have spun!" But she would give the host notes from interviews that weren't going to yield any actualities—factual information that might help flesh out the story.

Most important, Smith says, is to home in on actualities that will be riveting. If you've done your job right in the field, it shouldn't be hard to find. You'll want to avoid any cuts that are plodding, confusing, or obscure. "If there's something that needs to be explained, probably the reporter or the host can do it better. The tape has to be anecdotal, interesting, and engaging. Every cut should have 'tease tape' quality."

Almost all producers select more actualities than they expect to use in the final piece. If you can, it's best to sit with the reporter or host and explain why you think the structure you've come up with makes sense. Remember, your story is going to start with the intro, not with the first interview clip or first track; it will also need to have a conclusion, not just a sign-off pasted onto the last actuality. You may want to include a moment where we hear the reporter ask a question or respond to what a guest has said; be prepared to explain why you think that adds drama or texture to the story, or improves its pacing. Many field producers provide their colleagues with an outline that includes the location where each actuality was recorded and the name and title of the person who is speaking. Some go further and suggest a sentence to lead into each cut; they may include an explanation of why one scene transitions logically to the next, and write down facts that they've found through their own research and reporting.

Finally—and this can happen the same day the interviews are completed or weeks later—the reporter or host writes the script. Don't be surprised if he or she doesn't stick to your outline. "Sometimes Scott [Simon] will completely rearrange the piece," Matt Martinez says. "He did that when we reported a piece on New Jersey Democrats being threatened by the Republican machine for the first time in God knows how long. I'd set up the piece a certain way, and he completely changed it. He moved two scenes around, and I was really surprised. But it made it better." When the script is finished, run through it, and all the audio, with the reporter, to make sure you've covered everything you wanted to, and included all the voices and sound and ideas you discussed when you were on the assignment. "It's like the first edit," Jane Greenhalgh says. "The reporter will read it while playing the tape, I'll time it, and then we will both have our initial response to it—it's too long, this bit's dragging, whatever. We try to work out those initial obvious problems before we take it to the editor."

But your job is still not over. While the editor will be listening for logical inconsistencies, or confusing writing, or dull audio, he or she won't know the names of all the people you spoke to, or be able to verify the dozens of other facts in your piece. When you produce in the field, that's *your* responsibility. Lots of fact checking can be done simply by comparing what's in the script with what you have recorded. (For example, the reporter writes that Mrs. Reynolds has three children. But did she actually say that?) Many other facts require more energetic work on your part. When she was producing stories, Margaret Low Smith would underline every fact in the host's script and make calls to check every statement—"age, race, place, time, date, religion," she says. A lot of the time, she and the host would remember the same detail—like how long it took to drive from Birmingham to Montgomery—but she'd still confirm it. "By the time the words were recorded, I knew it was perfect."

Filing from the Field

Many field producers work furiously when they are on assignment, and then bring back all of their audio to assemble and mix their story in relative tranquility. But you should also know how to file your report from a remote location, since the most routine assignment can change suddenly when news breaks.[1] "I think of it sort of as a matrix," says Charlie Mayer, who has produced stories across the United States and abroad. "If you're

1. In September 2001, Matt Martinez and Robert Siegel went to New York to do several stories—a feature on the underground economy of illegal immigrants, a walking tour of the Five Points neighborhood, and a piece on people who commute to the city from Connecticut. But then terrorists attacked the World Trade Center, and Martinez and Siegel spent the week covering the biggest story of the careers.

by yourself, and you're bringing the tape home, that's the easiest kind of thing you could be doing. But if you are traveling and working with somebody else—a producer, a reporter, a host, an engineer—and you're filing from the field, that's the most complicated combination that you could have." For that reason, Mayer and other field producers say there are a few things you should always keep in mind, even if you file stories from the road only occasionally.

KNOW ALL OF YOUR OPTIONS FOR FILING. If you're working alongside other reporters and producers—at a political convention, for example, or at a big-time sporting event—there will probably be a filing center. If there is, find out in advance whether you'll have to send your piece during a "filing window"—a set amount of time when you'll have access to ISDN lines, the Internet, etc.

Some reporters and producers travel with satellite phones, which can allow them to send and receive real-time audio. On a breaking story when time is of the essence, or during a natural disaster when you may not be able to find a reliable source of power, a sat phone can be a lifesaver. "All sat phones are different," says Mayer, "but basically you point it at the sky, and when it sees a satellite it lights up and you can see the amount of signal strength you have, and then you go through a dialing process. And it's either connected or not connected."

For many reporters, filing from the field these days means uploading audio files to a server using FTP (file transfer protocol). To file by FTP, you may need to use a special program on your laptop (there are dozens available, and most are free), or access your news organization's Web site. The process is only a little more complicated than downloading a song or video, but you'll need to find out whether the place you're filing from has wireless Internet access—and if it doesn't, how else you can send your data.

If your computer has an EVDO[2] card—which gives you a broadband connection through a cell phone company's network—you may be able to file without going back to a hotel or searching for an alternative wireless hotspot. Determine ahead of time whether the place you'll be doing

2. "EVDO" is an acronym for "Evolution Data Optimized" or "Evolution Data Only"—neither of which is very meaningful to the layman.

your reporting has wireless coverage; you should be able to find a map online. "If you're going to a city, the answer is 'probably,'" says Mayer. "If you're going out in the country, the answer is 'maybe'—but it might not be very fast." Mayer offers one tip: If you are trying to use an EVDO card and can't reach the Internet, try driving to the on-ramp of the nearest interstate. "The cell phone companies have been pretty good about wiring up the highways. You can even file in motion if you're on an interstate," he says.

Finally, check that you can get through on a cell phone, or at least get to a landline. You may be able to play your sound over the phone by cupping your headphones over the mic of the phone's handset, though there's no guarantee the result will be airworthy. A reporter can always do a live interview—but only if the phone line is adequate and the connection is reliable.

KNOW THE TECHNICAL REQUIREMENTS OF YOUR NEWS ORGANIZATION. If you're filing by FTP, you'll need to ascertain the IP address of the server (given as a series of numbers separated by periods—e.g., 123.456.78.90), and whether it demands a specific user name and password.

Ask in advance what file format is preferred. Compressed files are smaller and much quicker to send, but they sacrifice some audio quality. If you send your acts and tracks as MP3's but WAV files are required, you'll have to refile—and may miss your deadline.

Also, make sure you've confirmed the directory, or folder, where you should upload your files. More than one story at NPR has not gotten on the air because the elements had been put in the wrong place.

TEST ALL YOUR EQUIPMENT. Evie Stone has produced stories in the field with more than twenty reporters—and despite all that experience, she has had a satellite phone stop working because a ninety-nine-cent cable had failed. She points out that when you're filing from the field, it's often because you're covering breaking news—a situation that is almost always stressful. Her advice: "Do everything that you're going to do later under duress when you're *not* under duress." If you plan on using a sat phone, she says, "go out on the front lawn, and dial Master Control, or whomever you're filing to. Make sure you can talk down the line and send tape down the line. Make sure all the cables are working, and that the sat phone has batteries and it's charged." If you have a power adapter that allows you to run your equipment off your car battery, test it out.

The same advice holds if you plan to file by FTP. Give it a try *before* you head out on your assignment. And if you think there's any possibility you'll need to download files as well as upload them, make sure you know how to do that as well.

DO A "PRACTICE FILE" FROM THE SITE. This is critical if you'll be filing live by sat phone. For example, while a reporter is in a courtroom waiting for a verdict to be handed down, you can be going through a dry run, Evie Stone says. "You need to know that you can get the connection—that you can get a strong enough signal, that you know where to point the dish—so that when the reporter comes running out of the courthouse and you're about to go live in five minutes, you're not saying, 'I wonder whether it's this way' or 'Oh, there's a building in the way!'" Stone says you should identify the precise location you where you plan to stand—"and perhaps you should even mark it on the ground with a little piece of tape, like producers do in TV for a live shot," she says. "Sometimes you're in some plaza, and there's only one place where there's a space between two buildings that happens to be where you can reach the satellite. When the moment comes, you have to be able to be cool, calm, and collected."

Test your cell phone from the same location. Even if you won't need to file your report by phone, you will probably need to communicate with a program producer or editor, and you don't want to discover at the last moment that there's no reception.

HAVE SEVERAL BACKUP PLANS. Plan for the worst, since eventually—no matter how carefully you prepare—you'll experience it. "No thing is too inconsequential or too mundane to go wrong," says Stone.

Just because you've been told there will be a filing center doesn't mean it will be appropriate for editing audio—much less for recording a reporter's voice tracks. "At a political event, sometimes there's blaring music," says Stone. "Then you may have to pick your cuts sitting on the floor of the bathroom. That's gross, but sometimes it has to be done!"

If you plan on sending your audio via FTP from your hotel, think about what you'll do if its Internet connection is sporadic—or non-existent. "I was in Miami with a reporter and we were in one of these great Art Deco hotels," says Stone, "and the hotel wireless was just not happening. So we did everything we could—we tracked the story and we had it all set up in [the digital audio program] CoolEdit—and we drove to a Starbucks, and parked in front, and filed from the car." Have a list of other places that provide wireless access to the Internet—office supply stores, public libraries, etc. And don't forget to bring an Ethernet cable along, just in case wireless isn't an option when you need it.

If your computer fails, or its battery dies and you can't get to a source of power, you may need to file by phone. Know where you can get a reliable cell phone connection. If the reception is spotty, locate a landline, and figure out in advance how long it will take you to get to it.

In general, you should have multiple backup strategies, to ensure that your story will get on the air one way or another.

IF YOU'RE FILING ELEMENTS FOR A STORY THAT WILL BE MIXED BY SOMEONE ELSE, GIVE PRECISE DIRECTIONS ON HOW YOU THINK IT SHOULD SOUND. "Filing from the field is very different from producing something of your own because it requires giving up a certain amount of control," says Evie Stone. You may work all day and night and upload your audio files at midnight for someone else to retrieve and mix the next day. So if you want the piece that airs to resemble the one you have constructed, leave as little to chance as possible. If the reporter has done several takes of the same track, delete all but the one you want to go on the air. If ambience needs to be mixed in a particular way, mark the script accordingly. "You want to be as clear as possible with your mixing instructions," Stone says, "down to which word you should start to hear the ambience at, where you want it faded out, that it should post [be brought up] for so many seconds, or for this phrase that the person is saying." If you can, find out who will be mixing your story—and call that person to make sure he or she has all your elements, and so you can answer any questions.

MAKE A LIST OF ALL OF THE EQUIPMENT YOU'RE TAKING ON YOUR TRIP. For one thing, it can serve as a final checklist, to help you pack everything you need. If you include the make and model number of each item, the list will also save you time if you lose something, or if a cable or connector breaks while you're on assignment. And it can be especially valuable if you go overseas. "Every time I travel abroad with a lot of gear," Charlie Mayer says, "I take the gear and the list down to the local customs office and I have them stamp it, so that the customs people on my return don't ask me if I bought all this stuff abroad and try to charge me duty on it." You can obtain a certificate of registration for this purpose from the Customs Service's Web site.

Taking Responsibility

Some of the things that go wrong when you're working in the field are more serious than others. You can overbook a trip, for example, or accidentally leave something off the agenda. "The first gig I went out on, with Robert Siegel, I forgot to include time for lunch," Matt Martinez recalls. Even when you've done everything right, there are always circumstances beyond your control, so you need to be ready for the unexpected. "You always have to have Plan A, B, and C," says Julia Buckley, "because people don't show up, or your car breaks down, or your guest doesn't want to talk to you—or for whatever reason, stories change. So you have to be able to plug in new people, or say, 'I foresaw that this was going to happen, and here's how we deal with that.'"

When Jane Greenhalgh was producing a story in Boston on diagnostic genetic testing, a critical interviewee seemed to have "disappeared." Greenhalgh had done a lot of advance work just to find a couple who would talk about their child's cystic fibrosis, and when she arrived at the airport she called them to say she and the reporter were on their way. But they weren't at home—and she had neglected to get a cell phone number. She finally tracked them down—in the emergency room. "At that point, as a producer—and this was uncomfortable for me, too, because I'm a mother as well—I had to ask them to still do the interview, but to do the interview at the hospital." It wasn't easy to make that request of parents who were already distraught, but Greenhalgh says changing plans ironically became "one of those sad, radio gifts." "It ended up being a much better scene for us," she says, "because we were interviewing these parents at the hospital, and they allowed us to go into the room, and talk to the little girl a bit, and we were able to get the doctors." On top of everything else a producer does in the field, she says, "You have to sometimes do some of the unpleasant work."

In short, a good field producer accepts praise when things turn out well, and responsibility when they don't. "Don't blame anybody," Margaret Low Smith says. "Don't blame the engineer, don't blame the reporter—don't blame *anybody*. Whether the hotel doesn't have a room, or the interview was lousy, just take ownership and solve it." The host or reporter is concerned about doing the interviews and making sure the story stands up, she says. Don't make him or her fret about logistics, or tape quality, or whether the next interviewee knows you're on the way. That's your responsibility, Smith says. "It's your job to worry."

Remember, even though a reporter or host will get to sign off at the end of the piece, it is your story, too. No matter who comes up with the original idea or who does the interviews, the reporting and production process is truly a joint effort.

CHAPTER 6

Story Editing

For some reporters and field producers, editors make an easy target—and it's almost always open season. If a story seems convoluted and poorly structured, they can blame the editor for insisting on adding material that wasn't in the first draft. If the piece is missing some key facts, the editor can take the blame for cutting them out or insisting the report be kept to a certain length. Sentences are too long? It must be the editor's fault. Reported the sidebar but somehow missed the real story? That was the assignment the editor made. And when editors do really extraordinary work, it often goes unrecognized. There are Pulitzer prizes for reporting, photography, commentary, and even cartooning—but not for editing.

In public radio, a single editor often does the jobs of several different editors at newspapers—he or she makes assignments, suggests sources, raises questions, helps structure the story, and revises copy. At the same time, a radio editor plays some roles that are unique to the medium, including coaching the reporter's delivery and writing story summaries for the DACS that goes out to stations.[1] A successful editor has to help the reporter see the big picture, but also needs to fret over details—and do it all, with few exceptions, without being credited on the air or on the Web.

The Editor's Role

Above all, public radio editors are responsible for making sure that reports are accurate and fair. This is not merely an ideal; it must be the

1. The DACS [pronounced "daks"] is the electronic system for distributing information across the public radio system. Among the many DACS messages are the NPR program lineups, which include summaries of all the pieces slated to air in upcoming shows.

foundation of every report that is broadcast, podcast, or posted online. Inaccuracies and bias reflect badly on the reporter, the editor, and the entire network. They break the bond of trust between journalist and audience.

As an editor, you're also responsible for ensuring that any given report is well-structured—that it has a beginning, a middle, and an end. The story should be focused; it must be clear from the start what it is about. And when the piece is over, it should be easy for the listener to recall key scenes and ideas.

Whether a story is memorable will depend in large part on the quality of the writing. A good radio report is visual and descriptive. It engages the listeners, and keeps them engaged throughout the piece. It neither presumes that listeners have extensive knowledge of the subject nor talks down to them. And it's written for radio—in short, clear, strong sentences that avoid jargon and journalese.

Public radio places a premium on sound. News reports must be well-delivered. If they use actualities and ambience, those elements should advance the story, not merely add production for its own sake. And while editors are not primarily voice coaches, they should recognize that their work is wasted if a piece is unlistenable, either because of the way the reporter reads or because of the poor quality of the audio.

The Ingredients of a Story

Ideally, the editor and reporter should collaborate from the very start—before the reporting has begun. One of the editor's key jobs is to help the reporter (and in some cases, the reporter and producer) focus the story; a reporter-editor conversation early on can save time when it comes to doing interviews and tracking down facts.

One way you can start that process, especially with less-experienced journalists, is to get the reporter to give you (or even write out) a "focus statement"—a one-sentence description of what the proposed story will be about. A good focus statement suggests the tension inherent in the story: "My story is about how teenagers with disabilities make the transition from public schools to private universities, and about the people and institutions that help them, or fail to help them, in their first year at college." Or "My story is about whether government-sponsored radio services, like the Voice of America and Radio Free Europe, reach their target audiences now that there are commercial satellite broadcasts, Internet access, and independent news media in many formerly closed societies." Note that a useful focus statement doesn't assume the *conclusion* of the reporting, though it may hint at a direction for the reporting to start; in these examples, the reporter doesn't presume that private schools are failing to help disabled freshmen make successful transitions to college, or that international government-sponsored broadcasts are a waste of money now that there are other media alternatives. The statement should be value-neutral.

As an editor, you should also assess whether the story is interesting enough to pursue. You should ask what is new about the proposed story, whether it is (or can be made to appear) relevant to people in other parts of the country or the world, and whether there's a news "hook" that makes the story worth doing now. That means you should be familiar with the beats and regions of the reporters you work with. As editor Ted Clark puts it, "The Victorian ideal of knowing something about everything and everything about something also defines the editor-reporter relationship." The editor needs to be able to determine whether news is incremental; whether it relates to events in other places; or whether a local story is simply so compelling that it should be heard by a national audience. Steve Drummond has fielded education stories from all over the country, and says local reporters sometimes don't have the same perspective he has. "What I see often are trends among cities, or among school districts," he says, "where the same problem that arises in the Cleveland public schools is right now being dealt with in San Francisco; and the thing that makes a story not just a Cleveland story is to say—somewhere in the story—this is also happening in San Francisco and Seattle."

When Alison Richards is editing science stories, she says the "so what?" factor poses a big challenge. "For most of our stories, it's going to be several years down the line before they make on impact on people's lives," she says. "So the editor's job is often to think about how you tell the story, or even whether you use the scientific discovery itself as a kind of springboard for moving off to explore different, bigger issues raised by an area of research." In short, one of your roles as an editor is to offer a perspective on the news that may be difficult for a reporter or field producer to achieve.

Part of the discussion of a story's focus may include whether the piece will be structured "vertically" or "horizontally." Horizontal stories often take a high and wide view of a news event, while vertical stories go deeply into a single idea or episode. Similar stories can be told in different ways. For instance, if a reporter wants to report on scientific advances in cloning animals, he may interview policy makers, ethicists, scientists, and businesspeople, so that different people who have a stake in cloning each tell a bit of the story. This "horizontal" story may also recount the history of cloning advances, told in chronological order. On the other hand, the reporter could focus on one scientist in one lab—on the moment when she painstakingly removes the genetic material from a cow's egg and replaces it with the DNA from another cow's skin cell—and use that scene as a doorway for information about why cloning is so difficult. That story would be structured "vertically." It takes a narrow view, but still has depth.

Once the editor and reporter agree on the focus of a story, they should make sure they have thought through the other essential elements. Almost every good story, from a children's fairy tale to an investigative piece for the *New York Times*, has a few basic ingredients. It has characters; it is set in some specific place; it has a clear beginning, middle, and end; and there is some sort of tension that makes the listener or reader eager to find out how things are resolved. Radio stories are no exception.

For instance, when a reporter is pitching a story, you may want to ask, "Who do you imagine will be the main characters in your piece?" This is the time when you can prod the reporter to think creatively. Can we try to do this story about U.S. policy *without* talking to any of the usual suspects at Washington think tanks? In this piece about school resegregation, did any of the teachers or other workers now at the school attend it *before* it was integrated? Former NPR editor Bebe Crouse says she always tried to ensure that each interview would help advance the story. "With some reporters, I'd be very specific about asking, 'Who are you going to talk to and what do you want from them?' Not 'What's the quote you want?' but 'What's the focus of your interview with them?' And that's always going to keep coming back to the focus of the story." It may turn out that the best "character" isn't even a person. If the proposed story is about limiting beach erosion in New England, the main character could be a lighthouse. If the piece describes new advances in photographic restoration technology, the main character could be a photograph—stained, torn, and faded at the start of the story, but revived and eye-catching at the end.

The reporter and editor should also discuss *where* the story should be reported. Sometimes the location is obvious—e.g., the story is about the New Jersey foster care system, or a plant species found primarily in

Mississippi, or changes to logging regulations in Alaska. But many other times, the story could be set in a number of locations. If a new study on illegal immigration from Mexico and Central America shows the numbers going up across the country, you *could* report the story from California, or Florida, or New York. But maybe you'd get the most interesting and surprising piece by going to Minnesota, or another state that has not historically had to deal with a large influx of Hispanic immigrants. If the reporter is planning a story on how some big cities are losing their Chinatowns to upscale development, you might ask if the story should focus on only one of those cities, or should compare two of them, or should take an overview of the situation across the country.

It's also worthwhile at this stage for you to ask if there are ways to tell the story through scenes and by using sound. Get the reporter to imagine the ideal scene or the best sound he or she might get—a construction site in Duluth where the Minnesota crew managers are trying to communicate with their Mexican workers, or a street protest by Chinese-Americans who are watching their community center be replaced by a new luxury apartment building. Then suggest ways to get those scenes, or similar ones, into the story. "A lot of times it's a brainstorming process," Steve Drummond says. "You're on the phone with the reporter, and you're doing a story about special education, and you say, 'I want to make sure that if we interview the parent, he's not just sitting on the couch in the living room talking to you. I want to hear the parent interacting with the child.' I'll suggest maybe hearing them doing homework around the kitchen table that night, or maybe doing dishes and the child is helping out and there's some interaction," he says. Once the reporter is actually on location or arranging interviews, events may not turn out as you'd planned. That's okay; the exercise won't be wasted. Thinking in advance about the scenes you'd *like* to have in the report will often point a reporter in the right direction.

Ideally, the reporter and editor should also discuss these same elements of the story while it is being reported, especially if the focus of the story evolves as more facts are uncovered, or if key sources can't be interviewed. It's your responsibility as the editor to make sure that the reporter has conclusively established essential facts, that he has solicited responses from people criticized by interviewees—even that he has contacted an expert who might debunk the thesis of the story. In the end, the report will be stronger if it can withstand these sorts of challenges. And the last thing you want is to promise a piece to a program—and then have to renege, when you discover the story was thin, or didn't hold together, or was wanting in some other way.

Structuring the Story

One of the main jobs of a good radio editor is to ensure that each story has a logical structure. The best stories are built around sound—either actualities or natural sound—but great audio doesn't necessarily mean a great story. All too often reporters assemble their pieces by collecting their best actualities, and then writing copy that simply gets them from one cut to another (usually ending with the cut that is most poignant, or emphatic, or forward-looking, or in some other way sounds like a conclusion).

In this sort of connect-the-dots structure, the copy blocks serve primarily to move the listener from one actuality or piece of sound to another.[2] The problem is that even if each copy-and-cut module of the story may make sense, the report as a whole may not cohere. A piece that is no more than a collection of good sound bites strung together by the reporter's voice tracks will be much less memorable than a story that unfolds in some systematic way.

Exactly *what* way will often depend on the story that's being told. For example, a reporter may be taking a listener on a tour of a place, in which case the piece may be structured geographically—as a trip down the river, a visit to the headquarters building, a survey of a battle site, and so on. Many reports describe events in chronological order. For instance, a piece explaining a detective's theory of how a crime was perpetrated may run through the sequence of events very straightforwardly. Or a report on a scientific discovery may begin with the question the scientist hoped to answer, and then take us through his research one step at a time. It's sometimes possible to invert the chronology, beginning with the latest developments and then telling us how we got to this point. For instance, a report might begin with a scene of community members holding a ceremony to remember a local activist—and then work backward through his career so that we learn why he made such an impression on people.

2. Often the disjointed nature of the report is cleverly obscured by the fact that the actuality and surrounding blocks of copy seem to dovetail. If the actuality ends with someone saying, "it's not the end of the world, you know," the next track may begin, "But for farmers, it *does* seem like the end of the world."

If you want to build suspense and keep people listening, you can suggest the reporter start telling the story close to the peak of the action, then halt the action to present needed facts and background information, and finally resume the story to resolve the tension. Let's say a reporter is working on a story about the dilemma of foster parents who have been separated from their children by a devastating hurricane; in the rush to evacuate people, the parents got on one bus and the kids on another. The reporter has interviewed a woman from New Orleans who was relocated to Texas, and who spent eight days looking for her two foster children before she found them at a shelter in Baltimore. He also has lots of other information from interviews with foster care experts and New Orleans officials. Using this sort of "interrupted action" structure, the reporter might begin his story of the foster mother, but *interrupt* it at the point that she gets separated from her children—that is, to set aside the end of her story for a few minutes. As listeners, we'll be hooked—we'll want to find out what happened to the children and whether the woman has been reunited with them. Now the reporter can tell us about the plight of *other* foster parents in New Orleans, the laws that apply to relocating children in other states, the problem of reconstructing legal documents lost in the storm, and so on. Since we want to hear how things turned out for the main character, the reporter just needs to find a way to segue back to the primary narrative. "For Wilma Mae Booth, things eventually turned out well. She found her two foster children in Baltimore."

Many news pieces can be framed more or less as an argument: the report may examine an issue by posing the problem at the beginning—e.g., whether curbside recycling actually saves communities money, whether building a new stadium for the local football team will revive the city's economy—and then letting people on various sides each have a say. A common—perhaps *too* common—variation on that structure starts with a case study to illustrate the problem: "Sylvia Clements gets up every morning at 4:30 so that she can feed her children. Then she takes a bus, to catch a train, to catch another bus, to get her to her job by 7:30—three hours of commuting every day that costs her more than ten dollars. That's nearly a fifth of what she earns each day."

The important job for you, as the editor, is to make certain that the structure serves the story well—that it makes the piece easier to remember. "You're trying to make sure the story is going to be interesting on the radio, that the listener will want to stick around for three or four or five minutes and listen to it," Steve Drummond says. "That's part of what you as an editor can do. The reporters have gathered the sound, they have all the facts and figures. You help them organize them in a way that has a narrative flow, and that you think will be interesting and exciting on the air."

Sometimes the structure will be obvious even before the reporting has begun, or will have been hammered out by the reporter and producer in the field. But frequently reporters come back with lots of good information and great sound, and while they know they have a gripping story, they're at a loss as to how to begin telling it. Or they're so determined to work in all the material they've gathered that they've lost their focus. So you can coach the reporter. Ask him or her to sketch out an outline of the story, and then see if that structure makes sense. Do all of the elements relate to the focus of the story? If you listen to the proposed actualities in order, does the sequence seem logical? (Many editors like to do this sort of "tape edit" before the script is written.) Has the reporter considered alternate ways of telling the same story? "The detective story is a metaphor we use a lot for science pieces," Alison Richards says. "Or we might make the story show something about the process of science. If the reporter has pre-interviewed the scientist and found they have this amazing story that you didn't quite expect, then you might decide, 'This becomes a profile.'" Taking the time to think about and discuss a story's structure—before the reporter starts writing—can actually be a way to *save* time during the edit.

Editing by Ear

Editing for radio involves much more than merely correcting and revising text. The editor serves as a surrogate for the listeners—a one-person focus group to determine whether a report "works" on the radio.

The most obvious difference between script editing in public radio and editing at magazines or newspapers is that the radio reporter is always expected to read his story *out loud* to his editor. Some editors will refuse even to look at a script ahead of time; they want to hear it "cold," as if they had just turned on the radio in time catch this story at the beginning. Other editors will ask for a script, but use it mainly as a notepad; they just glance down at the script as they're listening, and circle sections they have questions about, or which they think need extra work. A few editors do like to see the script before the formal edit, especially if they are concerned about the way the reporter has structured the story and think it might need major surgery. But all of them will end up either sitting face-to-face with the reporter or listening to him or her over the phone. They will edit by ear, to determine whether the story is intelligible, accessible, interesting, paced correctly, and generally ready for broadcast. Andrea de Leon—who has edited more than a thousand pieces from member station reporters—says it doesn't matter how a script *reads*; she only cares how it sounds. "Sometimes I'll hear a paragraph or hear a sound bite, and I'll think, 'Huh?' Then I'll look at the script, and say, 'That makes perfect sense.' But I didn't get it the first time."

For listeners, there is no second time. They get one chance to make sense of a story. So if a sentence doesn't work for the ear, it doesn't work—period.

As an editor, you should listen to all the sound in a report and judge whether it's airworthy—that is, whether it will be intelligible to someone listening in a car, or to a bedside radio with a small speaker, or in a kitchen while making dinner. Reporters can become overly familiar with and almost emotionally attached to their interviews; they can sometimes recite an actuality word for word. The editor helps provide a reality test, so that you don't end up with a report built on a foundation of incomprehensible actualities.[3]

A foreign accent on a studio-quality recording may just make an actuality hard to understand, but it may make the same actuality unintelligible when it's filed over the phone. Ambience that brings a piece to life when it's filed on an ISDN line may sound like so much noise when the

3. The NPR transcript archive has any number of pieces that include sections like this:

> REPORTER: Dr. Mohamed Khojandi says his research will be conducted carefully, to minimize any possible risks to the patients.
>
> DR. MOHAMED KHOJANDI: When it comes to (unintelligible), I have no doubts that we are (unintelligible). The government makes sure of that.
>
> REPORTER: But others disagree...

Presumably what is unintelligible to a transcriber listening carefully to a recording is unintelligible to a casual radio listener who is also driving in rush hour traffic.

report is sent over the telephone.[4] So you should think about the means of transmission, and edit accordingly. For example, one foreign editor knew that his reporter in the Mideast would be filing from Gaza by phone. So he told her to keep all of the actualities in Arabic, and then to translate them herself in the body of the piece. He knew the speakers' heavily accented English would be hard to understand. If you're editing over the phone, you may not be sure that an actuality is intelligible—but don't take any chances. "If I had a question about a piece of tape while I was editing," Bebe Crouse says, "I'd have the reporter feed it to me so I could listen to it before crunch time, just to make sure. You don't want to build your story around it and then lose it at the last minute."

Even editors who are usually hardnosed about making sure actualities are airworthy may be led astray if the cuts are transcribed in the script; if the editor is reading along while she's listening, it's easy to think a sound bite is much clearer than it really is. For that reason, and others, Alison Richards says she never wants to read an actuality before she hears it. "Once you've heard it you can never hear it cold again. And I'm listening for lots of things—obviously for content and for clarity, but I'm also listening for pace, whether it's grabbing me, whether it slows down, whether I feel I'm being clearly shown the way, whether there's a real sort of narrative momentum."

As the editor listens to the voice tracks, actualities, and sound, he or she will also time the piece, using a stopwatch. This is certainly not the most important part of the edit, but it's nonetheless a crucial one; most radio pieces are allotted a specific amount of time—four minutes would be average at NPR, although they may be much shorter elsewhere—so the reporter and editor need to know if the story is running long or coming in short. Timing a script is a skill, albeit not a very difficult one to acquire. You have to learn how to stop the clock if the reporter wants to reread a paragraph, or if an actuality doesn't play when it's supposed to, or if the phone rings and you're interrupted, and then to start it at precisely the *same* point when the edit resumes. (Many editors have been mortified to get through a lengthy edit only to realize they forgot to restart their stopwatch after an interruption, or never started it to begin with!) Since the time budgeted for any report includes the introduction, you should start

4. One reporter filing a piece from overseas by telephone once began a report with the sound of waves crashing on the shore—which sounded exactly like the hiss of the phone line, only louder.

timing the piece with the first words of the intro. And make sure the reporter doesn't rush through the story in order to squeeze a four-and-a-half-minute piece into a four-minute hole! "There's no one I know who reads faster when they're recording than they do in an edit," Bebe Crouse says. "They always read slower, so I would build that difference in when I timed a script." Steve Drummond says he is never surprised if the piece is too long after a first edit. "You're thinking already about what can be cut," he says. That's a lot easier if you've been helping the reporter focus the story from the start. "Having a concrete knowledge of what the story's about allows you to look at every paragraph and every piece of tape, and ask, 'Does this contribute to that larger story, or does it not?' If there's a piece of tape that's very interesting or lively or exciting, but doesn't help tell the story, then often it has to go," Drummond says.

By *listening* to the piece, rather than reading it, the editor ensures that it will sustain a listener's interest for its full length. As Alison Richards puts it, "You've got to be driving the audience's curiosity and engagement all the time." Bebe Crouse says she would sometimes write comments in the margin of a script, to remind herself, "I got lost, I started drifting, I got confused. I'd go back to those spots on the script, and say, 'You started losing me here. Let's look at why that was.'" Indeed, one of the harshest criticisms an editor can make of a radio report is, "My mind began to wander."

The Intro

The host intro is one of the most important—if not *the* most important—part of a radio story. It is the equivalent of a newspaper headline and lead paragraph rolled into one—the "hook" that is going to grab the listener's attention. It is the first thing a reporter should write, and the first part of a story the editor should listen to.

The intro has to whet the listener's appetite for the piece that's coming up, and explain why the news event or issue is worth paying attention to. It should set the scene for the upcoming report, by providing the time reference, the location, and perhaps even the main characters of the story. As much as possible, it should be visual; it should give the listener something to see in his mind's eye. It should make people care about the upcoming story.

And there are many things it should *not* do.

It should not paraphrase the entire report. Imagine a six-sentence intro that tells us that lawmakers are working on a new bill; that Democrats take the position the legislation is needed to correct some alleged abuse by big corporations; that Republicans say they think the law would hurt the economy; that the President says he'll veto the bill if it gets to his desk; that the GOP says there aren't enough votes to override a veto; and that NPR's Mark McMuffin has more. Why should anyone stick around for *more*?

An intro should not have to carry all the technical freight of the story. Some intros are filled with names and numbers—almost as if the reporter didn't want to clutter his piece with all the facts. But when we've got a piece on, say, a heated election in India, nothing could be duller than cramming the intro full of the candidates' names, their previous jobs in or out of government, the parties they belong to, and what each of those parties stands for. Listeners want to know what's at stake, and how it will affect them, directly or indirectly; you can't let those ideas get submerged under too many details.

An intro should lead logically into, but not echo the reporter's first voice track. It should avoid "layer-caking," where the host tells us what has happened and then the reporter tells it again in different words—sometimes more than once, depending on what's in the first actuality. And when the intro is part of a package of reports, it should not echo material in the preceding piece.

It should put the news at the top. A great many intros use a template that can be summarized: "It used to be that way, but now it's this way"—a pattern that almost always leaves what's new for the last line. Here's an example:

> One of the hallmarks of workplaces in the last decade was a marked change in dress codes. The vast majority of businesses relaxed their standards. Surveys indicate that about half of all office workers are now allowed to "dress down" at least once a week, and for some employees "casual Fridays" evolved into casual every day. But as N-P-R's—— reports, casual wear may be starting to fade.

There are lots of ways this intro could be rewritten, perhaps the easiest of which (and this is often the case) is to put the last sentence first:

> Casual wear may be starting to fade.

Of course, this means the writer has to follow it up with something more than a restatement of common knowledge. (And in this case, you have to wonder whether the typical public radio listener really needs to be told the history of casual Fridays.)

Common variations of this stock format include: "You probably assume it's this way, but it's really that way" (e.g., many people believe the folk music boom ended in the 1960s, but it's returning with a vengeance to college campuses); "Some people think this, but others think that" (e.g., hunters view rattlesnake roundups as a springtime tradition, but animal-rights groups say they're barbaric); and "*This* may be good, but *that* is bad" (e.g., waste-to-energy programs have helped keep thousands of tons of trash out of America's landfills, but they're also adding to air pollution). These sorts of boilerplate constructions are not only dull; they are also simplistic, squeezing any sort of nuance out of the intro. And they're confusing: putting too many "buts" in an intro leaves the listener unsure what the piece is about—we are in the process of absorbing one piece of information when we're presented with another that in effect undercuts it.[5] More important, it is a formula, and formulaic writing does not demand the listener's attention. A good intro has an element of surprise. Here are just a few examples:

> We've all heard that it's expensive to live in New York City. Well, it's not cheap to be dead there, either. New York's cemeteries are filling up, and the remaining lots are pricey. That's bad news for families that want to stay in New York forever, and challenging for cemeteries that will soon be out of the burying business. Here's N-P-R's Robert Smith.
>
> The Great Wall of China has survived assaults by Mongols, Manchus, and Chairman Mao, and now it's being threatened by modernization. The

5. And the confusion wreaked by "buts" seems to increase geometrically with their number. This intro feels at least four times as muddled as a straightforward lead:

> Lebanon's capital, Beirut, was once known as the Paris of the Middle East: a commercial and entertainment center for the entire Arab world. But then came fifteen years of warfare, and Beirut was left in ruins. Now, a little more than a decade after the war's end, the city is starting to sparkle again. The ravaged center of the city has been rebuilt. And new restaurants and nightclubs are hopping with young professionals. But scratch the surface, and a deep malaise becomes apparent. NPR's —— reports.

> wall is one of China's main tourist attractions for foreigners and the new Chinese middle class. Tourism means commercial development near the wall—which is raising concerns, as N-P-R's Rob Gifford reports.

> Not-so-recent graduates of North Carolina and Illinois are swarming Saint Louis this weekend, anxiously awaiting tomorrow night's final game in the N-C-Double-A basketball championship. They are the "loyal alumni," a term that really doesn't do them justice. N-P-R's Tom Goldman reports maybe "crazed" would be a better term for the middle-aged alums who still live and die with each game. He spent time with one man from Moline, Illinois.

An intro is a sort of promise to the listeners—it tells them what to expect to find out in the report that follows it. As an editor, you need to check that the reporter makes good on that promise. If an intro says, "Jim Rodino reports that federal documents show the coal mine has a record of serious safety violations," that's what he has to report. An intro should somehow let the listeners know if the report is an investigative piece, a profile, a chronology of events, etc. It needs to manage their expectations.

Because the intro is so important, the writing should shine—it should give the host an opportunity to connect with the audience and sell the reporter's story. As NPR Senior Vice President Jay Kernis puts it, "During a lead is when hosts become hosts.... Let them have their moment on the stage, in the best possible light, in front of the most captivating set."

The First Track

It isn't exactly the ancient chicken-and-egg conundrum, but radio journalists have their own mystery to grapple with—namely, if the story's news lead goes into the intro, what does the reporter say first?

One way around this issue has long been to start the piece with sound. The rationale for constructing a piece this way is that it sets the scene immediately, and allows the reporter to be (in the listener's mind) somewhere other than in a studio. The device *can* work. But after more than thirty years of reporters beginning their stories with sound, this has become a production cliché. So it should never be the automatic "default" choice; the sound should be as effective a follow-up to the intro as a well-written sentence. The sound of a passing car, or a ringing phone, or even a church bell ringing probably isn't good enough.

In general, as the editor, you should make sure there's some sort of logical connection between the intro and the start of the report itself, whether it begins with sound, an actuality, or a voice track. That means that the reporter should read his track as if he has just heard the intro—which, of course, is how it will be presented to the listener. Here's one example:

> INTRO: Human evolution is sometimes compressed to a kind of comic strip. It starts with a hairy apelike creature bent over and walking on his knuckles, and next to it are progressively more upright figures slouching along, and finally a fully bipedal, clean-shaven modern human. Some anthropologists hate this drawing, in part because there's been little evidence that human ancestors walked on their knuckles. Now it turns out that cartoon may be fairly accurate. N-P-R's -- reports.
>
> REPORTER: Brian Richmond is an anthropologist at George Washington University. He says he's tried to walk on his knuckles like a chimp or a gorilla, but he feels silly and his wrists hurt. He got to thinking about wrists a couple of years ago. Richmond was sitting in a small office at the Smithsonian's Natural History Museum...

Note that this is *almost* an intro of the "it used to be that way, but now it's this way" variety; you might call it a "veiled but" intro, since "but" is implied at the beginning of the sentence that starts, "Now it turns out." Yet this intro and voice track work well together. The intro begins with a strong visual image—the evolution cartoon that many of us have seen, and probably even seen parodied. It gives us some information we didn't know—that most anthropologists hate that cartoon. And it makes it clear what the piece is going to be about—namely, that new research shows human ancestors might indeed have walked on their knuckles.

Having put all of that in the intro, the reporter then begins his report with a new scene—a researcher, in his office, practicing walking on his knuckles. But the way he reads it makes it clear that this is a continuation of the ideas presented by the host. The reporter says, "Brian Richmond is an anthropologist at George Washington University. He says he's *tried* to walk on his knuckles like a chimp or a gorilla, but he feels silly and his wrists hurt." The reporter stresses the word "tried," not "knuckles"; he knows the host has already broached the subject of knuckle walking.

(As good as that *second* sentence is—and it usually provokes a laugh from anyone who hears it—the *first* one is awfully dull. The reporter could have postponed giving us the information about Richmond's title and affiliation, and just started with his name: "Anthropologist Brian Richmond says he's tried walking on his knuckles.")

The first line of the reporter's first track *can* be problematic. Since the news lead is in the intro, the goal of the first sentence is to advance the story. Reporters don't always meet that goal, as in this example:

HOST: The Bush administration will announce its decision this week on regulations to protect the confidentiality of medical records. The rules, issued by the Clinton administration, are the first federal protections for medical information, but they've been suspended for the past two months while privacy advocates and the health-care industry have fought over whether they go too far or not far enough. N-P-R's —— reports.

REPORTER: The medical privacy regulations issued last December would, for the first time, guarantee patients access to their own medical records.

There's not much in this first sentence that advances the intro, except perhaps the fact that the regulations were issued in December (although the intro does say they've been suspended for two months). And the vocabulary of both the intro and first line is technical, dry, policy-speak.

These sorts of policy stories can be hard to bring to life. But on the same day that the piece on medical privacy aired, several reporters working on similar stories came up with some more successful first lines. Among them:

If ever there was a situation ripe for political payback, this would be it.

Every year, Taipei makes requests to Washington for new weapons—but there are weapons, and there are *weapons*.

For the first few months of the Bush administration, there was a White House Office of National AIDS Policy—though no one seemed to be working there.

The Bush administration came into office talking tough about overthrowing Saddam Hussein, while uttering sympathetic words for the Iraqi people.

Even though these reports were about politics, weapons agreements, AIDS policy, and U.S.-Iraq relations, the reporters all managed to put something into their first sentences to catch the listener's attention. One of your jobs when you're editing is to make sure that attention never flags—and if it evaporates with the first line, you've lost your audience.

Of course, a tried-and-true way of grabbing the listener is to give a specific example of a person who is affected by a policy or an event. But this, too, can become a formula. For example, a quick check of NPR stories one afternoon found a number of scripts that started almost identically:

> When Tampa police officer Rick Dubinas gets a report of a stolen handgun, the first thing he wants to know is the gun's serial number.

> When Jay Schecter relaxes in the quiet of his home on Hannawa Pond in northern New York, there's one sound he can't stand.

> When Sarah Higley was a nine-year-old growing up in Glendora, California, she and her friends wanted be able to share secrets and pass notes.

In short, there's no recipe for writing a first track, just as there's no recipe for writing an intro. Ideally—and this has been used as a definition of good poetry—every word should seem unexpected and inevitable.

Copyediting

The best editors are good writers. They recognize disorganized or turgid prose, understand and practice good grammar, and know how to turn a phrase. As an editor, you should be familiar with the principles of broadcast writing outlined earlier in this guide, and keep an eye out for clichés, long sentences, odd vocabulary or syntax, and other common writing problems. In addition, you should be alert to other mistakes that are likely to occur in pieces that include actualities and sound. Among them:

ECHOES. We often hear echoes of thoughts and phrases between the intro and the piece itself, but echoes can also appear between the actualities and tracks. This is just one of many reasons why it's essential to listen to all the actualities in a piece, and not accept a transcript or a paraphrase from the reporter. Also, not all echoes involve repeating phrases word-for-word. If there's a first reference to a person or thing in the intro or at the top of a report, make sure there isn't another first reference later in the piece; you don't want the reporter to tell us twice, for instance, that "Jeremy Bentham is the manager at Kwikee Foods."

UNIDENTIFIED ACTUALITIES. If we've been hearing from Jeremy Bentham throughout a story, reporters sometimes assume we don't need to identify him when he speaks for the third or fourth time. But listeners aren't taking notes. We have to assume they're listening with half an ear most of the time, and so we should identify people concisely but frequently before each actuality: "When he leaves work each day, Bentham always turns out the lights." It never hurts to add some sort of ID—and almost always helps.

CONFUSING IDENTIFICATIONS. Often reporters come out of one actuality with the ID of the *next* person we're going to hear.

RONALD BABCOCK: ...So I think the government should use one tenth of one percent of the money it spends on roads to protect wilderness trails.

REPORTER: Richard Jenkins of the U-S Forest Service says many environmentalists are opposed to designating new trails because they fear they will only attract tourists.

RICHARD JENKINS: We're getting to the point with all of the sanctioned trails where it's costing millions of dollars just to pick up trash.

This placement of the identifier is really confusing, especially if the two people who are speaking have similar voices but different views; it would be natural for listeners to assume that the voice they just heard was Richard Jenkins, not Ronald Babcock. A simple way to clarify things is to contrast the two speakers in the voice track: "Richard Jenkins disagrees. He oversees the designation of trails for the federal government...." Or "Ronald Babcock is in the minority. Many environmentalists say they're concerned that designating new trails will create new problems. Richard Jenkins says..."

CONFUSING PRODUCTION DEVICES. Just as we don't need every piece to start with sound, not every actuality should begin with a blind tease of the speaker. Make sure that there's a reason behind the way each clip is used, and that the production device is effective. For example, reporters sometimes like to "butt-cut" (splice together, to use a metaphor from the analog days) the end of one actuality with the beginning of another, when the views expressed contrast well.

RONALD BABCOCK: . . . So I think the government should use one tenth of one percent of the money it spends on roads to protect wilderness trails.

RICHARD JENKINS: We're getting to the point with all of the sanctioned trails where it's costing millions of dollars just to pick up trash.

REPORTER: Richard Jenkins of the U-S Forest Service says many environmentalists are opposed to designating new trails because they fear they will only attract tourists.

RICHARD JENKINS: Every time you give a new trail a name and publish a guide book to it, you're inviting every backpacker in America to set up camp in the woods.

This can look fine on the page, but where the actualities are on the phone, and the speakers are of the same gender, it can sound confusing. So listen to the audio critically, as if it were on the radio. If you have any doubts about whether the listener will know what's going on, encourage the reporter to try a different approach.

There is only so much you can do as an editor to improve mediocre writing; you can remove the clichés, shorten the sentences, eliminate unnecessary numbers—and still have a dull script. So remind the reporter that he or she is *telling* a story—not presenting it as if it were a book report or lecture. Science editor Alison Richards says she makes a special effort to get her reporters to use everyday language. "You 'catch' language from the people you're talking to and whatever it is that you're reading," she points out. "So if you're talking and reading science all the time, there's a danger—as a reporter and as an editor—that you start to think in that kind of vocabulary. Very often the first draft of a science story will be completely lacking in feeling in just that kind of way that a scientific paper is."

You can also liven up the writing by asking the reporter a few, well-chosen questions. "If a reporter comes back and writes, 'Here we are in Mrs. Jones's class and the students are working on math problems,' I interview him," Steve Drummond says. "I ask, 'What kind of math problems?' And the reporter may say, 'Oh, they were doing story problems about two trains heading toward each other.' And I'll say, 'Let's put that in there. Were all the kids sitting at their desks, or were some of them at the chalkboard? Did they have their textbooks out and open, or were they all working on sheets of paper?' Those are the sorts of things many reporters don't think to put in or don't think they have time to put in," Drummond says. "By interviewing them, you can draw it out of them."

You should also urge the reporter to use strong, visual verbs. Target sentences that start with "There is" or "There are." The changes can be quite small and still make a difference. Instead of, "There were crowds of teenagers eager to buy tickets," you can encourage the reporter to describe what was actually happening: "The teenagers were elbowing one another aside as they pushed forward to buy tickets." Instead of "There was a small girl in front of the school," the reporter could say, "A small girl was fidgeting with her lunchbox as she stood in front of the school."

That means the reporter has to record those details—mentally, in a notebook, on his MiniDisc, with a camera, but somehow! Sometimes reporters are only thinking about sound, so they don't pay attention to the other aspects of a scene that can bring it alive. Andrea de Leon recalls the time a reporter spent a week at a jail that was running a summer camp for the children of inmates. "I wanted to know about the moment when these children first went in," she says, "and not only did I want to hear, but I wanted to *see* them waiting to be let in on their first morning. The reporter stood there with them, but she was so intent on her MiniDisc that she had to call the jail and ask what the door was made of. She couldn't tell me what color it was, if it had a window, or bars, or anything." De Leon's advice to her reporters: "Make sure that you report with *all* of your senses."

Sometimes—and that "sometimes" should probably be underlined—you can help a reporter write better by getting him or her to have a particular *perspective* on his story. This doesn't mean slanting the story; a reporter can have a point of view without showing an editorial bias. It's okay, for example, to describe what it *feels* like at the top of the dormant volcano where the observatory is located, or to make it clear that the sights at the refugee camp were appalling, or to reflect a sense of humor in a situation that is undeniably funny.

And at some point during this process, take a step back and ask yourself whether the reporter has succeeded in giving the listener something new. After all, one of the "core values" of public radio identified by the public radio program directors is "love of lifelong learning." The best news reports don't repackage ideas or points of view that are widely heard elsewhere; they tell us something we either didn't know or didn't remember. You don't want to get so caught up in the technical details of editing that you fail to make sure that the story rewards listeners for their time and attention.

The Ending

Just as you have to pay close attention to the intro and the start of the report when you're editing, you should make a habit of checking that the conclusion of the piece actually says something. Journalists joke about reports that conclude with the hackneyed phrase "only time will tell." But these are actual last lines from NPR pieces that all aired *on one day*, and they all say little more than "only time will tell":

> But as the fight over the medical privacy issue nears its five-year mark, it's clear that whatever the decision, it won't be the last word.
>
> The head of Gazprom's media arm, Alfred Kokh, says he has until tomorrow to respond to Turner's latest offer, but Kokh indicated a final deal could be months away. In the meantime, the test of wills continues.
>
> The debate over the budget and the tax cuts will likely get louder once lawmakers get back from spring recess.

Many radio stories conclude by saying, in effect, "Something is going to happen—but we can't say what." Some reports—like the last example here—make a vague prediction of what's going to take place, modified by "likely" and perhaps also by "may," to ensure that the reporter is not charged with speculating: "The President may push Congress to pass an even bigger tax cut—something Democrats are likely to oppose."

So what's an editor to do? Sometimes it is fair to ask the reporter to look to the future. In this conclusion, the reporter adds a new piece of information to her report on U.S. AIDS policy:

> All this is leading up to a key meeting in June in New York. For the first time, the United Nations General Assembly is convening specifically to address international AIDS issues.

Some reporters opt not to *have* a final track—ending their piece with an actuality, followed simply by the reporter sign-off. This is often just a cop-out; ending a report with an actuality is a device that should be used very rarely, and only when the last cut is, in effect, a conclusion. "The reporter is in control of the flow and the thought process you want your audience to be going through," says Andrea de Leon. "You told them this was the beginning, you told them this was the middle, and you should have the last word." De Leon and other editors also say that in reports on controversial subjects, ending the piece with one actuality may give the impression that the reporter sides with the speaker; it's as if the reporter were saying, "What can I possibly add to that?" In addition, an actuality that's more than, say, twenty seconds long may need to be reidentified, which means the reporter's final track will be something like, "Jim Smith, of the Chesapeake Waterman's Association. Jack Jones, NPR News, Annapolis"—and consecutive ID's hardly make an effective ending. Some editors oppose the conclusion-less report on production grounds, saying it sounds odd to have the reporter pop back in at the end of a report only to sign off. Despite all these objections, pieces lacking real conclusions air regularly on public radio, presumably because editors—at least under deadline pressure—are hard pressed to find something more for reporters to say that isn't platitudinous, speculative, or redundant.

Many reporters like to end their pieces with a quote, but *not* an actuality—that is, they tell us something one of the interviewees has said, just as you might in a newspaper or magazine story. For instance, a reporter ended a piece on Peru's presidential elections this way:

> "The process has changed 100 percent," Velasquez says. "Last time, it was totally fraudulent, but this time it's all very correct, and we hope that today's winner will be Peru."

Frequently the editor can help the reporter steal a quote, or idea, or fact from somewhere else in the piece. In fact, many of the best conclusions began their lives as intros.

When to Stop Editing

There's no rule about how much editing is *too* much. Some NPR editors say that if they've done their job right from the start—if they've talked the story idea over with the reporter, consulted while the reporting was still under way, and helped him or her structure the piece—then only one edit should be necessary. Others expect to have several edits, especially on a longer piece. "It's in the second and third edits that you polish the story," Steve Drummond says, "and deal with issues like: is the writing exciting and clear, does the reporter get into and out of tape well, is the tape exciting and useful? And also at these times, you're looking at the length," he says. "At a certain point, you take out the surgeon's knife and say, 'Let's cut this phrase.'"

However many times you go over a script, the process should involve *editing*, not rewriting. It's almost always better to steer reporters in the right direction and let them use their own words—words they're comfortable saying. One common technique is to get the reporter to tell part of the story *without* a script. Then the editor can note whether the phrases used in speech are better than the ones in the first draft of the piece: "See, you didn't call him 'the Wynton J. Rassias Endowed Chaired Professor of Psycholinguistics,' you just called him 'a linguist.' And it was much easier to follow you when you described how he figured out the ancient Greek text. All the detail in the script had me confused. Now go back and write it more or less the way you *told* it to me."

Similarly, a reporter should not expect his or her editor to boil an eight-minute piece down to four. If a story is twice as long as it ought to be, you might offer some general guidance about what needs to be cut; but you should leave it to the reporter to do most of the cutting.

Editors can, and often do, help out in the reporting of a story, by gathering material, lining up sources, isolating actualities—and even by conducting interviews. These are jobs usually carried out by the reporter, sometimes by an editorial assistant or producer; but editors should also be willing to lend a hand, especially when the reporter is on a deadline. Just make sure that it's clear what each person is trying to accomplish, so the two of you don't end up duplicating each other's work.

As an editor, you cannot know everything about every subject, but like a good field producer, you should know which facts need to be checked, and how to verify them. You want to make sure that names and dates and places are all correct (and don't forget that someone who was twenty-eight when he was interviewed last summer may be twenty-nine now). You should get in the habit of taking a hard look at all numbers ("half of all fourteen-year-olds have used marijuana"), at assertions from individuals or groups with political agendas ("the state government has made no effort to help single mothers find jobs after they've been kicked off welfare"), at historical interpretations that may require attribution ("the Japanese invasion of Manchuria unleashed forces which led ultimately to the attack on Pearl Harbor"), at anyone characterizing another person's motives ("Republican Congressman Jim Smith says Democrats are introducing the bill to embarrass the President"), and many other similar statements. They may not be wrong; but you need to know why the reporter felt comfortable making them.

When, despite your best efforts, a story is not coming together, you may also have to tell a reporter that his or her piece is not ready for broadcast. Especially if the subject is controversial and you are not convinced that the story is fair or that all the vital facts have been established, you have a responsibility to keep the story off the air. And on occasion, you may need to kill a piece, though it can take years of experience to know when to take a reporter off a story that is not materializing. In any event, remember that a five-minute decision on your part may undermine days or even weeks of work by the reporter—so don't make these sorts of decisions lightly. "You're a diplomat, you're a psychologist," Bebe Crouse says. "I always tried to remember what it was like being a reporter—being on the other side."

Finding New Approaches to Old Stories

One of the ironies of the news business is how much of what we report is not really new. At the beginning of each year, for example, most news organizations already know that there are certain stories that they will be covering during the next twelve months. Reporters overseas are sure to have pieces on new outbreaks of fighting in old trouble spots; arts reporters will do stories on annual awards ceremonies, like the Grammy and Academy Awards; national reporters will have pieces on the start of the school year, on big moves in the financial markets, on the Super Bowl and the World Series; science reporters will have scores of stories based on research journals. Every "routine" story poses the challenge of presenting it in an unusual way.

After hearing these sorts of reports year after year, most editors can predict what the next incarnation will sound like. Try it yourself. Imagine a day on which the big news story concerned a hurricane heading toward Florida. Forecasters say that it will be the biggest storm in twenty years, that it's heading straight for a city on the Atlantic coast, and that it's expected to make landfall in forty-eight hours. A reporter has been assigned to cover the story today for *All Things Considered.* What would you expect to hear?

When we asked different groups of public radio editors that question, they independently said almost the identical thing:

- The piece would start with the sound of people preparing for the storm—probably hammering up plywood to cover the windows.
- It would include a scene somewhere of people stockpiling supplies—water, milk, toilet paper, etc.
- It might also have an actuality of, or at least describe, the people evacuating, taking the one road out of town. (For some reason, one editor noted, there always seems to be just one road out of town!)
- It would have to include an actuality of an expert—e.g., someone at the National Hurricane Center, describing the path of the storm and why it's to be feared. (This cut would probably be on the phone, while the others would probably have come from in-person interviews.)
- It would conclude with an old-timer who is staying put because he's been through big storms before.

No doubt you've heard pieces like this, and seen very similar stories on television. Such a story might well meet most of the tests for a good piece of reporting. It would certainly be timely. It would have lots of relevant sound, and only one cut taken off the phone. It would be well-focused and well-structured (we'd probably hear from people who are getting ready to leave, who are leaving, and who are staying put, in that order). A reporter who produced this story on deadline would probably feel pretty good about having covered all the bases. But the piece would be almost comically dull—because it would be so similar to the dozens of hurricane stories reporters have produced over the years.

Now, if there really were an oncoming storm that threatened to take people's lives and destroy huge amounts of property, *All Things Considered*—or any other program—would be justified in wanting some sort of coverage. So we asked editors to take just a couple of minutes to think up some alternative ways to get at the same story. Here are some of the things they came up with:

- Have a reporter ride along with a policeman conducting the evacuation.
- Have a program host check in repeatedly throughout the day with one person to get a sense of the growing crisis—a storeowner, a policeman, the person running a shelter, and so on. Or ask a reporter to do the same thing.
- Talk to a meteorologist about the science of predicting when and where a hurricane will make landfall.
- Talk to a TV weather channel producer who thinks the storm is exciting.
- Have a reporter or host collect advice about coping with a hurricane from people who have been through such monster storms in the past.
- Have a reporter stake out a route *into* the city to find out why people would be entering a place from which others are being evacuated.
- Have a reporter talk to insurers gearing up for the aftermath of the hurricane before the storm has even struck.
- Have a business reporter find out who profits from a destructive hurricane (e.g., lumberyards, homebuilders, landscapers, etc.).

Not all of these ideas would pan out, and the success of some would depend on the skills of the reporter. Others might sound better a day or two later. The point is that most editors can generate four or five alternative ways to get at a story in just a few minutes of brainstorming. So it's important to take those few minutes to try to conceive an approach that doesn't fit one of the old templates.

It's not as hard as it may seem. Here's how one science reporter began a story about an article in a journal:

> A scientist returns from an exotic place every now and then and describes something that's stranger than fiction. A biologist has done just that in Costa Rica. N-P-R's Chris Joyce has a story about death and transfiguration in the rainforest.
>
> JOYCE: When you're crawling around a tropical forest looking for unusual forms of life, there's a simple rule of thumb to remember: one thing leads to the next. For William Eberhard, who works at the University of Costa Rica and for the Smithsonian Institution, the trail began with a spider:
>
> EBERHARD: It's sort of a medium-sized spider. It's really beautiful, its silver abdomen with black streaks and patches of gold and green and red.

JOYCE: What really intrigued Eberhard wasn't the spiders, though; it was their webs. These were orb spinners. Normally, they make oval latticeworked webs. But many of these webs were weird looking, just a big X with a sticky tangle at the center. And hanging from this tangle was a single cocoon, one Eberhard recognized as belonging to a small, stinging wasp.

EBERHARD: I could see that these webs were very strange, and that they somehow or other were associated with the presence of the wasps. And that set me going.

JOYCE: Eberhard didn't have to go far. He found spiders in the act of weaving these strange webs. Each had the larva of the wasps stuck to its abdomen. They wove the basic superstructure of a web, but didn't finish. Eberhard watched these spiders for hours.

Look what the reporter has done. He's written an intro that tells you you're about to hear a story, not about some newly published research, but about something that's "stranger than fiction" and deals with "death and transfiguration." He's told the story as a sort of mystery, where "one thing leads to the next": a scientist found an unusual spider, then he noticed the something was wrong with the web it was spinning; and so on. When editors were asked what this piece was about—an important question for every editor to ask of every reporter—they almost always responded "nature is amazing" or something of that sort. In other words, the piece—nominally about how certain wasps can get certain spiders to reprogram their web-weaving—was really about something larger, and considerably more interesting, especially to people who are not usually science junkies.

This is an unconventional piece for other reasons that may not be apparent in this excerpt. The journal in which the research was published isn't mentioned in the intro; in fact, it's only cited at the end of the piece. The intro doesn't include the usual pro forma statement that "while the results are preliminary, the research could eventually lead to new treatments for [some disease]." And the piece is built around a single interview with the researcher. (The reporter conducted other interviews, but decided that the researcher was the right person to tell the story.)

Not every reporter will be able to think up a new way to get at a conventional story. But editors can always challenge them to try. "It's a little bit like a dance," Bebe Crouse says. "You're both coming to it with something and don't want to step on each other's toes, but you work it out together."

After all, as the editor, you are the critical middleman between the reporter and the audience. Your role may often be out of sight, and your responsibilities unappreciated—until something goes wrong. Then the first question asked by executives and listeners alike is "Who *edited* that?"

CHAPTER 7

The Reporter-Host Two-Way

Every day, public radio hosts conduct interviews with staff reporters. These two-ways (or two-way reports) get news on the air quickly, albeit without the actualities and natural sound that are public radio's stock in trade.

However, a good two-way can give a program a sense of immediacy that's hard to duplicate in a highly-produced piece. Consider this excerpt from *All Things Considered*, which is pretty typical of the genre:

> HOST: And now to reporter Rusty Jacobs of member station W-U-N-C, who is in Washington, North Carolina; that's in Beaufort County. And, Rusty, I guess the eye of Isabel crossed over Washington earlier today.
>
> JACOBS: It did, and bringing along in its trail was the back side of the storm, which is really causing some problems now. I'm standing on the back porch of a downtown business in Washington. And if I took two steps down to street level, I'd be knee deep in the waters of the Pamlico River, which have overspilled their banks a while ago. And this is the dreaded storm surge that officials here were most concerned about. That's where the hurricane just churns up the waters of the Pamlico Sound, pushing it back up the river over the banks and plaguing communities likes Washington, Belhaven, and the other riverside communities in eastern North Carolina.
>
> HOST: These riverside communities have a history. They've had a history of past storms flooding their area, haven't they?
>
> JACOBS: Indeed. In fact, you know, I visited people in the last couple of days to see how they were preparing for the storm, putting up boards on their windows, taping the windows. And I've been to at least a handful of places where the owners marked the water lines from all the past hurricanes. And I met one gentleman who rigged up a special hoist so he could raise all his expensive machinery in the garage off the ground whenever a hurricane rolls through...

HOST: Rusty, we heard earlier at the beginning of this segment from a couple of folks in Virginia Beach who'd actually gone out to watch the storm. Is there any of that going on in Washington, North Carolina?

JACOBS: Not right now. A little earlier today, there were indeed people, especially when the eye was passing over, the wind had died down, even the rain had diminished to almost a drizzle. And there were a few hardy teenagers out in their shorts wading in the waters, walking out on the piers, which are invisible to the eye, because they're submerged under one or two feet of water...

Even though this sort of live interview lacks the careful writing and the polished production of a conventional piece, it conveys a lot you don't get when the reporter goes back to his station to write a script and record it in a studio. You know from the start that he's on the scene, that he has personally observed (and is still watching) the news event, and you even get a peek at some of the reporting he has done. (Later in the interview, Jacobs mentions talking to emergency operations workers.) From the program's point of view, the two-way may have been a necessary compromise, perhaps because the reporter didn't have the equipment or opportunity to produce a full-fledged story. From the listener's perspective, it can make the news more vivid, and convey a greater sense of urgency, than many of the other pieces in the show.

There are other legitimate reasons for a reporter doing a two-way. Sometimes the deadline alone makes writing a script impossible: A jury verdict is announced at 11:45 a.m., and you have to be on the air at 12:06. Sometimes there is little or no audio to build a radio report around: microphones weren't allowed in the courtroom, and none of the lawyers has spoken publicly yet, or the people who have spoken to you have refused to let you record them. And some stories lend themselves to reporter two-ways because the reporter can touch on a wide number of related points: a reporter debunking a supposed "cure" for cancer can discuss why the media pumped the story up, the business impact, what doctors tell patients who get their hopes up, and so on.

The Problem with Scripted Two-Ways

The behind-the-scenes glimpse at how the news is gathered was one of the original goals of the reporter two-way, according to *All Things Considered* host Robert Siegel. Back in 1968, during a newspaper strike in San Francisco, KQED-TV launched a public TV news program called *Newsroom.* The reporters who were not working for their newspapers would cover stories each day and then be "debriefed" on the air. "The conceit was that the conversation that a reporter has with an editor in the newsroom, telling what he or she has got, being questioned about how he or she knows this or that, is more interesting than the story that often results. So, the process of a newspaper city room was unveiled to the audience," Siegel says. The result was TV news reporting without any of the usual video or graphics that are considered essential for television. "It gave reporters who had no film a format for presenting a story. Other reporters were present and could ask questions. The spontaneity—and the fallibility—of the reporter were essential."

Today, many reporters—perhaps knowing they're fallible—often fear that spontaneity.

Unlike the other guests interviewed on public radio every day—experts, authors, regular folks, and journalists from other news organizations—many radio reporters expect to have a say in the questions they are asked, or at least to know the questions in advance. Especially when they're going to be reporting live, they don't want to be blindsided by a question they can't answer, or suddenly at a loss for words because they don't have an important fact at hand. This has led some reporters to script not only the questions, but also the *answers* for their interviews, so the interview becomes a sort of radio play—as in this exchange:

HOST: What have you learned about how Corporal Hassoun ended up in Lebanon from Iraq?

REPORTER: Nothing. There's no word on that from the Defense or State Departments or from Hassoun's family in Lebanon or here in the United States. What we do know is that somehow Hassoun and U-S embassy officials in Beirut came into contact and arranged a pickup. Hassoun was taken to the embassy and is expected to stay the night. The embassy spokeswoman told me a short time ago that Hassoun is there voluntarily and can leave at any time. He has not been questioned or debriefed and won't be until the appropriate officials from the Defense Department show up to do that. That's the Defense Department's job, the spokeswoman told me.

HOST: It appears, then, that Hassoun is not in custody. He's not being detained.

REPORTER: That's right. He's free to go, the embassy spokeswoman said. That's an indication that the State and Defense Departments are moving very carefully here, despite an investigation by the Naval Criminal Investigative Service. A spokesman for the Marines confirmed the existence of that probe yesterday, noting that the investigation includes the possibility that Hassoun's captivity was a hoax.

HOST: What's the basis for thinking Hassoun's captivity might have been a ruse?

REPORTER: Remember that when Hassoun first disappeared, the Marines classified him as on unauthorized leave. Even a week later, when Hassoun appeared blindfolded in a video, with a sword hanging over his head, and a death threat issued by apparent captors, it was a day or so before the Marines changed Hassoun's status to captive...

There are many reasons not to script two-ways in this fashion, not the least of which is that it usually *sounds* awful; when the reporter and host are both reading from a script, the result is stilted and artificial. Because reporters aren't used to writing dialog, they frequently come up with sentences whose syntax works better on the page than on the radio—sentences that don't sound like anything a person would utter in normal discourse. In the example above, it's hard to believe a reporter would spontaneously say, "He's free to go, the embassy spokeswoman said" (as opposed to "The embassy spokeswoman told me he's free to go"). And it strains credibility to think that a reporter would spontaneously say, "Even a week later, when Hassoun appeared blindfolded in a video, with a sword hanging over his head, and a death threat issued by apparent captors, it was a day or so before the Marines changed Hassoun's status to captive." Few people talk that way. So—because most of us are neither actors nor playwrights—we're left with a mediocre performance of an amateur radio play.

Hosts don't like these sorts of orchestrated two-ways because the scripts force them to ask questions they might not normally ask. (They sometimes call them the "dumb host" questions.) "When we do two-ways with our own people, the entire network has an investment in the host looking like he or she has some command of the story, some knowledge of the story," host Scott Simon says. His advice: If anyone reading a news crawl on cable TV would know something, assume the host does too. If you, as a reporter or editor, feel the need to get certain background information into the story, don't suggest the host ask a ridiculously naïve question like, "Now, this is not the first time the tobacco companies have been sued, is it?"

A variation on this is the question that makes it appear that the host has suddenly forgotten the introduction he has just read—that amnesia set in immediately after the reporter was introduced. For example, look at this intro and the questions that were suggested to follow it:

> In Los Angeles today, the Metropolitan Transit Authority and bus mechanics announced a tentative deal to end a strike that has idled public transportation in that city for more than a month. Most issues that divided the union and the M-T-A were resolved over the weekend, with just one major dispute unresolved—health care. Both sides agreed to non-binding arbitration on that issue. The buses should start running tonight. Joining us now to talk about the deal is ——.
>
> Did today's announcement come as a surprise?
>
> What kinds of problems or hardships has the strike caused?
>
> Is the agreement official?
>
> There's still a big sticking point. What is it?
>
> When will the buses start running?

At least two of the five suggested questions—on the sticking point and when buses will start running—are *explicitly answered in the intro.* (That the agreement is not official is also hinted at by the deal being described as "tentative" and the statement that both sides have agreed to go to non-binding arbitration) True, the host is always free to change the questions; but when the reporter has written out her answers, she is likely to be thrown off balance by suddenly hearing different questions, or even similar questions posed in a different order.

Even if you are prepared to answer questions off the cuff, poorly drafted questions can get the interview off to a bad start. Reporters and editors preparing for two-ways often suggest questions that require the host to pretend to misunderstand something—which hosts resist, with good reason. The rationale for this is usually to correct a misapprehension that the reporter or editor assumes may be widespread among listeners. Then the conversation usually sounds something like this:

> HOST: But a high-fat diet is *always* unhealthy, isn't it?
> REPORTER: Not necessarily. It depends on the kind of fats you eat.

If they spot them in advance, hosts often turn these questions around so that they reflect their own knowledge: "A high-fat diet isn't always unhealthy, is it?"

Reporters and editors can also go to the other extreme, drafting leading questions that suggest the hosts know more than they do. No one is fooled by questions like "Does chlorophyll play a part in this process?" or "I understand that some of the Duma members are not happy with the Russian president's press conference. What are they saying?" They sound like the setups they are.

Finally, hosts never want to serve simply as props to help you get to your next point. Two-ways structured around questions like "How many people were there?," "And then what happened?," "What else did he say?," and "What happens next?" are really just monologues in disguise. Noah Adams has been on the receiving end of this sort of two-way when he has filled in as a host. "After you come out of a highly scripted reporter two-way," he says, "you feel just a little bit like a prostitute."

In their defense, reporters point out that they suggest questions mainly because they know more about their stories than the hosts in the studio; they are in the best position to know which questions will elicit productive answers. Also, agreeing on questions ahead of time can avoid the situation where you have to correct a host on the air:

HOST: The car bomb went off outside the police station. Why were there so many people in front of the building?

REPORTER: Actually, it wasn't a police station—it was an army recruiting office.

This can happen when show editors or hosts are depending on news wires for the facts—and that information becomes outdated, or conflicts with what the reporter has found out firsthand. Sometimes the wire reports are accurate, but the reporter is simply not in a position to confirm them, so the questions can end up missing the mark. For instance, AP or Reuters (or both) may be reporting that someone has been killed at a huge anti-government demonstration in Moscow. Even if the reporter has spent the entire day with the protesters, she may not be aware of the shooting; all she really knows about is what she has observed, or what she's been able to confirm through her own interviews with protesters, police, and other people on the scene. In that case, a question drawn from information on the wires ("We know that a twelve-year-old boy has been killed—what have you been able to find out about that?") will leave the reporter dumbfounded.

Going Scriptless

The best compromise between ad libbing the questions and answers and writing them all out verbatim is for the host and reporter to confer ahead of time on the topics to be covered in the interview. *All Things Considered* host Melissa Block says she tries to read up on the subject so she can come up with her own questions. "I know I'm sitting here reading the Associated Press, which may be 180 degrees apart from what the reporter in the field is seeing. But let's say I've read some stories and there are a couple things that have jumped out at me. Maybe the reporter and editor have sent questions in. I'll look at them and say, 'Gosh, they never mentioned Yassir Hamdi [a U.S. citizen held as an enemy combatant]. Shouldn't we have a Hamdi question?'"

If logistics make that host-reporter conversation difficult or impossible, their proxies (i.e., their editors) can have the discussion, though if the reporter has been incommunicado—inside a courtroom, for instance, or out of cell phone range at the site of the forest fire—an editor may not know what questions to ask. (When you're reporting, you don't want your editor to suggest that the host ask, "And what did the defense have to say today?" if the trial adjourned before the defense had an opportunity to make an opening statement.)

Then don't write out your entire answers; just jot down a few words for each of the key points you want to make. That way, Scott Simon says, "you won't have a sense that you've got to race to complete that answer and get into the next thought." By providing the show with an intro and an outline, you can suggest the shape of the conversation without insisting that every word be spoken as it is written.

Here's a sample of one reporter's suggested questions and his notes:

It appears that the negotiations between Europe and Iran over Iran's nuclear activities are near collapse.

For days, Iranian officials have been openly suggesting that Iran is about to restart some of the nuclear activities it suspended last year. Yesterday, the foreign ministers of Great Britain, France, and Germany sent a letter to the Iranian government warning that such a step would bring the negotiating process to an end and have negative consequences for Iran. N-P-R's Mike Shuster was in Tehran recently and he joins us now.

1. What exactly are the Iranians threatening to do?
   - uranium conversion, a precursor step to uranium enrichment.

2. Why are they saying they will restart uranium conversion? What's so pressing about this for them?
   - they are impatient with the negotiating process.
   - they suspect the Europeans under US pressure will string out the talks indefinitely.
   - hardliners inside Iran were never happy with the process in the first place.
   - presidential election politics.
   - could all be a negotiating tactic.

3. Is there a compromise here?
   - hard to see one.
   - Iranians want to enrich uranium under monitoring of I-A-E-A.
   - U-S firmly opposed to that outcome.
   - Europeans say, Iran can acquire nuclear fuel from others.
   - so far not enough incentives to Iran to abandon its position.

4. What will happen if Iran does go ahead and restarts some of these activities?
   - possible emergency meeting of the I-A-E-A in Vienna and referral to U-N Security Council.

5. The Bush administration is also facing nuclear trouble from North Korea right now. What's going on with that?
   - possible preparations for a nuclear test.
   - possible removal of fuel rods from a nuclear reactor, leading to increase in plutonium stockpile.

6. Is it just a coincidence that the Iran and North Korean issues are intensifying at the same time?
   - doubt it's a coincidence.
   - each country watches the US and its actions on both fronts carefully.
   - each could be reasoning, two crises too much for the US to deal with.

Many of these notes anticipate that the host will ask the suggested questions. And the interview that aired was indeed similar to the one sketched out here—but only *similar*. The first answer alone shows how much the reporter was prepared to expand on his notes:

> Well, what the Iranians are threatening to do is *not* to start uranium enrichment. Sometimes people get confused about that. There are a number of technological and engineering steps before uranium enrichment, one of which is called uranium conversion, and that's taking the relatively

raw ore called yellowcake and turning it into a gas that then later on can be put through centrifuges and turned into enriched uranium—which, of course, at a high enough level, can be used in a nuclear weapon. So they're *not* threatening uranium enrichment, but they *are* threatening to start uranium conversion.

Also, the host inserted a question (between numbers 2 and 3) that wasn't in the script, and the reporter answered without hesitation:

HOST: Mike, take us back here. What is the ostensible point of these negotiations? What is it that the Europeans are trying to get the Iranians to do? What might they be offering Iran in those negotiations in exchange for that?

SHUSTER: Well, the Europeans, with the backing of the United States, want to see Iran not only suspend uranium enrichment, but forgo it all together. They don't want Iran to enrich uranium. The Iranians want ultimately to be allowed to enrich uranium, but under the supervision of the International Atomic Energy Agency in order to create nuclear fuel for an eventual civilian nuclear power industry. How to bridge this gap is a tough one.

The Europeans had hoped that they could offer the Iranians economic incentives, and they've discussed that, including a clearer way to membership in the World Trade Organization. The United States has also agreed to waive some sanctions in order to allow Iran to purchase spare parts for some of its airliners. The Iranians want more, clearly, and haven't gotten it from the Europeans, and that's what they're complaining about in public. So it's difficult for me to see how this gap can be bridged right now even in further talks, or further immediate talks.

Because the notes were there, the host knew how much latitude he had to bring up related issues. The extra background information helped clarify the suggested "compromise" question. And it's worth noting that the host changed that question slightly, to make it transparent that he *knew* the reporter did not see a compromise:

HOST: You don't see the possibility of a compromise—or, more to the point, the parties don't seem to see the possibility of a compromise?

SHUSTER: Well, when I was in Tehran a couple of weeks ago, and I talked to officials and semi-officials about that, they believe that they have the right to enrich uranium...

The result is a two-way with a reporter that sounds as conversational as most other interviews, but has the virtue of condensing a maximum amount of information into a minimum of time. In this case, the reporter knew about the subject from his travel to Iran and from having covered nuclear weapons and arms control issues in the past. If you're a reporter doing a two-way, the best preparation is to know your subject cold. But regardless of your level of experience, you're probably going to be called upon at one time or another to field questions on the air—perhaps live—about a story you have reported. "This should be a minimal requirement," says Simon. "I'm not saying this of some twenty-five-year-old field producer who happens to be the only person on the scene of a news event—but a staff reporter, a staff correspondent, ought to be able to go live."

Two-Way Tips

These tips have been suggested by reporters who frequently do two-ways. They aren't hard-and-fast rules. But if you keep them in mind, they may help you overcome any hesitation you may have about going live—and make your interview much more intelligible and interesting for listeners, whether it's live or recorded.

LET THE SHOW KNOW WHERE YOU'VE BEEN AND WHAT YOU'VE SEEN. If you've been stuck in one corner of a giant demonstration, chances are you won't be able to talk about the skirmish that happened half a mile away. The host needs to know that.

SUGGEST QUESTIONS, OR AT LEAST TOPICS THE HOST SHOULD ASK ABOUT. But you might also solicit questions from the show; that way you're not likely to be surprised on the air.

LET THE SHOW KNOW WHAT YOUR ANSWERS WILL COVER—especially when a suggested question includes a reference whose antecedent won't be obvious to the host (e.g., "Why is this meeting so important?" when previous questions don't *mention* a meeting).

FEEL FREE TO SUGGEST OR PROVIDE AN ACTUALITY TO GO IN THE HOST'S QUESTIONS. It can sound smart for a host to say, "When the President spoke to the striking miners, he made a point of mentioning national security. Let's hear a little of that speech." Then the sound plays, and the host can ask you, "How did the miners react to that appeal?"

IF YOU'RE INCLINED TO USE AN ACTUALITY IN YOUR ANSWERS, THINK ABOUT HOW IT WILL SOUND TO THE LISTENER. You don't want to evoke the scene in the film *Annie Hall*, where Woody Allen just *happens* to have philosopher and critic Marshall McLuhan at hand. If it does make sense to use an actuality as part of your response, introduce it conversationally: "I've been talking to people all day, and they seemed pretty angry. Here's a bit of what one of the miners told me..."

IF YOU WRITE THE INTRO, DON'T SUGGEST A FIRST QUESTION THAT ECHOES FACTS YOU'VE ALREADY INTRODUCED. This makes hosts sound like they're not listening to themselves. If the intro says the President addressed the United Nations to drum up support for rebuilding Iraq, you can't have a first question that says, "Why was the President addressing the U-N?"

DON'T WRITE QUESTIONS THAT FORCE YOU TO CORRECT THE HOST—that in effect imply "your assumption was wrong; let me set you straight." This is especially annoying to hosts when they would never have asked that question except for your having suggested it.

USE SPECIFIC, VISUAL EXAMPLES AS MUCH AS POSSIBLE. This reminds us where you are, and what you're reporting about.

HAVE NOTES HANDY, ESPECIALLY IF THERE ARE FACTS YOU COULD EASILY GET SCRAMBLED. When you're under the gun, it's easy to say "twenty-five million" when you mean "twenty-five *billion*," or to say "1968" when you meant "1986."

IF YOU DON'T HAVE INFORMATION A HOST ASKED ABOUT, ADMIT IT. You can often deflect a question that you can't answer directly. ("I don't know how *many* tiles are on the space shuttle, but I can tell you they should be able to withstand a temperature of about three thousand degrees Fahrenheit...")

DON'T INTRODUCE A SUBJECT THAT HAS BEEN ON THE AIR ALL DAY AS IF THIS IS THE FIRST TIME ANYONE HAS HEARD OF IT. If a topic or event has been covered in newscasts or on the Internet or on cable TV, you can start with something between a first and second reference to it: "The President's address at the United Nations this morning was meant to drum up support for the reconstruction of Iraq."

RUN A CLOCK, AND KNOW WHEN YOUR TIME IS RUNNING OUT. You may want to ask the program producer or studio director ahead of time whether there's wiggle room—i.e., whether you can run ten or fifteen seconds past your allotted time—but don't push your luck or you'll be cut off.

CUT YOURSELF SOME SLACK; THIS IS A CONVERSATION, NOT A POLISHED SCRIPT. You're allowed to think about an answer for a second before you start talking, especially if you weren't ready for it. If you misspeak or stutter, you can correct yourself.

DON'T INADVERTENTLY ENCOURAGE LISTENERS TO TURN OFF THE RADIO AND WATCH TV. When cable networks are broadcasting riveting video of the same event you're describing, it can be hard to stop yourself from saying, "We're seeing just incredible TV pictures right now taken from helicopters that show how huge the crowds are at this demonstration!" But a comment like that will make a lot people stop listening to *you* and turn on the television. Try to think in advance whether there is some other was to relay any information you're gleaning from TV that supplements your firsthand reporting—for instance, "I've seen pictures that show the crowd stretching from the Capitol all the way to the Washington Monument."

DON'T RESPOND TO A QUESTION THAT YOU'VE SUGGESTED TO A HOST BY SAYING, "THAT'S A GOOD QUESTION!" The audience might not realize what's going on, but your colleagues will always have a laugh at your expense if you praise your own questions on the air.

CHAPTER 8

Reading on the Air

Lots of people on the radio talk without any scripts. DJs identify songs and artists and make jokes, sports announcers give the play-by-play of football games, talk show hosts field phone calls, pundits discuss political issues—all without reading from prepared texts. (People who worked with the late humorist and storyteller Jean Shepherd say he used to come to the studio armed with only a few notes, a newspaper article, or a kazoo; without ever once using a prepared script, he somehow managed to talk for a total of about five thousand hours over the years.) There will be some occasions—during fundraisers, on talk shows, during live events or when there's late-breaking news—when public radio reporters and hosts also will have to ad lib, but most of the time they will be *reading* from scripts they have written themselves, or that others have written for them. So if you are on the air, or hope to be, you should be able to read your own writing effectively.

"You can't get around the fact that if you're a radio reporter, there is a performance aspect to the job," says Sora Newman, who has been a trainer at NPR for more than fifteen years. "If you don't want to perform—at least a little bit—you're probably in the wrong business."

That doesn't mean you have to have an operatic voice to be on the radio. "I have heard people with really beautiful voices, and you can get lulled to sleep by the beauty of it," Newman says. On the other hand, "We have some people on our air who, in the heyday of radio, would not have gotten jobs—they don't have pretty voices—but they write for their own voices in a way that makes them good storytellers."

And it doesn't mean you necessarily have to become a master of elocution. Most people listening to NPR are native English speakers; over-enunciating is not the best way to get them to pay attention. Unless you have a speech impediment or an unusually harsh voice, you can probably be on the radio.

"Probably"—because there is no substitute for talent. Some people seem to have been born with the "performance gene," and sound like old hands almost as soon as they start reporting. Others appear to struggle with their delivery; they may always retain some fear of being on the radio, consistently stutter or make other mistakes, and never become satisfied with the way they sound. There are limits to how much you can learn about delivery by reading a guide like this; for many people, the best training will come from spending some time behind a microphone. But there a few important goals to keep in mind, and some tips that can keep you on track toward reaching those goals.

Sounding Conversational

The principal thing to remember when you're on the air—just as when you write your scripts—is that each listener feels like you're talking directly to him or her. You are not giving a lecture; in fact, as far as that listener is concerned, you're not even reading a script. You're just talking. So the first step toward reading well is writing well. "You can take something that's written conversationally and read it badly," Newman says. "But it's really hard to take something that's not written conversationally, and make it sound conversational."

When Newman is coaching reporters or commentators, she asks them to picture the specific person who's listening—a relative, friend, or colleague—and to imagine where that person is, what he or she is doing, even what they're wearing. If she's coaching someone who's working out of a home studio, she may suggest the reporter look at a photograph. Newman says, "When the person starts practicing their script, I'll ask, 'Are you *really* talking to your Mom?' And often what I get is, 'Well, I wouldn't talk to my Mom like that.' So I'll ask, 'How *would* you say it?'" Once you know who you're talking to, you can imagine *how* you would talk—and edit your script accordingly. Much of that editing will take place in the studio, as you're reading your script out loud.

"What I'm actually trying to do is get people—it seems so obvious—to *think* about what they're saying," Newman says. When we're chatting with friends at a bar or talking on the phone with relatives, most of us naturally speak with expression, energy, and interest. But something changes when we are reading from a script. Many people start to read the words on the page without reflecting on their meaning, the way children sometimes recite poems they don't understand. We may speak in a monotone, or with a sing-song pattern to our voice. So Newman says—in addition to writing conversationally and imagining one actual person to talk to—a good reader needs to "inhabit" the copy. "What that means is that you're not just reading the words—blah, blah, blah—but you're *really* thinking and feeling what you're saying. You're trying to communicate."

For instance, if your news report includes a description of a place—say, the ruins of a church that had been burned down overnight—you need to remember what that scene *looked* like as you read what you've written about it. If you're paraphrasing what someone told you—"Mary Jordan says she'd been coming to that same church since 1970; now she says she feels lost—and for the first time since her husband died ten years ago, completely alone"—remember what Mary Jordan looked and sounded like as she told you those things, and how you felt when you heard them. Your script doesn't have to be laden with emotion for you to get involved in it. If you're explaining how a new kind of fuel cell works, think about how the inventor or marketer described it to you. *All Things Considered* host Michele Norris says, "I try somehow to have a visual image of what I'm talking about, whether it's some new microbial discovery that a science reporter is telling us about, or the space shuttle—I try to have some sort of image in my head so I can get inside the words."

Here Norris is talking about reading intros written by other reporters and editors; her experience of the subject is secondhand. If you're reading your own report, chances are it will be easier for you to recapture some of the enthusiasm you felt when you pitched the story, or were reporting it, or were writing the script.

Thinking about and feeling what you're reading is not just an exercise; it's the sine qua non of being on the radio. "The example I use in my delivery workshop is 'the broccoli conversation,'" Sora Newman says. "Remember, people are driving down the street. Some idiot is cutting in front of them. The kids are in the back seat making noise. And you're giving your little speech, you're reading your copy. Now, even though you may be reading perfectly well, there are all these things competing for the listeners' attention, and they sort of drift away and end up thinking, 'You know I have some broccoli I need to use up. So I could maybe pick up some fish, and get some carrots . . .' And they're gone! They're not listening to you talk!"

This doesn't mean a good radio reporter has to be an actor. An actor assumes the character of another person; radio reporters simply need to be themselves, and to read with the same intensity and cadence and music in their voices that they exhibit outside the studio.

Marking Up a Script

Encouraged to put a little life into their words, some people overdo it.

One mistake many newcomers to radio make is to stress too many words. If you listen to the way people talk when they're *not* on the air, you'll find that most of the time, they vary their pitch to emphasize one or two key words in each sentence. In contrast, reporters and commentators—perhaps because they want to sound authoritative, perhaps because they feel so much of what they've written merits attention—frequently stress five or six words in every sentence of their scripts. As a result, their pitch falls into an unnaturally regular pattern.

For instance, here's one way a person might say the sentence above:

> If you listen to the way people talk when they're not on the air, you'll find that most of the time, they vary their pitch to emphasize one or two key words in each sentence.

There are certainly other ways to read the same sentence; the point is that the stressed words don't fall into any predictable pattern. In contrast, a new radio reporter might read the sentence more like this:

> If you listen to the way people talk when they're not on the air, you'll find that most of the time, they vary their pitch to emphasize one or two key words in each sentence.

Besides sounding mannered, the problem with reading this way is that it doesn't help the listener focus on what's important and what isn't. The abundance of stresses, and their regularity—coming every third or fourth word—can make the reporter sound like he's reading a bedtime story to a child.

So read your story out loud before you mark it up in any way, and listen to how you are speaking. As you "inhabit your copy," just underline the one or two new ideas in each sentence. And then make sure the underlining is only a mnemonic. The goal of marking up a script is to remind you how you would say the sentence if it were not written down at all, not to force you to read it with artificial emphases.

There may even be times when you want to remind yourself what word *not* to stress. For instance, many reporters inappropriately emphasize the word "people" in phrases like "a demonstration by about two thousand *people.*" This always sounds like the speaker was concerned we might think the demonstration was staged by armadillos, and just wanted to clear up any possible confusion. If you have a tendency to fall into that trap, or similar ones, you may want to indicate to yourself in some way—maybe by putting a squiggly line under the word—that "people" *shouldn't* be stressed.

News anchors and hosts who read other people's copy may want to mark it up still other ways. Michele Norris says, "I know that there are certain words that are going to trip me up, so I have to 'mouth edit' everything. I'll read it, and I'll figure out where I need to slow myself down. Sometimes I'll put a 'yield' sign before I get to a complicated word—I'll make a mark to myself, sort of like, 'Hairpin turn coming up! Slow yourself down here.'"

Norris also writes in the margins. "I tend to talk fast and I tend to read fast. On a long piece, I'll pick up momentum as I'm reading, so sometimes in the margins I'll have to write—literally write to myself—'Slow down.'"

Some people like to put a vertical line everywhere they would naturally pause. "In some ways, pausing is more critical than emphasizing a particular word," says Sora Newman. "Pausing lets the listener have a second to take in what somebody's saying, particularly if it's complicated, and it lets you, the reader, breathe and stay aware of what you're talking about." Even if you don't want to use underlines, and slashes, and yield signs, you should at least heed your own punctuation. Many people use commas or dashes when they write, and then ignore them when they read—plowing through their scripts headlong, with scarcely a break. Many experienced broadcasters don't even use commas in their scripts . . . they use three dots instead . . . simply to remind themselves where to pause.

Remember, not every sentence—or every part of a sentence—has to be read at the same speed. You don't want to race through a phrase so quickly that listeners can't keep up with you, but you can adopt the same varied pace in your on-air delivery that you have when you speak. If you're explaining something technical, or describing something sad, you may want to slow down. If you're repeating the name or title of someone who's been identified clearly and repeatedly earlier in your report, you can speed up a bit.

If your sentences are short enough, you may not want or need to mark up your scripts at all. "People generally—not all of us, but most of us, especially if you wind up in this business—will emphasize the right words, and pause at the right places," Newman says. "It's just a natural thing that happens. But you have to get into the idea of *communicating*, and thinking about what you're saying."

Getting Physical

Reading on the air is a *physical* process. It requires you to use your lungs and diaphragm, your vocal cords, and the resonance of your entire upper body—including your chest, pharynx, mouth, and upper head cavity. Anything that limits or constricts or compresses any one of those things is going to keep you from sounding as good as you can.

The greatest obstacle many people will face is their own nervousness. When you feel tense, your vocal cords get tight—that raises the pitch of your voice and limits your entire vocal range. (If everything you say is higher than when you normally talk, you don't have as much "head room" if you want to vary your pitch.) Most people will overcome "mic fright" with time—but not everyone will. "It's one of these Catch-22 situations," says Sora Newman. "If you're really worried about your delivery, then you're consciously trying not to create a problem—which means you get tense, which causes a problem!" It's a lot easier to be relaxed if you haven't stayed up all night finishing your story, and aren't rushing into the studio to track it just a few minutes before it goes on the air. But even if that's how you work, there are a few things you can do to "tune up" your vocal instrument—and to keep it in tune.

LOOSEN UP YOUR BACK, NECK, AND JAW MUSCLES. "People come to the studio with their teeth clenched, and their shoulders are up at their ears—you just can't sound good if your body is so tight," says Newman. "Drop your shoulders—twist them around a little and move your neck—and get your mouth moving."

BREATHE DEEPLY A COUPLE OF TIMES BEFORE YOU START TO READ. "I suggest that when they get in the studio, they take a deep sigh—a deep breath—and then vocalize," Newman says. "Make a sort of sigh with sound."

COUNT DOWN FROM THREE—LOWERING YOUR VOICE WITH EACH NUMBER. You want to set a starting pitch for the first words of your story that's more or less in the same register you usually use for speaking. If anxiety is pushing you into a higher key, counting down can sometimes bring your voice back where it belongs.

SIT UP STRAIGHT—OR TRY STANDING. There's a reason you don't see opera singers sitting down when they perform. You want to use the resonance of your whole body if you can. Newman says, "Standing up works for a lot of people, especially women with more shallow voices. Or anybody who's tired, because if you're tired you tend to fall into the microphone!" (When one reporter was cajoled to keep her head up, to sit up straight, and to breathe deeply, she quipped, "If you tell me to get married, you'll sound just like my mother!")

AVOID CAFFEINE AND CIGARETTES. Most reporters drink lots of coffee when they're trying to complete their stories, which Newman says tends to tighten up their vocal cords. And smoking dries them out. "It's bad, bad, bad, if you're going to be on the air," she says.

Newman and other voice coaches will often suggest various vocal exercises, designed to address specific problems—a lack of breath, or sibilance, or mushiness of enunciation. Like most regimens that demand self-discipline, vocal exercises are followed much less often than they are prescribed. But they can pay off. "It's not like other exercises, where your cholesterol may go down but you don't look any different," Newman says. "You can really hear the difference."

The Need for Feedback

Many reporters like to have someone in the control room to coach them as they read their scripts. A good coach can remind you to relax, to sit up, and to get involved with your story. But in the long run, developing an effective delivery on the radio requires having a good ear, as well as a good voice. You need to be able to listen to yourself as you're reading, and to recognize when you're sounding interested and energetic, when you're varying your pace and pitch naturally—and when you're not.

Start by listening to a report or commentary that you've recorded. Don't read the script at the same time—just *listen.* Do you sound like you're talking, or reading? Do you seem to be stressing the key words? Are you reading more slowly or giving more emphasis to the ideas that need to stand out? Does your pitch seem natural—or does each sentence seem to have the same "melody"?

Most important, can you hear what needs to change? And if you *make* the changes, can you hear the improvement? "People who cannot tell if they're doing one take better than another will have a harder time improving," Sora Newman says. "It's like anything else. If you're doing a tennis serve and you can't tell that your arm is in the wrong place, you can't get better."

If you're not satisfied with your performance but unsure what you're doing wrong, ask someone to interview you about the same subject you cover in your script—and record that interview. Then compare your *scripted* version with the *unscripted* one. Chances are you'll hear some obvious differences. You may discover that you always stress the last word in every sentence when you read your script, or that you have trouble with certain sounds, or that you read in a monotone. You can't eliminate flaws you don't hear, Sora Newman points out. "Some people who can't change right away will say, 'Oh, I did that wrong again.' That's the first step toward developing a feedback system—recognizing what's wrong."

One of the hardest things to recognize is whether you're reading at a comfortable pace—comfortable for someone who's trying to take in everything by ear. Some of us naturally talk very quickly or very slowly, and we may need to change our usual pace so that listeners are not racing to keep up with us—or impatient that we don't get to the point quicker. "People who have trouble pacing themselves may really need someone else to give them feedback," Newman says. "Fast talkers who slow down think they're crawling, and they're still probably too fast." Word processing programs used in broadcasting sometimes will estimate the amount of time it will take you to read a piece of text. If pace is a problem for you—if you read what should be three minutes of copy in two minutes and forty seconds—don't adjust the "read rate" so the word processor jibes with your delivery; instead, adjust your delivery so that you begin to match the word processor. Or identify someone whose pace you'd like to emulate, get a script and recording of his or her report, and read along as you listen. Try to get a sense of what it *feels* like to read at a more leisurely pace.

In fact, Sora Newman says even people who can't *hear* a difference between a good read and a bad one may be able to feel it. "They'll say, 'I felt a little more animated, or I felt my body perk up.' Some people are going to get that feedback physically, rather than through their ears," she says. "What matters is that they get it *somehow*."

CHAPTER 9

Hosting

When Michele Norris arrived at *All Things Considered* in 2002, she turned to veteran host Susan Stamberg for advice. Norris had worked in TV and at newspapers, and was hoping to get a sense of her new role as an NPR host from someone who had defined the job for many years. She remembers that Stamberg used a metaphor to give her the big picture. "She said hosting is like sitting at the head of the table at this big dinner party, with all these varied guests—who don't know each other, who don't necessarily speak the same language, who don't necessarily even like each other! And your job, as the host," Norris says, "is to make sure everyone has a good time at the party, that everyone is enlightened when they leave, that you keep the conversation going."

The people who present public radio programs do have a number of traits in common with good hosts *off* the air. They are consummate storytellers who have a keen sense of what needs to be explained, what facts can be left out, and how to keep people waiting for the story's conclusion. They make clear and concise introductions—to interviewees, to reporters, and to commentators. When they are speaking with a guest, they don't dominate the discussion or draw undue attention to themselves; they keep the focus on what the interviewee has to say. They also listen attentively and know how to draw people out by asking thoughtful questions. They speak fluently, and their voices are pleasant to listen to. And they maintain—or seem to maintain—a high level of energy and concentration, even when they have gone many hours without sleep or food.

Admittedly, the comparison only goes so far. In public radio, the hosts of news programs are also seasoned journalists. Their interviews are intimate and engaging in many circumstances—but tough and adversarial in others. They play critical editorial roles on their respective shows, helping to decide what stories will be covered that day, that week, and over the course of the next few months. They personify the network and give it voice—so that listeners rely on them when the news is fast-breaking, catastrophic, or sorrowful. As *Weekend Edition Saturday* host Scott Simon says, "They expect us in a sense to register the depth and the gravity and the consequences of what's happened in public along with them. They turn to us in moments of need—less for solace than for evidence that the world is still spinning." The public radio audience expects hosts to be honest, credible, versatile, quick-witted, articulate, and indefatigable, at times tough or soothing or whimsical or funny—and always trustworthy.

The Host's Day

Just by listening to the radio, it would be hard to tell exactly what the hosts *do* over the course of a day or a week. Listeners hear them conduct interviews, but don't know if they come up with their own questions. They hear the hosts read intros to news reports and commentaries, but don't have any idea where those intros come from. Considering how close many regular listeners feel to the hosts, it's sometimes startling to discover how little they know about their jobs.

A host, Michele Norris says, "is in some ways the compass and soul of the show"—an integral part of the program's editorial team, along with the senior producer and senior editor, and on many shows, the executive producer. That role is built into the job. In a memo defining the roles of an NPR News host, Senior Vice President for Programming Jay Kernis specifically highlighted "editorial leadership":

> They begin each day with stories they hunger to get, and questions they want to ask. They have ideas on how stories and interviews should be focused and on what items might be included in that day's line-up. They argue effectively for stories that will mean a lot to listeners or stories that need more attention or stories that bear repeating or that need updating. They know the value of strong editing.

Hosts contribute their own story ideas, and they help sift through the other people's pitches. If someone suggests doing an interview about a town in Oklahoma that has tried to outlaw rap music—this is a fictitious example, by the way—the host would subject the pitch to the same vetting process as any other story. Why would people elsewhere care? Who would we talk to? Is there another side? And since the host will be the one conducting the interview, especially if she's not thrilled with the story, she may ask, "What do you hear as the second and third questions?"

Hosts say they need to be persuaded of the viability—and focus—of any story they'll be pursuing; they don't want to end up in the studio wondering why they're doing the interview. But they recognize that they work in concert with the editors and producers, not in opposition to them. They shouldn't have veto power over story ideas, says Noah Adams, who hosted *All Things Considered* for more than twenty years. "I think you can assume if the program [producer] wants it done, and there's no moral objection to it, that any host is going to go in and do their very best," he says. "And you're not going to be able to tell if they liked the idea or not—or *shouldn't* be able to tell."

In public radio, host interviews can deal with nearly any topic. As Kernis writes, "Hosts are extremely versatile . . . They can converse with the shopper at the mall, the scientist in the lab, or the President in the White House." For proof, look at almost any news magazine. On a single *Weekend Edition Saturday* program, for instance, Scott Simon conducted interviews on Israel's planned withdrawal from Gaza; on why a celery farmer there did not plan to move out; on the obstacles to passage of an Iraqi constitution; on the legal ramifications of racial and religious profiling to combat terrorism (which was a three-way interview, with two guests); on blues legend "Little Milton" Campbell; on Dreamworks' plan to divide itself into an animation and a live-action production company; on some of the oddest and most annoying interpretations of Elvis Presley's songs; on news of the week in review (which included anti-war protests, the nuclear ambitions of Iran and North Korea, and the Gaza and Iraq stories covered in other interviews); on research connecting mathematical game theory to romance; on the White Sox winning streak; on a protected habitat or "fish hotel" beneath the Chicago River; and on guitarist Eric Johnson's blend of country, rock, and jazz (which was a "performance chat," where the guitarist played excerpts from his music). Taking on so many diverse subjects—which may sound effortless on the radio—involves a lot of work. A successful host has to be as credible talking about nuclear weapons or racial profiling as he is when he's chatting about jazz or baseball. And hosts have to be able to modulate their on-air performance accordingly, without sounding as if they are acting.

The key to hosting such a heterogeneous news magazine is preparation—and for Michele Norris, it begins hours before she goes on the air. "You start by sitting down and figuring out what's the toughest interview you think you have that day—toughest maybe in terms of talking to someone who's not forthcoming, or a subject you're not familiar with, or the thing that is changing most during the course of the day—and prepare for that first," she says. "Make sure that you arm yourself with whatever you need for that interview, and then work backwards." For longtime NPR host Liane Hansen, hard news interviews often require the most planning and study. "When I know I'm doing an interview on a certain subject—because it can range from North Korea, to weapons of mass destruction, to agriculture in Brazil, to whatever—I try to read up on it, to read all the wire stories, and all the articles in the newspaper." And if she has enough time, she often talks with a reporter or editor whose beat includes the subject of the conversation. When your day includes such a varied array of topics, it's smart to get assistance wherever you can find it—from news wires, producers, editors, reporters, or librarians. "I need as many safety nets as I can get."

It also makes sense to check with the senior editor and senior producer to be sure everyone agrees on the focus of the conversation—that they're looking for analysis and not "color," or that they want a personal story from the rancher in Wyoming to flesh out a report from Washington on new environmental regulations. If you have several interviews scheduled back to back—which happens often on daily shows—you may want to take a moment to "cleanse your mental palate," says Norris. "You just stop, and try to think, again, about what am I trying to achieve in each interview? What's the tone? When you're talking to Cal Ripkin about baseball, it's very different from when you're talking to a U.S. ambassador about the Gaza Strip or someone at *Jane's Defence Weekly* about a Chinese aircraft carrier."

Many of the interviews on public radio programs are recorded and edited, rather than conducted live—and some hosts want to be involved in how their interviews are edited. *Morning Edition* host Steve Inskeep says, "Sometimes I argue for more time for the interviews, and sometimes I argue to cram less in." He gives the example of his interview with Republican Senator Rick Santorum of Pennsylvania. "We had a really, really absorbing thirty-minute discussion, which would have been nice on talk radio," he says. "And when I talked to the editor, I said, 'It's important to me that we not try to cram in so many sections that we don't have a back-and-forth.' It was both for substance and style, in my opinion, because I felt we had a really interesting conversation, as well as a really substantive discussion that called for push-back." Scott Simon says he usually leaves the cutting of his recorded interviews up to the producers and editor—but he says if he has asked a question that would occur to most listeners, he expects it to stay in the interview—even if a producer thinks the answer wasn't informative or substantial. "The classic example to me years ago was when we were interviewing a young woman who was on her high school wrestling team," he recalls. "In those days, she was the only young woman on a high school varsity team who was competing against young men. And I asked what I think is the first question on almost everybody's mind, which is, 'Are your opponents sometimes reluctant to touch you because you're a girl?' And her answer to that was, 'No, I haven't noticed that.' So it didn't get in the edited version! And it sounded like we didn't ask it."

When you're hosting, you may also want to listen to a recorded interview after it has been cut but before it airs. *Morning Edition* host Renee Montagne says she often acts "as a sort of first-line editor," suggesting a change if she's not pleased with the way the interview has been edited the first time around.

When you host a public radio news magazine, you will usually write the intros for your interviews. Depending on the format of the show, you may also write an essay; the billboards and returns that start each hour or half hour; and the script for a story reported in the field. Even when you're just reading introductions to other people's reports or commentaries, to a greater or lesser extent you'll probably tailor the scripts you read to suit your own style of speech. Much of this editing takes place in the studio, sometimes just moments before it's time to read the script on the air.

In other words, what the audience hears each morning or afternoon—the hour or two of "hosting" the news program—is really the culmination of a whole day's effort, or for the weekend shows, a whole week's work. On public radio, a host is not just the "talent." Even when the live program is over, you may still need to stick around for updates. Keeping the shows current for refeeds to different time zones can require rereading intros, or redoing interviews.

For many hosts, it all adds up to a ten- or twelve-hour workday. Renee Montagne says that as a reporter, she'd sometimes work on a story for eighteen hours in a row; reporting is "really tough work, no question about it." But hosting *Morning Edition* is harder, especially when the show includes big stories that require constant updates. "When you're out reporting, stories have an arc, and then they're over. But what happens to us, in these situations [i.e., with continuing and changing stories], is it just goes on. You get to the weekend, and you get to read all those books for the interviews that start on Monday! There are times I've been so busy that people had to bring me food in the studio. The hardest thing is the relentless quality of it."

The Host Interview

A radio reporter usually does an interview to get material for a story; a ten- or fifteen-minute conversation may end up as two or three actualities in the final piece, totaling perhaps forty-five seconds or a minute. In contrast, a program host assumes that his or her interview will stand on its own—that it will *be* the story. That difference affects almost every aspect of the interview, and calls on a different set of journalistic and performance skills from those needed to do a good job as a reporter. In his memo defining the attributes of NPR News hosts, Jay Kernis lays out the standard for host interviews:

> They know how to quickly establish rapport—in person or over a satellite; how to gain an interviewee's trust; how to guide a conversation; how to ask provocative and challenging questions without offending the interviewee (although this doesn't necessarily bother them) and the audience; how to add information and not interrupt the flow of the interview; how to elicit new information, insights and memorable stories; how to take what is essentially an artificial situation and make it sound urgent and real; how to serve as advocates for the listener.

GETTING READY

Most hosts will spend time getting their guests comfortable before the interview begins—especially if they're going to be speaking with people who are not used to being on the radio. When he was doing phone interviews as the host of *All Things Considered,* Noah Adams even placed the call himself so he wouldn't come off as officious; he didn't want a producer to tell the guest—in effect, if not in so many words—"Stand by for Mr. Adams." Once the guest is on the phone (or the connection is made between the host's studio and the one the interviewee is in), you may want to engage the guest informally while the engineer gets a voice level.

"I just start talking with them about stuff that might have nothing to do with the interview," Liane Hansen says, "or say, 'Hi, how are you?'—and try to strike up a conversation, so they're used to talking to me before we begin to record." Adams would often talk to people for five or ten minutes before the real interview began. "What I'm trying to do there is teach them how we're going to do the interview," he says. "This is the 'key' we're going to be in, this is the pace, this is the amount of friendliness we're going to have, and I'm going to tell you a tiny little story about when I was in Minnesota—that's to show you that you can tell *me* stories, too." The whole point of the warm-up is to put the guest at ease, so the interview you have planned for begins as seamlessly as possible.

"Sometimes it's *too* seamless," Adams says, "and they'll ask, 'Have we started yet?' Which means you've done something wrong." As a host (or when you're out reporting, for that matter), you should always make sure the guest knows when you're recording. Most hosts have this down to a formula—a sort of Miranda warning for recorded interviews that also suggests how much they might be edited. You might say, "We're going to talk now for about ten or fifteen minutes, and then our conversation will be edited down to four and a half minutes for tomorrow morning's program. That means you can repeat an answer if you don't like the way you said it the first time—but the more concise both of us are right now, the less we'll have to be edited. Okay?" On the other hand, when time is really tight—if you only have four or five minutes to record a four-minute interview, and especially if it is being broadcast live—you should let the guest know, so he or she will be succinct.

How long the interview actually takes, whether it coheres or rambles, will depend on the topic, the guest, and the amount of time you have had to prepare. Steve Inskeep says the best producers and bookers pass along not only information about the guest, but also anything they've found out by talking to other people about the topic of the interview. He says he was lucky they'd done just that when he was about to talk to Chris Hill, the chief negotiator with North Korea, who was representing the U.S. in talks on nuclear weapons. "It's ordinarily a subject I would have spent considerable time myself prepping for, but I had no time," he says. "But there was an excellent pass-off [that included comments from] an arms control expert whom we interviewed just a few months ago about something else. And I guess [the editorial assistant] basically asked, 'If you had a chance to ask Chris Hill a question, what would you ask him?'" In ten minutes, Inskeep was able to read through the booker's notes and get ready for the interview. When there is more time, most hosts make an effort to study up on their own. "You just have to educate yourself a bit more," Michele Norris says, "by doing the research, and by talking to the person who originally pitched the story, plus one of the head producers or one of the line producers, to understand what are the two central questions that we're trying to answer here."

BEYOND A LIST OF QUESTIONS

The most obvious way host interviews differ from reporter interviews is that your questions are more likely to be heard on the air when you're hosting a show. (Occasionally a reporter will include a question as part of an actuality.) Much of what makes a host interview memorable is the nature and quality of those questions—whether they are insightful, informative, skeptical, clever, and so on.

You may want to sketch out your questions ahead of time, especially if there's an important idea or turn of phrase you want to remember. But don't let your inclination to follow your own script supersede what Robert Siegel calls the number one rule of interviewing: "Listen *very carefully* to what the person you're interviewing is saying." Even as you're framing your next question in your head, you have to be tuning into the guest's answer to your *last* question—and following up, if necessary. "Rather than move onto the question you've placed on your page, be sure that if you hear that person saying something that leads you elsewhere, you go *there*," Siegel says.

That's what Siegel did when he interviewed Homeland Security Secretary Michael Chertoff in the aftermath of Hurricane Katrina in 2005. Tens of thousands of people had sought refuge for days in the New Orleans Superdome and at the city's convention center, but they lacked food, water, beds, and toilets. So Siegel asked Chertoff, "How many days before your operation finds these people, brings them at least food, water, medical supplies, if not gets them out of there?" Chertoff replied with an answer about the situation in the Superdome. So Siegel pointed out, "But this is the convention center. These are people who are not allowed inside the Superdome." Chertoff spoke again about the Superdome, adding, "The limiting factor here has not been that we don't have enough supplies. The factor is that we really had a double catastrophe. We not only had a hurricane; we had a second catastrophe, which was a flood." After another question and answer, Siegel and Chertoff had this exchange:

SIEGEL: We are hearing from our reporter—and he's on another line right now—thousands of people at the convention center in New Orleans with no food, zero.

CHERTOFF: As I say, I'm telling you that we are getting food and water to areas where people are staging. And, you know, the one thing about an episode like this is if you talk to someone and you get a rumor or you get someone's anecdotal version of something, I think it's dangerous to extrapolate it all over the place. The limitation here on getting food and water to people is the condition on the ground. And as soon as we can physically move through the ground with these assets, we're going to do that. So . . .

SIEGEL: But, Mr. Secretary, when you say that there is—we shouldn't listen to rumors, these are things coming from reporters who have not only covered many, many other hurricanes; they've covered wars and refugee camps. These aren't rumors. They're seeing thousands of people there.

CHERTOFF: Well, I would be—actually I have not heard a report of thousands of people in the convention center who don't have food and water . . .

This was not what Siegel had actually planned to talk to the homeland security secretary about, he says. But because he went where the interview took him, he got the secretary to make an admission that made news around the country—namely, that this was the first he had heard of the dreadful conditions at the convention center. "Very often in an interview the best-laid plans are abandoned for good reason," Siegel says.

One reason not to adhere to a list of questions—particularly if the guest is not a political figure—is to make sure the interview sounds like a *conversation* and not an interrogation. "I think particularly with artists, with musicians, with actors, with directors, with authors, the conversational aspect is something that's important," Scott Simon says. And one of the ways Simon keeps things sounding natural is not to ask questions at all.[1] "You sometimes don't get a response by asking a question the same way you do by making a statement."

Here are a statement—not put as a question—that Scott Simon put to John Madden, director of the film *Proof*, and Madden's response:

1. Many conversations in daily life are actually interlaced monologues. When we talk to our friends and spouses—even for hours at a time—we ask relatively few questions, but allow our statements to dovetail loosely.

SIMON: The underlying anxiety that animates many of the characters in the film seems to be, and I don't mean to fill in the blanks too much on this, but, of course, the whole idea is that a proof is something that is certifiable. A proof is something that works or it doesn't. And in a sense, that's the anxiety that Gwyneth Paltrow's character and others have about their mental state—either this is valid or it's not.

MADDEN: That's right. The story contrasts the kind of concrete realities and the certainties that are available in math—for those people who don't know, a proof, a mathematical proof, is a way of validating a hypothesis; some very simple things are very, very hard to prove, and that sort of Holy Grail of mathematics is to find a proof to some very simple conjectures—so it contrasts that sort of certainty with the kind of lack of certainty that we experience in human discourse and human experience . . .

Here Simon is jump-starting the response—putting a premise before the director, and letting him agree or disagree with it. This is in fact the way we tend to get people to talk about their work when we're *not* on the air.

Here's another example, this time on a news topic. Simon spoke to Tariq Modood, an author and sociology professor at the University of Bristol, about the issue of multiculturalism in Britain. The conversation took place after four British citizens of Pakistani descent carried out a terrorist attack on the London Underground system. The prime minister and others had questions about whether immigrants were being assimilated into British society.

SIMON: I think a great many Americans, at least, saw a distinct difference between, let's say, at least British and French approaches to multiculturalism in the head scarf ban of last year.

MODOOD: Yeah.

SIMON: We should explain—let's refresh our audience's recollection—that the government of France said that in the, at least in the state schools, that Muslim girls could not wear head scarves and, for that matter, Orthodox Jews could not wear skullcaps.

MODOOD: That's right. There is a general philosophical principle in France that they refer to as "laïcité"—their understanding of secularism, of the separation of state and religion. And they saw this as a necessary way of entrenching secularism from Muslim pressures and encroachments.

> In Britain, we do have similar debates, but we don't have a problem with head scarves in schools. We see this as part of the natural diversity of society . . .

In this nonquestion, Simon is also performing another critical role of the host during an interview—namely providing "drop-in context," so that the conversation doesn't rely on references that might not be obvious to listeners. This is especially important when you're interviewing someone on a foreign affairs issue, or a technical topic, or any other subject that would usually demand a paragraph or two of information. The challenge is to provide the background facts without making the interview sound like a lesson.

Steve Inskeep faced that challenge when he was talking to U.S. negotiator Chris Hill about North Korea's plans to develop nuclear power—or nuclear weapons:

> HILL: What has been agreed is that at an appropriate time, meaning that when North Korea is back in the NPT in good standing—
>
> INSKEEP: Back in the Nuclear Non-Proliferation Treaty.
>
> HILL: —back in the Non-Proliferation Treaty, that they withdrew from in order to start making nuclear bombs. But when they're back in it, in good standing, the participants have agreed to have a discussion about a light water reactor.

Note how Inskeep interrupts to define an initialism that would have gone over the heads of most listeners. And, at the end of this exchange, Inskeep integrates the mention of the Clinton administration's deal with North Korea and the Bush administration's criticism of that deal. It's appropriate—and necessary—at this point in the interview, because he's asking whether willingness to talk to the North Koreans represents a change of policy for the U.S.

> INSKEEP: Now already I think we have a difference of opinion here because the North Koreans appear to be saying in a public statement, 'First, we get a nuclear reactor, and then we'll dismantle our nuclear program.'
>
> HILL: . . . What I can assure you of is the North Koreans understand full well that they can't have a discussion about a nuclear reactor until they're back in the N-P-T.

INSKEEP: Isn't this a dramatic change? Because administration officials came in—and you were there in the early parts of this administration—saying that it was a terrible mistake to negotiate with North Korea. During the Clinton administration in 1994, an agreement was reached to provide the North Koreans with peaceful use of nuclear energy and other concessions as well, and the North Koreans agreed to dismantle their nuclear program, and didn't.

HILL: Well, you recall what was done there, was there was a freeze . . .

The secret to success for all of these interviews is to be fully prepared—but not to let that preparation get in the way of an animated and organic conversation. You have to be ready to explain what the NPT is, or to mention the year that the Clinton administration came to an agreement with North Korea, or to summarize France's policy on head scarves and yarmulkes; but you don't want to condescend or sound professorial.

Some of the best facts to use in an interview may come from other conversations you have had, either on or off the air. When *Day to Day* host Alex Chadwick interviewed U.S. Commerce Secretary Carlos Gutierrez about China's imposing "internal tariffs" in response to American quotas on Chinese textiles, he referred to what he'd learned from someone else:

CHADWICK: A textile lobbyist told me yesterday we've lost two million manufacturing jobs in the United States just during the Bush administration. When are those jobs going to come back?

GUTIERREZ: Well, let me just talk to you about the facts as I see them. Manufacturing employment has been on the rise for the past two years. And instead of picking out one number from our overall economy, we should step back, look at the economy at large and look at the bigger picture . . .

One of the most important roles of a program host is to guide people through all the topics and bits of information—as Steve Inskeep puts it, "to connect pieces on the same subject that develop over time." Many subjects are simply too big to be presented in one program; they get covered facet by facet over weeks, or even months—and the hosts can help stitch the pieces together for the audience. "When people speak in the abstract about a host bringing a certain sensibility to the show, I think that's something you can do. You can say, 'This is something we're following, and we're learning about it together, you and me.'"

As a host, you can also accelerate the pace of an interview by summarizing events. Look how Michele Norris advances the story as she is interviewing foreign policy analyst Ellen Bork about the plight of two ethnic Uighurs held at Guantanamo Bay:

NORRIS: Ellen, if we could just take a step back, what were the circumstances of their incarceration?

BORK: . . . A lot of my information comes from a declaration made to the court by their lawyer. He recently had the opportunity to interview them, and they told him their stories, in which they say they were arrested in Pakistan, that they were trying to get to Turkey, which has a substantial Uighur population, where they could work. But through a complicated itinerary and refusal of visas, they were unable to get there, and did end up in Pakistan. And there are any number of other Uighurs also whose transcripts suggest similar kinds of cases of people finding themselves in Pakistan or, indeed, in Afghanistan, which makes it very interesting.

NORRIS: A military tribunal has looked at their case. They found that the men were wrongly incarcerated. They ordered—the tribunal ordered them released. What would happen to them if they returned to China?

BORK: Based on the Chinese campaign against the Uighurs and based on some of the men's comments about China, I think we could fear the worst, that they might be imprisoned, they might be executed . . .

By encapsulating a series of events in her second question, Norris puts the interview into "fast forward" to get us to the next interesting idea—namely, the likely fate of the men.

You may want to do this explicitly, as Melissa Block did when she interviewed *Washington Post* reporter Walter Pincus in 2005 about the leaking of CIA officer Valerie Plame's identity. She was asking him about an internal administration memo in which Plame's name appeared. "I was very mindful of the fact that I needed to—in a quick way, in the questions—establish facts, rather than establish them through a question and a long answer," Block says. So she gave listeners a timeline of the case:

BLOCK: Let's set up just a bit of a timeline here. This memo from the State Department was on Air Force One on a trip that President Bush and then-Secretary of State Colin Powell took to Africa on July 7th, 2003.

PINCUS: It was a month after the date of the memo. But what happened was that Ambassador Wilson sort of did his opening blast, so to speak, on Sunday, July 6th.

"In fifteen seconds," Block says, "you've given the backbone of the stuff you need to know next." Hosts need to develop the ability to pace the interview through the questions—and it's not a skill that comes naturally, even if, like Block, you've got years of reporting under your belt. "It's something I hadn't paid attention to before, until I started thinking about the architecture of the host interview."

GETTING TOUGH

When Renee Montagne made the transition from reporter to host, she had to learn how to do a different kind of interview. As a reporter, she would try to gain the trust of the person she was interviewing—even if she personally loathed his position or politics. "When I was in South Africa, you would interview a racist Afrikaner farmer, and there's no payoff to challenging him. The payoff—and I'm making up this particular example—is when he says what might be astonishingly awful things, and he hangs himself. Or the opposite payoff might be he turns out to be a person whose ideas you don't approve of, but you come to a realization of his humanity and why he thinks what he thinks." When she was reporting—and even now, when she's on the road doing stories—Montagne says she ensures balance by pitting one person's statements against those of his opponents. "You don't have to be the person doing the challenging."

That changed when she became a program host, and realized there would be times when she would have to take on a guest and do an aggressive, probing interview. "Then the question is how to create the right touch, because you still are essentially a surrogate for the listener—you're still in that person's home, and the sense is you have to be a little polite," Montagne says.

Despite public radio's reputation for cerebral or genteel interviews, newsmakers shouldn't get a free ride—and they usually don't. Politicians and high government officials know that they may face some tough and embarrassing questions when they agree to be on the air—so as a host, you should recognize that they do know that. "They have been to the same school you have," Scott Simon says. They are prepared to stay on message, or to deflect difficult questions, or to equivocate just enough that it will appear they've answered when they really haven't. "I think with a public figure is where you have to listen carefully. Who cares ultimately if you follow up or not on a question with a novelist? But with a public figure, you have to be particularly eager to listen to what they have to say and to look at their own words they choose to explain something, and be able to speak in terms that are clearly responsive to the statement that they had. Otherwise, what you're going to get is just a prepared list of responses from them."

As a case in point, look how long it took Robert Siegel to get an answer from presidential candidate John Kerry about whether he would ask for more troops for the war in Iraq:

> SIEGEL: . . . What do you do if you ask the Joint Chiefs of Staff what they need to achieve their mission in Iraq, and they say, "We need a lot more troops"? Do you escalate the troop levels, or do you plan for a quick or a constant exit instead?
>
> KERRY: You have to support our troops, and you have to do what's necessary to try to make this mission successful. But they have not asked for that, and I have to wait until I'm President and sit down with them and see where we are.
>
> SIEGEL: But you yourself have pointed out that General [Eric] Shinseki, the former Army chief of staff, said there should be hundreds of thousands of troops in Iraq, and you say he was fired for saying that. What if you get now the real story, as you would say it—the Army speaking candidly to you?
>
> KERRY: I'll have to make a decision when I get there as to what the probabilities are. I can't hypothesize as to what I'm going to find on January 20th, whether I'm going to find a Lebanon or whether I'm going to find a country that's moving towards an election. That depends on what the President does now. I think the leadership has been arrogant and disastrous.
>
> SIEGEL: But should either you or whoever is President next year consider the possibility of an increase in troops? Is that even a consideration, or should it be completely off the table?
>
> KERRY: I do not intend to increase troops.

Whether you see this as dogged questioning or harassment of the guest may depend on your own politics; in the end, Siegel got a firm and unambiguous reply. From the host's point of view, this sort of back-and-forth illustrates a dilemma—especially when the politician or his staff has set a time limit for the interview, and there's a possibility you'll never get a candid answer. According to Siegel, "The pressure in those interviews is, what if I come away with a recorded interview in which we have a sterile exchange between me and some public official yielding no answer, over three exchanges, that will make somebody in the audience satisfied that we're being tough and persistent with the official? Is the theatrical civic event alone worth our putting on the air?"

Remember, you're conducting the interview not simply to put a public figure on the spot, but to elicit his or her positions, or opinions, or plans. "Doing an interview where you sound tough and getting the person to say something may be two completely different things," Steve Inskeep says. So you have to know when to give up. "There are times where it sounds like we're waiting for an expected answer, and we'll keep hammering the subject over the head with the question until we get the answer that we want," Liane Hansen says. "I figure if I ask the same question three times—different variations of the question—and the answer is not forthcoming, you're not going to get it. And what it's going to sound like on the air is that you're badgering."

Here again, though, to do a proper job as an interviewer, the host needs to know—to use the example above—what Shinseki said about troop strength in Iraq and what Kerry said about Shinseki. "How tough you can be depends completely on how prepared you are," Melissa Block says.

You can ask tough questions of people even if they're not politicians, diplomats, and cabinet secretaries. Years ago, Noah Adams interviewed a doctor whose company was using genetic engineering to try to create a cat that would not trigger an allergic reaction in humans. Adams began the conversation with the simple question: "What's in it for the cat?" After the doctor discussed the science of what his company was doing, Adams went back to where he had begun:

ADAMS: Yes, but to go back to the question, what's in it for the cat?

AVNER: Really, there's—all we're doing is we're turning off a protein that does not perform any significant biological function in the animal.

ADAMS: That's suggesting there's no benefit to the cat—obviously to people, but none for the cat. But are you sure there would be no damage, are you sure in the course of evolution that the cats don't need this particular allergen gene for some reason?

AVNER: Right. This particular protein is considered a redundant protein and it's a protein that's secreted by the cat's sebaceous glands and when the protein is secreted it covers the fur, it covers the skin, and then when people come into contact with this allergen, that's when they exhibit the traditional allergic symptoms that we're used to seeing, such as red eyes, itchy, scratchy throats, difficulty breathing.

ADAMS: You know, the reason I ask you that question is that Dr. Jerry Yang at the University of Connecticut, where the research would be done for your company—

AVNER: Yes.

ADAMS: —was asked, "Will knocking out the gene hurt the animal?" and he replied "We hope not." It doesn't sound like he's very sure about it.

AVNER: Right. Well, you would never know for sure until you actually do it.

Here the host could have saved the "tough" question for last, but by getting the guest to grapple with it right from the start he raises an issue we might not have thought of—and makes us want to stick around to hear the answer.

A certain amount of skepticism can be a useful ingredient in almost any interview. It certainly enlivened Michele Norris's conversation with moviemaker Morgan Spurlock, who ate only high-calorie fast food from McDonald's for thirty days for his 2004 hit documentary *Super Size Me*. Spurlock was trying to show the excesses of a fast-food culture, and Norris asked "if the message got lost in what some would see as a stunt."

NORRIS: Do you think that there's someone in America—

SPURLOCK: Yeah.

NORRIS: —that follows your "McDiet"?

SPURLOCK: I think there's people—I don't think—

NORRIS: Is there a single person in America that—

SPURLOCK: —that eats this for breakfast, lunch, and dinner for thirty days straight? No, God, I hope not, you know, it's like—'cause it's a terrible, terrible way to live. You know, it's a terrible diet. You know, this film is just an example. And is it unrealistic that people eat there for breakfast, lunch, and dinner for thirty days straight? Of course . . .

Norris says it felt a bit like they were having "a scrum in the studio." But the guest jumped right in—"and we wound up having this very robust and yeasty debate." In interviews of this sort—where the guest is not a politician, though his work may have a political motive—you have to know how far to push a line of questioning before the interviewee clams up or turns hostile. Norris says, "It's something I learned at the dinner table growing up—how to question authority and not get sent to your room!"

EDITING IN YOUR HEAD

Whether it's extensively edited or done live, a successful host interview will have a beginning, a middle, and an end—and you'll want to think up questions that will help give the conversation that sort of narrative structure. For that reason, whenever you host a program, even as you listen to the guest and think of your next question, you also need to keep in mind whether—and how—the interview will be edited.

If it becomes clear during a recorded interview that an answer will not survive the editing process, you may need to reask the question to solicit an answer that will. But you have to do it in a way that doesn't suggest that you thought the guest's answer was muddled, or boring, or a waste of time. "You have to do it somehow so the guy doesn't quite realize what you're doing," Steve Inskeep says. "The guy will spend three minutes describing all of Washington, so you'll start the next question saying, 'Washington has a troubled history as a city. What are you going to do about it?' That's a way you can kind of edit as you go."

Here's a real example, from an *All Things Considered* interview with a farmer on Martha's Vineyard who used DNA testing to discover the identity of the dog that had been killing his chickens. The interview started straightforwardly enough—the man described where he lived and how many chickens he'd lost to dogs recently—but it began to run aground on the key question. This is what the unedited version sounded like:

SIEGEL: Well, the reason we're calling you is because of what you did to track down the culprit from among some neighborhood dogs.

FARMER: Right.

SIEGEL: Why don't you tell us about it?

FARMER: Well, it started—let me see. It was in the early part of March, during . . . My daughter and I went out to put the chickens away for the night. It was snowing—big snow drifts. And we found a chicken that had fallen off the roost and fallen into the snow. And my daughter picked it up and brought it back, and I noticed it was pretty cold—snow wouldn't even melt under its wings. And we put it by the woodstove. And it slowly came back to life a little. And then she went on to the chat line and talked to different vets, what she could do, and she nursed it back to health. This is on her, whatever, spring break from college.

By this point, it was probably becoming clear to Siegel, the producer, and the editor that this Q&A would be cut from the interview. So Siegel leaves the daughter and the hens behind, and tries another tack. Here's how the replacement Q&A came out in the edited version:

SIEGEL: Well, at some point, you figured out that the problem had to be one of some dogs that lived nearby. But what evidence did you have of what dog it might be?

FARMER: Yeah. Well, I mean, I had a hair from the chicken house—that a dog went into the chicken house. And in the process of getting through the door—which is just made for chickens—he squeezed through and some hair got stuck in some splinters. And so I hooked up with the animal control officer, and she collected hair from suspect dogs and I sent them off.

SIEGEL: You sent them off?

FARMER: Well, there's a service where you can look at hair through a microscope and compare certain features, but that doesn't identify a particular dog. And so I kept getting the advice, "Well, you know, to do this, you're going to have to do the DNA." And I called up the hair and fiber department of the FBI, and they wanted $1,300 per hair. But they gave me some leads, and I talked to other people and those people gave me more leads. And so eventually, I found somebody that was very reasonable and interested in the case and did the work.

Here Siegel advances and refocuses the interview, and in the process allows for a very long and rambling section to be edited out.

Here are a few other situations when you'll want to reask a question:

- **IT TAKES THE GUEST A WHILE TO GET TO THE CRUX OF THE ANSWER.** "If I'm thinking of cutting [an answer] in the middle," Renee Montagne says, "I'll just say, 'Let me just ask that in a simpler way. You can answer it, or we can use part of your previous answer.' So then if I focus it down to the part that I want, they'll pick up the answer naturally."
- **AFTER ANSWERING YOUR QUESTION, THE GUEST STARTS TO RAMBLE.** Then—especially if time is tight or the interview is airing live—you'll want to interrupt somehow. You may need to verbally

elbow your way in with a phrase that will cut the guest off—"Mr. Jones, let me ask you this"—so that your question is in the clear. When an interview is being recorded, you can be a little more subtle, Steve Inskeep says. "Frequently what I'll do is deliberately stutter: 'Do you— Do you have to . . .' That way, a producer has the option of cutting it off there."

- **YOUR QUESTION IS MARRED BY OTHER NOISE.** You'll want to repeat a question if, say, the guest coughs while you're talking, or there is some loud background sound that's irrelevant to the story (such as construction next door to the office where the guest is speaking on the phone).

And you may well want to repeat part of an *answer* if the phone line is marginal, or the guest has a strong accent, or both:

GUEST: . . . and all around us, vee could see MAH-rins, viz zair MAH-shin guns.
HOST: You could see Marines all around you, with machine guns.

Remember, although you're probably doing the interview in a studio and are wearing headphones, the listener might be in a car, with a bunch of hungry, hyperactive children in the back seat. So the audience will be grateful for anything you do to make a conversation more intelligible.

Noah Adams says his early experience as a tape editor helps him as a host—and he tries to subtly communicate with the producer about which Q&As should be cut. When he's doing an interview, he's thinking to himself, "I'm going to make a little edit here in my mind, and you're going to hear it in a little inflection change in the question. Or I'm going to ask the next question with a neutral inflection that doesn't tie it to the previous answer, in a way that will let you drop seven minutes, and just pick it back up to another place." Producers know they can sometimes stitch together pieces of an interview that occur several minutes apart, but there's little they can do to cover up a sudden shift in the tone of the host's questions.

Whether an interview is live or recorded, it will sound abruptly cut off if you don't end with a concluding question and answer; just as there are times at a cocktail party when you can't gracefully end a conversation, there are times in an interview when it will sound very odd simply to say, "Thank you very much, Ms. Richards." Even when they know the interview will be edited, hosts try to come up with an "ender"—a question that yields an answer that brings the conversation to an elegant close. "There's always got to be an end," Montagne says. "Even if it's an interview with an expert about automobile sales or the price of oil, you still want them to round it out somehow."

Reading Scripts Other People Write

Fifty years ago, it was common in radio broadcasting for one person to write the news and another one to deliver it. The announcer was hired more for his (and it usually was *his*, as opposed to her) attractive voice and acting ability than for his journalistic skills. As a rule, the news reader wasn't really invested in what he read; he just had to make it sound good.

Radio news anchors and hosts generally still need to have pleasant voices and know how to bring a text to life, but they are much more than just readers; they are also reporters and editors, even when they are hosting their shows. From the very start of National Public Radio, Robert Siegel says, "we were integrating journalism and presenting the material into a single role. People did write things for the hosts, but the hosts put their own imprint on that script that was written for them." Giving a script that "imprint" may involve as little as changing a word or two—or rewriting the whole thing. But it means that the host has a stake in what he or she reads—and that almost every intro represents a collaboration between the host and a reporter, editor, or producer. No host thinks of himself as merely a voice.

At one extreme, the host may get the chance to write—or at least edit—everything he reads on the air. For instance, Scott Simon says he usually has a chance to put his own stamp on all the scripts for his program; and producers who write for the show say they expect him to "Simonize" their copy. As a result, Simon doesn't have to figure out how to interpret the script, because he has more or less written it himself. All he needs is the raw material. "As a rule, what I'm looking for are the facts that we need to know to begin the interview—that they're indisputable, as far as we can make them, that they're in the proper order, and that it's apparent why we're doing the interview."

Other hosts may not have the time or opportunity to craft each introduction; in fact, they may have only a minute or two to make sense of an intro—especially if the piece has just been added to the lineup. In a situation like that, Michele Norris says, the first thing she tries to do is just grasp the main point of the copy—it's not always obvious. For instance, if she is handed the intro to a commentary, she may not know if the lead is intended to be read ironically or straightforwardly; to get it right, she'll ask an editor or producer. "They have heard the commentary, so they understand how it works. So I'll call and say, 'What exactly are we trying to say?' If I need to, I'll rewrite it."

Frequently a host may understand the *point* of the script, but not be sure of the *tone* that's appropriate. Since the program producer and editor have presumably tried to make the show have a logical lineup, one of the ways you, as the host, can help the material flow naturally is simply to be aware of the preceding story when you read an intro. Melissa Block says that this requires "listening and trying to match the inflection or reference point of what you just heard." Listening to the show as it airs isn't always as easy as it may seem, since hosts also have to get instructions from the studio director, field calls from the editor, and sort through various iterations of the script and roadmap as changes are made. Fortunately, with digital audio, you don't even have to wait until a story is over to hear how it ends; you can ask the studio director or engineer to let you listen to the conclusion of a piece even as it's on the air, so you can be sure to start the next introduction with the appropriate tone.

Matching your voice to the preceding item gets trickier when you're recording an intro hours (or sometimes days) before a show is broadcast. Often it's clear just from the way a host emphasizes certain words that he or she had no idea what the preceding item was going to be. (For example, you will read the line "New York's mayor is looking for a few good men" one way if the preceding item ends ". . . local police will be discussing the problem this weekend at a conference in New York," and a different way if it ends ". . . another challenge for Chicago's new mayor." Only in the second instance would you put any stress on the words "New York.") So, whenever possible, when you're recording an intro or other script, try to find out where it's going to be in the show's lineup. "I used to drive people crazy when they wanted me to record an intro for the show ahead of time, because I'd say, 'Let me hear what comes before it' or 'Let me hear how it starts,'" Liane Hansen recalls. "They'd ask, 'Why?' And I'd say, 'I want to match my voice to what I did then. I'd like to see what the tone is.'" Hansen says she always tries to remember that she is hosting an entire program, even when she's just recording a thirty-second intro. "Ultimately it goes on the air as a whole, even though it's made up of separate pieces."

Finally, there may be times when—for one reason or another—you are not comfortable reading something another person has written. Some hosts don't want to make puns on the air, since they don't make them any other times. Others feel they don't sound like themselves when they're reading someone else's attempt at humor. You may believe that an introduction implies your endorsement of a book or CD you haven't heard, or suggests you interviewed someone when the audio was obtained by a producer, or in some other way puts you on the spot. Often these problems will crop up at the last minute, as you're reading the script in the studio. Because the host is the very last editor of the program, it's always your right to raise questions about what you say on the air. (The listeners don't make a distinction between what you say ad lib, what you write, and what you read that others have written.) But whenever possible, it's a good idea to let the show editor, producer, and the rest of the staff know ahead of time where you draw the line.

Hosting Live Programming

When it comes to getting the news to people, one of the great advantages of broadcasting is its *immediacy*. NPR runs newscasts every hour (and every half hour, during some parts of the day), so listeners never have to wait very long to find out about breaking news. And when the news event is of great national importance, NPR and its member stations—like commercial news broadcasters—will provide "special coverage." That can mean anything from adding a live news insert into a recorded program to sustaining "rolling coverage" uninterrupted for hours (or even for days).

Hosting this sort of unscripted program calls for a greater degree of flexibility and versatility than is generally called for when you host a news magazine. But since one NPR program, *Talk of the Nation*, is always live, understanding the host's role on that show is a good first step to appreciating what you'll need to do if you're asked to host a live special report.

TALK OF THE NATION

Neal Conan, the host of *Talk of the Nation*, works with a team of producers who help generate story ideas, line up guests, write segment scripts, and suggest questions. Most but not all of the programs also include calls from listeners, and the producer for any segment usually also acts as a call screener. But once the mic light turns on, the host takes over; Conan says he makes many of the decisions that would fall to a senior producer or senior editor on a news magazine. He goes into the studio with a roadmap and a script, like other hosts, but he knows that the lineup may change as soon as a guest starts to talk, or when he takes the first call. "It's like the von Clausewitz idea that no plan survives contact with the enemy. Well, no plan ever survives contact with the audience."

In contrast to recorded interviews, which can be edited by a producer under the supervision of an editor, interviews on a talk show have to be "edited" by the host as they're under way. "A lot of it's for pacing: if *you're* bored, then the listener's bored," Conan says. For that reason he may politely cut off a guest—or suggest that he or she condense what looks to be a long response. "When somebody says, 'I have three points I want to make,' the first thing you say is, 'We may only have time to get to one, so what's the most important?'" If you're nearing the end of a segment, you can also ask them to answer "briefly," or "in the little time we have left." (As on other talk shows, guests are warned that time is running out if they hear the theme music playing; but Conan says guests on the phone can't hear it, so he has to actively manage the time from the studio.) On the other hand, as a talk show host, you may decide to extend a conversation beyond the budgeted time—to keep someone there after the station break, for instance, even though the next segment was meant to start with callers or a different guest. And when you make those decisions, you won't be able to communicate orally with your studio director if the guest is in the studio. "It's almost always done with hand signals," Conan says. With a sweep of the hand in a rainbow arc, he can tell the studio director or engineer in the control room that he wants the interview to span the station break. A wave goodbye means the interview is about to end.

The host also rearranges the lineup every time he selects calls from listeners. Because screeners provide capsule descriptions of each caller's question or comment ("Sally from Topeka, KA—Why should lawmakers get to decide whether I can smoke?"), Conan can use the calls to move a conversation in one direction or another, or to provide an opposing point of view. Sometimes the callers have thought of something else they want to ask by the time they're called on. "A lot of the time, you choose them to say a certain thing at a certain time in the program and they say something else," Conan says. "What I will do is rephrase their question to encapsulate it and get back to the idea that we were going to—because I did call on them for a reason."

Many of the other techniques used on *Talk of the Nation* apply to other live programming as well.

For example, when you have more than one guest, you have to help radio listeners tell them apart by reintroducing them: "James Richardson—what do you think about Ms. Rosenberg's assertion that the government is wasting billions of dollars on fruitless research?" Helping the listener keep straight who is talking can be especially difficult if the guests are the same gender. Conan advises, "Never introduce two people at the same time and say, 'And welcome to you both,' because they'll talk over each other and say 'Hi!'" Instead, he says, give the audience an opportunity to match the voice with the name. "You say, 'Joanne Smith, nice to have you on the program,' so we can hear Joanne Smith's voice. 'And Leslie Jones...' and you can hear Leslie Jones's voice."

Assertions made during live programming can't be fact-checked the way they can when an interview is recorded. So as a host, you may be obliged to aggressively challenge some controversial or inflammatory statements—asking for specific evidence, or pushing a guest to explain where he got his facts or figures. Sometimes the best you can do is make it clear you can't vouch for a caller's statement—for instance, when someone asserts that "not a single state has rescinded its concealed carry gun laws" or that "more African Americans are now infected with HIV than all other racial and ethnic groups combined." "You don't try to get into an argument over facts," Conan says. "You'll burn up a lot of time, and what's the point? Nobody wants to hear it." You can still use the claim as the basis for discussion without accepting it as accurate: "Mr. Fiore, if it's true, as Susan Ramsey says, that no state has rescinded its concealed carry laws, do you take that as a sign that the laws are working?" One way or another, Conan always tries to indicate that he has not confirmed the statement. "You say, 'That may or may not be true, but certainly he's not alone in that opinion.' Or, when you know that something's in dispute, you can say, 'Well, other people would argue about that, but let's respond to his point.'" The goal is just to make clear that a speaker's "facts" have not been corroborated. "I want to at least create the opening in the listener's mind that there's more than one interpretation of this."

Conan's other tips about hosting *Talk of the Nation* (and any other program relying on live interviews) are what he describes as "interview basics": Ask one question at a time; organize your questions so they make logical and narrative sense; and *listen* to your guest. "That is the key element," he says. "I used to be terrified about what I was going to ask next, without realizing that if I just listened to what the guest was saying, they'd *tell* me what to ask next!" Perhaps most important, know your subject. You never know when the phone line will break up, or the ISDN line drop out, or the guest just won't get to the studio in time, and you will find yourself very much alone in the studio. "If I'm at all prepared, I can natter on about something for a couple minutes, while I look at my line producer and look at my director to find out what's going on." The ability to "natter on"—to ad lib confidently—depends on your being steeped in the subject ahead of time. "Preparation is the absolute sine qua non of live radio. Preparation is everything."

SPECIAL EVENTS

A host may have time to prepare for some live events much as he would for the daily topic on *Talk of the Nation*. It's not that difficult to bone up for congressional elections, or the President's State of the Union address, or confirmation hearings for a Supreme Court nominee. Like regular news magazines, many special reports will incorporate interviews with analysts or reporters—although they'll usually be live, and often involve two or three guests at a time. But whatever the format, hosting a live event requires an extra measure of flexibility. "It calls for a somewhat different set of skills, because you can't rely so much on highly edited intros and other scripts reaching you," says Robert Siegel, who has anchored many special broadcasts over the years. "A lot of what you're going to do is just speak spontaneously, so you have to feel comfortable about that."

Often the only parts of the program that will be written in advance are the beginning and end. When Neal Conan anchors a White House news conference, for instance, he says he likes to have a script to get the show started, even though he will be improvising from then on. "What I try to do is write a minute of copy. Normally you get a two-minute warning when the White House does these things. If I have a two-minute warning, I can vamp the extra minute."

Some events don't start so close to the top of the hour, which means the host may have to hold the air for five or ten minutes. Siegel gives the example of the President's annual State of the Union address, which begins with the President walking slowly down the aisle of the House chamber and shaking hands with members of Congress—shown at length on television. That raises the question of what we should do on the *radio* during those first minutes of the broadcast, he says. "What we do when we've got it right is we talk to a news analyst or political reporter or with our White House correspondent about the State of the Union. We *don't* narrate the video, because it really isn't that important," he says. Siegel says the sound of the hubbub in the House chamber is generally left in the background. "If something actually takes place—obviously the knock on the door of the chamber announcing the President of the United States, or the President reaching the podium—you can acknowledge it and then continue with what we're there for, which is to discuss the speech."

Siegel says hosting some live events is a lot like doing play-by-play at a baseball game—especially one without much action or tension. Neal Conan, who has actually *been* a baseball play-by-play announcer, agrees that you can learn a lot about anchoring live news events by calling a ballgame. "A lot of it is improvising story lines. There's the narrative of the game, which is compressed into however many moments there are of every pitch and every play—the 'and then' moments, when something happens and you have to describe it. But between those times you have all these spaces, where you have to weave other kinds of stories, which depend on the game, what the score is, what the nature of the game is." In one game, you may talk about tactics; in another, a pitcher's recovery from an injury; in another, the team's odds of breaking a losing streak. In a similar way, the interviews a host conducts during live coverage will be suited to what's happening outside the studio. Sometimes you'll want to hear the biography of the nominee; sometimes you'll ask about the political implications of a phrase used in a speech; sometimes you'll want to put the event in some historical context. And—just as at a baseball game—you have to find a way to squeeze these conversations into whatever time you've got, without cutting someone short. "They have to be telescopic stories—accordion-like."

Once the speech (or news conference, or hearing) begins, the host usually recedes into the background. "The program is not about you!" Conan emphasizes. "If there's something going on, we want to hear it. If it's not clear what's going on, you can describe it—but not at the cost of hearing what it is." For many events, that means the host provides the equivalent of audio captions. He may simply identify the speakers at a hearing ("That's Democratic Senator Joe Jacobs of New York") or briefly describe what's happening at other events ("The casket is being carried down the steps of the Capitol and placed in a hearse"). Sometimes the identifications are self-evident, Conan says. "When President Bush was introducing his nominee to be Chairman of the Federal Reserve, Mr. Bernanke, the President finished, and then we heard another voice. It was pretty obvious who the next voice was going to be. I didn't have to say anything."

As the event is under way, Conan advises, you should listen and take notes—"not on everything everybody said, but on points of interest. Don't put them in the form of questions, but take notes to list the bullet points that emerge from what we just heard." These notes will lead you to logical follow-up questions when the event itself ends and you turn to your "buddies"—the reporters or analysts on hand to provide background information and perspective.

Conan emphasizes that during any special event, the host is responsible for hitting all the time posts accurately. "Don't leave it to your buddies, no matter how polished they may be, to do that last minute of a segment. That's *your* job." His advice whenever you're reading live to a clock counting down to the end of a segment: "Know what you want to say, get ahead of yourself, and slow down." It always sounds better if you hit a post by slowing down than if you do it by racing to cram in another sentence.

OTHER LIVE COVERAGE

Special coverage may focus on planned events, such as the swearing in of a political leader or an important news conference; but frequently it stems from news that isn't in anyone's daybook. The death of the pope, the outbreak of war, the attacks on the World Trade Center, and many other big stories all precipitated live "rolling" coverage on public radio. To make these sorts of broadcasts a success, producers, bookers, and directors have to work in concert. And they place a special burden on hosts.

For one thing, you will be working with much less of a roadmap than usual. In fact, you may only have a few seconds warning from a director that a guest is on the line ("We've got Senator Jacobs!"), and your script may consist of an ID for the guest and perhaps a single suggested question. For much of the broadcast, you may feel like you're flying by the seat of your pants. But this is one case where it may not hurt to let the seams show a bit. In 2003, when the space shuttle Columbia burned up on reentry, Scott Simon found himself hosting live coverage nonstop for several hours. And throughout the broadcast, he provided a sort or running billboard to the rest of the program—communicating at the same time to his director, to NPR member stations, and to listeners:

> We are anticipating a briefing from NASA officials in Houston in about eleven minutes at the top of the hour. We also anticipate some kind of appearance and presentation from President Bush. President Bush was at Camp David this weekend. He was meeting with Prime Minister Tony Blair over a matter of great consequence—what seems to be the imminence of U-S and British military action in Iraq . . .

"I think we should feel looser about telling the audience this isn't magic, what's happening now," Robert Siegel says. "It is a broadcast. We're sitting in a studio, and you're hearing the sound of this event unfold, and here's what we're planning."

Another role of the host during this sort of unscheduled special coverage is to recap events frequently and succinctly. "You have to find some way to reintroduce the story to people without implying that it's time for all the people who have been listening for an hour to leave," Siegel says. "I think part of the secret when we have to go to that coverage is to use phrases like, 'For those of you just joining us . . . '" Here's how Scott Simon handled that chore when he anchored the coverage of the shuttle disaster:

> It is now just after 12:45 in the Eastern time zone, 9:45 on the Pacific Coast, where we imagine a number of listeners might just be joining us. And I am afraid we have to welcome you to the day, of what promised to be a momentous week, with the news of a national tragedy—in fact, an international tragedy that involves at least the United States, Israel, India, and, to be sure, each and every other nation around the world. The space shuttle Columbia has broken up in the atmosphere as it sought to return to Earth shortly after 8 AM Central time this morning.

Even as you recap events, you should stress that the story is developing —that death tolls may change, or statements from eyewitnesses may contradict one another. As Simon says of the shuttle broadcast, "When someone tells you about something they saw in the sky, you have to be careful not to say, 'Well that might have been the tail!' What do we know? For all I know it was a mail plane from St. Louis! On the other hand, you don't want to keep people like that off the air. This is, in a sense, an ongoing drama." So you need to help listeners sort things out, even as you summarize events. "You will very often get confusing counterclaims during live coverage," Robert Siegel notes, "so it's important very often to recap what we know, and what we *don't* know, about the story that we're covering."

Sometimes during live coverage (as on a talk show), the host has to serve as producer—and even director. For instance, you may need to catch the engineer's eye to indicate that you want your mic on or off, or that you want sound brought up or faded out under you. "When everything goes crazy, you're just communicating with the tech," Neal Conan says. "Sometimes we will have a deadroll [i.e., music to end a segment] planned, and I'll tell him with a hand gesture, 'Stop—don't do it!' Or during the National Memorial Service after 9/11, we were taking music down and up as it was coming in live for various descriptions of what we were doing. I had to cue the fades. An intermediary just messes it up."

Rolling coverage can be physically hard on the hosts. The programming may run uninterrupted for hours at a time—making it hard for a host to get food and water and go to the bathroom. Conan says the stamina he acquired announcing baseball games definitely comes in useful during live events. "You do a double header by yourself and you're talking for a good long time. And that was tremendously helpful in 9/11, when we were on the air for hour after hour after hour."

But often you'll be energized by the excitement of live coverage—by the fact that you're in the middle of the story—reporting it, as well as presenting it. "I find it gratifying to be learning things along with the audience, and to somehow try to give things some kind of coherence and some kind of form," Scott Simon says. "If you're in the news business, you welcome those chances not just to analyze, not just to reflect, but to actually bring the news to people in real time, and to let them know this is happening now."

CHAPTER 10

Newscasting

The most popular show on National Public Radio isn't really a show at all.

Many public radio listeners aren't getting their news from one of the flagship programs, like *All Things Considered*, but from the network's five-minute hourly (and sometimes twice-hourly) newscasts. According to NPR's Office of Audience and Corporate Research, more than twenty-five million people tune into NPR newscasts each week.[1] That's millions more than will hear *Morning Edition*, the news magazine with the biggest audience. (The difference comes in part because newscasts run throughout the day—when stations are providing local programming—and they may run on stations that mostly broadcast music.) While public radio reporters and editors make a big distinction between a short forty-five-second "spot" intended for the hourly newscast and a longer "piece" aimed at *All Things Considered* or one of the other programs, listeners hear them all as "news reports."

In fact, when news breaks, the Newscast Unit often has to be the first to mobilize. "We're on a different body rhythm than most of the rest of the building," longtime producer Carol Anne Clark Kelly says. People at NPR often talk about an important news piece going to "the next show up"—at midnight Tuesday, they're referring to *Morning Edition*, on Friday evening it would be *Weekend Edition Saturday*. But from the listener's perspective, newscasts are almost *always* the next show up.

What Is News?

Even on a slow news day, the AP and Reuters wires may turn out six or seven stories a minute. That can add up to hundreds of items an hour—or

1. Based on audience research from spring 2007.

thousands of stories a day! Admittedly, a lot of these are one-line alerts or revisions of previous stories. But the sheer number of items underscores the fact that one of the primary jobs of a newscast unit—at NPR or anywhere else—is to sort through the available material to decide what's worth putting on the air in the next hour (or whenever the next newscast or headline summary comes around).

Alas, there is no single rule to apply when making that decision. As newscaster Paul Brown points out, people listen at different times and for different reasons. "There are many people tuning in who have not heard any news that day, and there are also 'news junkies' "—so in a way, he's got at least two different audiences every hour. In deciding which stories to include, Brown and other newscasters take a whole series of factors into consideration each hour, some of them quite subjective:

IT WILL HAVE A GREAT IMPACT ON PEOPLE. A story is newsworthy when it affects people in a dramatic or consequential way. That doesn't necessarily mean it has to affect a large number of people directly. It's bigger news when the U.S. decides to keep American troops fighting overseas another six months than when the White House announces it supports a 3 percent cost-of-living increase for members of the armed services, even though more people might be affected by the latter announcement.

IT'S UNUSUAL OR UNEXPECTED. Listeners are less surprised when a Democratic candidate holds a news conference to criticize a Republican administration than when a Republican senator does—especially if the senator had previously been a supporter of the White House. Unpredictability helps elevate the status of a news story.

IT'S THE FIRST OF A KIND. It may seem tautological to say an item is news because it's new, but something that has never happened before (e.g., the first person to hit seventy-five home runs in a season, the first Democrat elected in a Republican state in sixty years) usually merits coverage, even if it fails the impact test.

IT'S TIMELY. The very fact that something is taking place right now or has just happened can move it to the top of a newscast (e.g., "At this hour, the Dow Jones Industrial Average is down more than five hundred points—or more than 4 percent"). That immediacy is one of the strengths of radio news.

IT'S CONTROVERSIAL. It can sometimes be difficult to determine whether a story is controversial chiefly *because* the news media are paying attention to it. To be sure, Americans in 2003 *were* sharply divided over whether to go to war in Iraq; there's no debating whether that was a hot topic. But were they really arguing over the dinner table about whether to add a prescription drug benefit to Medicare? The point is that newscasters have to make sure the controversy is not taking place only on cable television or inside the Washington beltway.

IT INVOLVES PROMINENT PEOPLE. It's more of a news story if an annual physical shows that the President has heart disease than if the same problem is diagnosed in a member of a second-tier '60s rock band.

IT DEALS WITH DEATH OR TRAGEDY. Public radio generally ignores the old newsroom maxim that if it bleeds, it leads. But in most cases a newscaster would be remiss not to report on the death of a prominent sports figure or politician, the collapse of an apartment building, the crash of a jet, or similar events.

IT HAS TO DO WITH THE U.S. The unwritten but widely accepted rule in many newsrooms is that the more local a story is, the more it's news. Even though public radio usually takes more of a global view than commercial TV networks, a flood, riot, or plane crash happening in *this* country will be more newsworthy than the same event taking place five thousand miles away.

IT CONCERNS AN IMPORTANT ISSUE. There are stories that may have no *immediate* impact, but which nonetheless raise or help define issues of great importance to the country. Oral arguments before the Supreme Court, for example, often draw attention to weighty social, legal, or economic issues, though they may deal with events that took place a long time ago, and the Court may not hand down a decision for months.

IT IS OF HUMAN INTEREST. From a medical perspective, it may not be remarkable that a team of twenty doctors is spending twenty-four hours to try to separate conjoined twins, but there's a side of that story that goes deeper than science—that has to do with what it means to live an independent life. Many stories have a compelling human-interest component. Newscaster Craig Windham says when people talk to him about something they've heard on the air, they usually *don't* cite the kinds of stories that make newspaper headlines. "It's almost never something about hard news. It's something about human interest that caught their attention."

IT'S USEFUL. There is a category of news story that's worth putting on the air mainly because it would be a public service to broadcast it: a carmaker is recalling vehicles with faulty brakes, the government has held a news conference to outline simple ways to avoid contracting West Nile virus, etc.

IT'S "OUT THERE." Occasionally there's a story that needs to be included in a newscast simply because it's been so constantly reported *elsewhere* that it's generated widespread interest. Public radio wouldn't normally report on TV programs, for instance, but if one-third of the country stays up late to see who wins the *American Idol* singing competition, there's *something* going on that probably is worth mentioning on the air. One newscast producer calls this a "didja hear?" story.

Obviously, newscast producers and anchors will use different criteria when considering different types of story. The important thing is to judge each story on its own merits, so that every item really deserves to be included. There are usually only seven or eight discrete stories in a five-minute newscast, so each one has to compete for that valuable airtime. The Associated Press may run a story about the Austrian national airline's decision to lay off four hundred workers, and no doubt there are Austrians (at least those who work in the airline business) who care; but that's not reason enough to include it in an hourly newscast in the U.S.

Public Radio Newscast Values and Principles

The criteria listed above might help you winnow a day's wire stories down from, say, ten thousand to a thousand or fewer. But the process of selecting stories is much less scientific and much more intuitive than such a list might suggest. One thing that can make the job easier is understanding the values and principles that inform public radio newscasts.

THE NEWS IS ACCURATE. News wires are in the business of getting the news out as quickly as possible, and as a result they often have to make corrections. (Search the wires for the word "correction" on any given day and you'll see how many mistakes they have to fix.) The wires place the highest value on *speed.* NPR places it on *accuracy*—even if that sometimes means holding a story for an hour until we get a second source to verify it. "The big difference here [in public radio] is the amount of editing we do," says Paul Brown. "We try to make sure we have something confirmed, and that it is true, based on a combination of our own reporter's work and the fact that we have access to a number of news sources."

THE NEWS IS UP-TO-DATE. NPR may not put speed at the very top of our list, but it's a close second. Once you have verified that you have the facts about a story—but no sooner!—you should get it on the air.

THE NEWS STORIES ARE EASY TO UNDERSTAND. It's possible to write a news story that is entirely factual and yet almost incomprehensible. Responsible newscasters value clarity—in the way they select facts, structure stories, and write copy. That simply means that at the end of a news story, it should be possible to ask a listener, "What was that story about?" and get a reasonably accurate reply.

THE NEWS STORIES PROVIDE CONTEXT. Context would seem to be a luxury in a newscast, where the longest report may be just forty-five seconds and many stories run just half a minute. But not all listeners have been keeping up with, say, new Medicare legislation working its way through Congress, or the growth of the federal deficit. As a newscaster, you need to help listeners make sense of the news by providing background information with an extra line saying, "The House and Senate have each passed different bills to add prescription drug coverage" or "This is the third time this year the administration has raised its estimate of the budget deficit."

THE NEWS IS BALANCED. People on all sides of an issue can draw the attention of the news media by issuing a report, holding a press conference, or staging a demonstration. You should approach all such groups and individuals skeptically and seek out opposing points of view.

THE STORIES ARE WELL WRITTEN. That means they are written for the ear and incorporate the key elements of strong broadcast writing.

THE NEWS STORIES AVOID HYPE. Just because a drug company describes something as a "breakthrough" doesn't mean that it really is, and not all administration policy announcements are "major." Newscast writers avoid characterizing the news, especially when its importance is open to debate. Corey Flintoff says the same goes for the way a story is read on the air. "I try not to be sensational, not to unnecessarily pump a story up, or give it a sense of urgency that's not there."

THE NEWSCAST TAKES A GLOBAL PERSPECTIVE. That doesn't mean that a labor dispute in, say, the Philippines is as important to listeners as a similar event in the United States. But it does mean you shouldn't confine yourself to American news.

NEWSCASTERS DON'T LET THEMSELVES BECOME PART OF THE PUBLIC RELATIONS MACHINE. Most companies and large organizations have whole divisions devoted to churning out news releases that show them in a favorable light. Newscast producers should be careful not to do their PR work for them unwittingly. "We're in a culture now where news is so carefully packaged, and given to us, that I think we have to be careful whether we parrot it back or not, and how we present it," Craig Windham says. When Crayola decides to add a new color to its box of sixteen crayons, or the Mars candy company says it will replace tan M&M's with blue ones, or Mattel announces that Barbie's proportions will be changing, some news organizations relay the "news" unthinkingly. Don't confuse PR with news.

Composing the 'Cast

Composing a newscast, like most other broadcast news jobs, is a mixture of art and science. But if people understand and agree on the criteria listed at the top of this chapter, they should end up with similar news story lineups at any given hour.

So we performed a little experiment.

We gave eighteen wire stories of different lengths and subjects to each producer, editor, and anchor in the NPR Newscast Unit. All of the stories had run within an hour or so on a given day.[2] The assignment was to sift through the various items and decide which ones would warrant inclusion in an early-afternoon public radio newscast—and why. (For the purpose of the exercise, we didn't insist on a second source for each story.) We also asked participants to justify the story's *placement* in the newscast—to be able to explain why something was a lead, or why two stories were paired together, or why a story was an "ender." Finally, we asked them to say which stories were "spot-worthy" (that is, which ones would be better told by a reporter within the newscast) and which would work best if we could add a bit of sound or an actuality. We emphasized that this was to be an *idealized* newscast—one in which reporters were always on the scene and available to file, where actualities and natural sound were plentiful and on hand, and where there were no disputes about the facts.

2. This exercise was conducted in 2003, so the references in some of the stories may now be obscure. "SARS," for instance, stands for severe acute respiratory syndrome, which had struck Asia, North America, and Europe.

The exercise had two main goals: to get newscast producers and anchors to verbalize a process that is usually intuitive, and to find out why actual newscasts often don't come close to the ideal.

What we found was that when newscasters are freed of the reality of working around time posts and available audio, their proposed newscasts are remarkably similar to one another (and sometimes identical). Here's a typical lineup that people came up with:

BUSH-TAXES with reporter Q&A (President signs tax-cut bill at 2:15pm)
MIDEAST-BUSH TRIP with actuality of President or spokesman (President will hold summit with Israeli, Palestinian leaders)
MIDEAST-HAMAS reader[3] (Hamas leader says peace roadmap a "trick")
SAUDI-BLASTS-ARREST with reporter spot (Saudis arrest mastermind of suicide bombings)
MOROCCO BLASTS reader (Morocco captures leader of Casablanca bombings)
**** Time post ***
SARS VIRUS with reporter spot (Beijing officials impose new restrictions to fight epidemic)
SARS-HONGKONG-NOSE reader (scientists say they may be able to block transmission of disease with nasal spray)
ABUSE CHAPEL reader (L.A. cardinal opens chapel to sex abuse victims but doesn't invite victims to opening)
EVEREST-50th anniversary with actuality (Sir Edmund Hillary complains about legions of Everest climbers)

Among the thirty people who went through this exercise, there were some variations on this theme. A few paired the tax-signing lead with a short story about a Democratic senator criticizing the President for not doing enough to get Americans back to work. Some added one more domestic story about the arrest outside Atlanta of an alleged serial killer. A few led with the arrests of the alleged terrorist leaders in the Mideast, then segued to the President's plan to hold a meeting with Israelis and Palestinians, and *then* went to the tax-cut story.

3. A "reader" is simply a story read by a news anchor (as opposed to one filed by a reporter, or a story that includes actualities).

In any case, almost everyone could make a good case for including each item in his or her lineup. The tax-cut story is worth leading with, they said, because it's going to affect millions of Americans and because it's just about to happen (since we said this was an early afternoon newscast). The Mideast news is unexpected—two countries arrest two alleged terrorists—and follows naturally from the announcement that the President will be holding a summit with Israeli and Palestinian leaders. That, in turn, comes logically out of the other White House story. The SARS items update an ongoing story. And so on.[4] Producer Rob Schaefer says that the goal of any lineup is to make a newscast memorable. "Ideally, you're telling the most important things first, followed by the less important stories." But, he adds, "You are writing for the ear—and sometimes, given the mind of the listener, if things are grouped together by category it's easier for people to take things in."

So, what would keep you from putting such a newscast on the air?

The most common reason, by far: A lack of reports—or news "spots"—filed from people on the scene. Five-minute public radio newscasts generally include three spots, but the ones that are available at any given hour often don't deal with the top news stories. If the lineup above were for a real newscast, it might not be possible to get a report out of Saudi Arabia or Beijing—or, for that matter, an actuality of Edmund Hillary. Instead, the most recent spots might deal with continuing western wildfires, the latest earnings data from a Fortune 500 company, and a United Nations report on poverty in sub-Saharan Africa. These are not insignificant stories, but they're certainly not in the top tier of important news. Building a newscast around such reports just pushes other, more pressing stories out of the news.

So don't let your news judgment be clouded by the availability of sound; make editorial decisions first, and production decisions second. Sometimes there are reporters or stringers who can do Q&As on breaking news; sometimes a producer can place a call and get an actuality to flesh out a story; but sometimes the anchor will just have to read copy. Corey Flintoff says that can be a *good* thing. "With digital editing, there's a big tempta-

4. Commercial broadcasting shares some, but not all, of the news values that distinguish public radio. The AP news summary for that hour began with the Bush summit announcement and the tax-cut signing, but then went to the arrest of the serial killer and a story about a woman in California who put her two-year-old in a laundromat washer.

tion to cram a newscast full of sound bites. And the trouble is that sound bites don't always tell the story that well. So if it would be better for me, and clearer, and simpler, *not* to include a sound bite, then that's the way I'll go." In other words, when news spots are in short supply, the burden shifts to producers and editors to make sure the top stories are covered.

There are other factors that keep actual newscasts from matching the ideal. An obvious one is the time post that comes in the middle of the newscast—a fact of life at most broadcast networks—to allow stations to cut away from the broadcast. There's no guarantee that listeners will ever hear the items at the end of the newscast; individual stations may substitute their own local stories, or insert recorded features, traffic and weather reports, or anything else.

But whatever the obstacles, you have to put together the best newscast—ideally one with what Corey Flintoff calls "narrative flow." "I used to do music shows when I started out in radio, and one of the best compliments someone gave me was, 'Your music show has continuity. It has a beginning, a middle, and an end. It tells me a story.' I try to do that with newscasts. I try to give it some sort of narrative arc."

Newscast Writing

Public radio broadcasters disdain newscasters who just "rip and read"—reading wire copy unedited on the air. They know it's wrong—but they usually don't say why. As it turns out, there are lots of good reasons.

For one thing, your newscast should be your own work, not the work of a writer at Reuters or the Associated Press. You should meet your own standards and have your own style, and you are ignoring them when you lift a wire service story verbatim.

Also, the wires are sometimes wrong. In their rush to get news out, they make mistakes. Some are small (they place Rockaway Beach in the wrong borough of New York), some are significant (they report a sale of fifteen million shares of a company's stock, when it was really five million), and some are critical (they misidentify the victim of a shooting, mistakenly announce the arrest of a terrorist, and so on). That's why it's important to verify the facts in each story against a second source—ideally a primary source, but at the very least a second independent wire.

You should also take care to check where the wires are getting *their* information. Wire stories may quote "sources" whose identities are not specified. When you read such a story verbatim, you imply that you have obtained the information yourself. Since you can't vouch for the reliability of sources that you can't name, you should always attribute such stories to the AP or other wire. Some wires source their news bulletins to a cable television network, which makes their "alerts" secondhand information at best. And even when the source seems reliable, it pays to be skeptical, and to verify the facts if you can. Long before he came to NPR, Rob Schaefer was producing a newscast when two wires ran alerts saying comedian Bob Hope had died. The stories were based on comments made by an Arizona congressman on the floor of the House. Schaefer recalls, "When the news broke, I said, 'This is odd. It's coming from the House floor, from a member of Congress. Why would that be?' " Schaefer's news organization happened to have Bob Hope's home phone number, and so they called it to confirm the news. "His daughter answered and said, 'No—he's sitting right here!' Other networks went with that story and had to retract it."

Even when the wires get all the facts straight, their notion of the lead—of what's most important in the story—may differ from yours. They may elevate the findings of scientific research to the status of a "breakthrough," or describe a demonstration as "unprecedented" or "massive." Those adjectives may turn out to be accurate, but there's no way that an anchor or editor working from the wires would be able to determine that. So read all the way to the bottom of the story to make sure you're familiar with all the facts, and then decide on the lead yourself. As Corey Flintoff puts it, "You have to think about the story—you have to digest the story yourself, and then tell it in your own words."

Many wire stories are written for print, not for radio, so they sound odd or confusing when read aloud. For example, because they start with datelines, wire service stories often don't clearly state where the news is coming from:

> LOS ANGELES—Police have asked the FBI to review forensic evidence that helped convict a man of killing his father 18 years ago, but officials say they still believe he is guilty.

Here only a *reader*—as opposed to a listener—would infer that the police were located in Los Angeles. A radio version of the same item might begin, "Los Angeles police are asking the FBI to review forensic evidence..."

There are many other reasons not to just copy text from wires and paste it into your scripts. Wire stories are often much longer than radio news items; many AP stories run three minutes or more when they're read out loud. On the other hand, the wires may summarize *too* much for your purposes. An item on a three-hundred-page human rights study may focus on one aspect of the report—the treatment of U.S. prisoners, for example—even though that may not be the main thrust of the original paper. Each wire story is written to stand on its own, so when related stories are inserted in a news broadcast, they may repeat some of the same facts. Finally, a wire story is likely to be written in the past tense, so it will have less impact when it's read out loud than a present-tense story written expressly for broadcast.

Unfortunately, writing original, accurate news copy is easier said than done, given the deadline pressure of putting a fresh newscast on the air every hour—and more often on some news cycles. It's also hard because newscast stories have to be so short. If a typical five-minute newscast includes three forty-five-second spots, with perhaps an actuality or reporter Q&A, it may end up about half actuality and half copy. That puts a premium on every word. Another way to look at it: newscasters have to abide by the same principles of broadcast writing as everyone else—only *more* so.

"Be direct—don't use excess baggage in your writing," Craig Windham says—and that means writing in simple, declarative sentences. An intro for a spot in a newscast may only be ten or twelve seconds long, compared to half a minute or forty-five seconds for the intro to a news *piece* headed for a news magazine. So present the facts as straightforwardly as possible. "It's always going to be subject-verb-object." Try to put just one or two ideas in each sentence: "Wholesale prices rose last month," "Strong storms in France have killed several people and left thousands without power," "In Baghdad today, Iraqis fired a rocket-propelled grenade at an American convoy." The cardinal rule for news writers is not just "keep it simple," but "make it even simpler."

And make sure those simple sentences actually include the news. One news spot dealt with a city's preparations for a possible protest against a meeting of agribusiness executives by people who oppose the globalization of agriculture. It mentioned that hundreds of extra city police had been told they'd be working overtime to make sure things didn't get out of hand. The story itself was quite engaging. But the *intro* began by saying only that the U.S. Department of Agriculture was going to host more than a hundred agricultural delegations from around the world "to introduce them to the latest in farming technologies." That was true, certainly, but it sounded like a government press release. You had to dig into the spot to find the real news.

Similarly, how much news does the following intro convey?

> A team from the U-S Army Central Identification Lab in Hawaii is deploying for Alaska this week. From member station K-H-P-R in Hawaii,—— reports.

Curiously, the spot itself began with a *second* reference to news that should have been in the intro: "The recovery effort will be centered on the Aleutian island of Kiska." What recovery effort? The intro clearly lacked some critical information.

When you're writing an intro for a news spot, ask yourself how you'd summarize a story if you had only a few seconds to tell it to someone—go for the "blurt factor," as one newscast producer calls it. That's likely to be the lead, or close to it.

Most newscasters agree that stories should be written in the present tense, especially when they deal with policies or opinions—things that remain in effect after they are articulated. "The President says the United States will contribute troops to an international peacekeeping force" sounds more current—and grabs the listener's attention more effectively—than "The President said this morning the United States will contribute troops to an international peacekeeping force." It's often possible to inject energy in news copy just by changing the verbs from the past tense to the present: "A new study is raising questions," "California's governor is facing a recall motion," "An Israeli soldier is free at this hour."

Also, look for ways to write spots and other copy in the active voice. (This is easier when you are a reporter covering an event than a newscaster working from the wires or other sources.) Consider the difference between these two possible lead sentences: "In New York, all-night talks are being held to find a way to avert a teachers strike" and "In New York, city executives and union leaders are meeting through the night to find a way to avert a teachers strike." When you read or hear the second sentence, you can easily imagine the negotiators staying late to come up with some sort of resolution. (You can almost picture them with their sleeves rolled up and their coffee cups out on the table!) Good broadcast writing is *visual*, whenever it can be, and creates a picture in the listener's mind. Writing in the active voice goes a long way toward painting that picture.

Keep numbers and dates in news stories to a minimum, and make them as easy to understand as possible. Because newscast stories are so concise, you can't risk losing listeners' attention for even a few seconds while they make sense of what they're hearing. Do everything you can to keep people from having to work too hard. It's often better to say something happened "nine months ago" than to say it took place "in December of last year" or "in December of 2007"; those phrases force listeners to do the math. If 300 of the 1,200 people surveyed opposed the referendum, you'll want to change it to "one out of four." And you should consider the level of precision that's really necessary for a particular story. You may in fact want to say that the gross domestic product grew "at a three-point-four percent annual rate," because a change of just two-tenths of a percent can indicate what's happening in the economy. But you can probably drop the digits to the right of the decimal point when you're dealing with election results, corporate earnings, and so on.

Finally, there are many things to *avoid* if your writing is to be clear, concise, visual, and easy to grasp.

Don't write in the shopworn idiom of newscasting. Because so many news stories are similar to one another—Republicans and Democrats always disagree about issues, rebels always oppose governments, stocks will continue to go up and down—it's easy to use the same boilerplate language to describe today's events that you used yesterday, the day before, last week, and last month. In a lot of news copy, debates are always "heated," deals are "hammered out," politicians in trouble are "beleaguered," and so on. The result of this sort of writing is to obscure what is actually new and interesting about the current story. Stories that begin "Republicans and Democrats are facing off over Medicare," or "Stocks dipped today on profit taking," or "Firefighters continue to battle the wildfires in Arizona" could be much more effective if they had stronger verbs, less common language, or a few real details. Will the Republicans introduce their own, competing bill? Are the firefighters going to drop chemicals from helicopters to try to put out the fires? Do we know that "profit taking" really explains why the stock market went down? (And if it doesn't, why are we *saying* it?)

Keep your ear tuned to what might be thought of as instant, short-term, or "nonce" clichés—phrases that enter the news lexicon when a story is hot, and then somehow become hard to avoid. When Congress was considering impeachment of President Clinton, one of the key figures—according to news reports—was "presidential friend Vernon Jordan."

(This led some people at NPR to wonder aloud whether that's what it said on his business card.) Before that, the Los Angeles Police Department was under investigation for the beating of "black motorist Rodney King." (Even many years later, if you asked someone to fill in the blank after the words "black motorist," they'd often say "Rodney King.") Here again, the challenge is to keep from parroting certain phrases and idioms just because they are used in wire stories and by other media, and thus come to mind first. With a little effort, you can often come up with something that is fresher, more accurate, and more picturesque.

Remember, your goal is not only to write news that's up-to-date, accurate, and balanced; it's to make sure people pay attention. Newscaster Paul Brown says he always assumes his listeners are distracted. "They're driving a car, they're getting their kids to school, they're shaving, they're taking the dog out, they're late to work, they're trying to get to Boy Scouts—who knows what." So he writes with the intention of drawing their attention away from the *other* thing they're doing. "They only get one shot at understanding the news, and my job is to make sure they understand it even if they aren't paying close attention. That's a tall order, but I think that you can satisfy that order a lot of the time by writing short, declarative sentences, with clear references, and one idea at a time."

The NPR News Spot

Public radio news spots are usually no more than forty-five seconds long, not counting the intro; those on commercial radio are often even shorter. So the reporter and the newscast editor have to be concerned with keeping the spot within the time limit; if it is too long, it has to be trimmed. But in many ways, that's the easiest part of the job.

When you are filing a spot for a five-minute newscast, make sure your story has a clear focus; because news spots are so brief, they usually can't be about more than one development. If there's too much information to put in a single spot, you can always file two—for example, one on the meeting between the rebels and the government officials, and a *second* spot on the emergence of the new rebel leader.

And remember, it has to begin with the *news*. "Every news story must answer who, what, when, where, how, and why," Rob Schaefer reminds us. "The lead—the anchor intro—should tell the listener *why*, the main reason they want to know about this story, or the latest development."

Newscast producers will rewrite reporters' intros if they begin in the past: "Two years ago, Congress passed the 'Free to Be You and Me' bill," or "Last week, the Administration . . . ," or "It has been six months since John Jones was convicted of paper rustling." Starting with this sort of background information just postpones giving people the news. "I don't want someone to turn the radio off thinking this is *yesterday*'s news," Carol Anne Clark Kelly says. "I want to hear what's new *now*, and then we can usually logically work backwards for people who haven't been following the story." So make sure the intro starts with the latest developments—e.g., Congress is passing another bill, the administration has a new proposal, Jones's lawyer says he was framed—whatever. It's usually easy to put the historical material in the body of the spot.

While you're writing (or editing) the intro, make sure it contains any date references or other information that might change—the number of confirmed dead in a plane crash, for instance. You don't want a spot to perish just because it's running at 12:01 a.m., and "tomorrow" has just become "today," or because one more critically injured person has just passed away.

You should also check that the intro doesn't summarize the entire story—it's the *lead*, the way *into* the story, and its purpose is to give us the top of the news as quickly and vividly as possible. The following intro, which is about half a minute long, is *too* complete for a newscast. It's really a whole news story itself:

> In Liberia, the defense minister says more than six hundred people have been killed in fighting, as rebel forces try for a third time to take control of the capital city, Monrovia. Aid workers and hospital officials say more than 90 have died in the violence. Marines have evacuated by helicopter about two dozen foreign aid workers and journalists from the U-S embassy compound. West African nations are meeting to discuss a plan to send in peacekeeping forces. From Monrovia, ——reports:

A good intro should help focus the story; at the end of this one, we don't know whether the reporter is going to report on the differing death counts, the evacuation of U.S. personnel, the imminent arrival of peacekeepers—or all three. (The answer: the African peacekeepers.)

During the edit—and all spots should be edited before they go on the air—the reporter should read the entire spot out loud. Most people read faster during an edit than when they're recording and have to enunciate, so the editor should insist that the pace of the delivery be realistic. Otherwise a spot that's forty-five seconds long during the edit will be fifty-five when it's recorded. If you're the editor, encourage the reporter to read naturally, and not in a singsong fashion; listen for tongue twisters or rhymes or other things that will trip up the reporter when he's filing, or sound odd on the air.

As the editor, you'll want to check the reporter's first line to see that it dovetails with the anchor's lead-in. If the reporter has written the intro first, it should—but often you'll find echoes, or first references that can be second references (e.g., the reporter will repeat the title and full name of someone already identified in the intro). The intro and spot should fit as well together as if one person read them both.

These are mostly copyediting issues. But one of the hardest tasks for any editor—especially one dealing with breaking news on a rapidly approaching deadline—is making sure that the reporter has all the facts straight. He or she is on the scene; the editor isn't. Reporters have beats, or at least areas of expertise; chances are the editor doesn't. Yet the editor is the one who's supposed to check for accuracy. So a good spot editor should become familiar with the story before he starts editing. "I'm always concerned about attribution," Carol Anne Clark Kelly says. If the news wires have reported the same news, she says she tries to have the wire version handy when she's editing a spot—to make sure, for instance, "that we're not alleging someone has committed a crime they haven't actually been charged with." Checking other sources also allows an editor to raise questions about any apparent discrepancies—you can point out to the reporter, "The AP says the ship is called 'The Hot Potato,' but you're calling it 'The Potato' " and give him a chance to explain how he got his information.

If you're editing spots on deadline, you need to have both a keen ear and ready access to lots of information. Know where to find atlases, gazetteers, and other reference books that will help you assess what you're hearing. Beware of superlatives—listeners will rush to correct you if you wrongly claim that something is the first, the biggest, the only, the oldest, etc. And always take a hard look at very big numbers.

For instance, can you trust the numbers in this news spot about a Texas Transportation Institute study on traffic congestion?

> The study says Americans wasted 3.7-billion hours stuck in traffic during 2003—a big jump from the year before. The report also says Americans

> wasted 2.3-billion gallons of fuel—another big increase. No surprise, Los Angeles-area commuters averaged a league-leading 93 hours a year stuck in traffic. That's actually a slight improvement. San Francisco drivers lost an average 72 hours, with Washington D-C folks close behind. The Institute says the U-S is gaining cars more rapidly than new miles of highway. The report came out just as the U-S Senate is poised to consider a new 284-million-dollar highway spending bill—one that experts say won't do nearly enough.

Unless you have a lot of facts at your fingertips and are a wiz with a calculator, you aren't likely to be able to figure out if Americans could even *use*, much less waste, 2.3 billion gallons of gas every year. (Ironically, one figure was wrong in this spot—by a factor of 1000! The price tag for the highway bill was actually 284 *billion* dollars—a mistake that slipped by the reporter, producer, editor, and news anchor.) Most of us would be instantly suspicious if someone described a "seven-hundred-car accident" or said the lottery winner got "three billion dollars." But an editor is not likely to have enough personal experience of federal spending—or corporate lawsuits or necktie sales—to know whether big numbers in stories on any of those subjects are reasonable, so it's incumbent on him or her to ask. Does the reporter really mean that the U.S. spends seventeen *billion* dollars on that program? Can there really have been 1,500 civil suits against that one company? Are you sure the factory sews a thousand ties an *hour*? Just raising the question will usually help the reporter catch a mistake.

Fact checking is much harder if the reporter is the first to file on a particular event. If you're editing in a situation like that, you have to be rigorous about asking how the reporter came by his or her information, whether the reporter has confirmed it (and with whom)—and if the facts are in doubt, you have no alternative but to demand additional reporting. In some situations, it's best to ask a reporter to rely on things he has personally seen. It may be too soon to report authoritatively whether anyone died in the train crash, but the reporter *can* say that three train cars were derailed, that six ambulances are on the scene, that medics are removing injured people from the train cars, and so on.

Like other radio editors, newscast producers and editors act as surrogates for the listeners; if the editor can't follow what the reporter's saying, it's a good bet the audience won't either. If you are editing news spots, take care to shorten long or convoluted sentences; weed out jargon, journalese, Briticisms, and clichés; and listen for awkward sentence construction. If there's any part of a spot you don't understand, have the reporter rewrite it. Since U.S.-based reporters often write their spots for their local station first and then file for the national network, their reports may include references that are meaningless to many listeners—"The twenty-five-car pileup happened at the intersection of highways 81 and 50," for example. Ask the reporter to insert geographical references, if they're needed.

If the spot uses an actuality, the editor should listen carefully to see whether it really advances the story, or whether it simply provides factual information the reporter could give more succinctly. "It's not enough if a policeman tells us a plane crashed in the neighborhood and there's nobody dead," says Rob Schaefer. "The anchor can say that and probably add a few more facts in the same amount of time. But if the policeman says, 'Clearly the pilot was trying to miss the homes. He landed on the street, and by God, he didn't kill anybody but himself and his copilot. It's a miracle!' *That's* the sound bite we should use, because it adds analysis by somebody who was there and knows what happened and has some emotion in it. It's more than my anchor could ever say—and it's subjective, so it's more than he *should* say." If an actuality in a spot seems weak or off the mark, Carol Anne Clark Kelly encourages reporters to look through their out-takes for better ones. "Chances are they may have a piece of tape they may not have realized was better, and I can coax them into using it." But you won't know if you don't ask.

In fact, one of the most important roles of the editor—in a newscast unit or anywhere else—is just to be actively curious. Anticipate the questions a listener might ask: Why is that company interested in merging? Who paid for that research? Isn't this the opposite of what that politician said earlier? The editor has to listen hard for what's *not* in the spot, and then get the reporter to flesh out the story.

Finally, if the answers to your questions don't elicit something worthwhile, know when to turn down the spot altogether. There are times when the news is so incremental that it can be handled in a few lines of copy—when the reporter isn't saying anything an anchor can't say as well or as authoritatively himself. Spots usually add a lot to a newscast, but if they don't include much news, they do little more than soak up forty-five seconds that could be better used for other stories.

The Reporter Q&A

Sometimes the news is so pressing that it's more important to get *something* on the air than to wait for the reporter to write a polished spot and subject it to the usual editing process. Even when the anchor could deliver the news himself, we may want to hear directly from the reporter on the scene—for example, when a court in New York has just found a business leader guilty of fraud, or a passenger plane has crashed in South Dakota, or rebels have begun lobbing mortars at the U.S. embassy in Liberia. In those cases, and many others, the reporter will need to ad lib—compressing the latest developments into a few sentences. Ideally, the reporter would file live into the newscast, in which case the anchor's lead-in may just be a sentence to sum up what's happened and a line saying, "Reporter Justin Case has the latest." But if the reporter's filing opportunity doesn't coincide with the next newscast, an anchor or producer will have to do a Q&A and excerpt the reporter's answers for use on the air. (This is just one of the ways this sort of Q&A for a short newscast differs from a reporter two-way on a news magazine.) The goal is to get the person on the scene to report on what's happening as succinctly and colorfully as possible.

"Succinct" is the operative word. We rarely speak as concisely as we can write; a reporter may answer the question, "Bring us up to date on what's happening" with a seventy-second monologue. So if you're doing a Q&A with a reporter, make it clear that you're looking for a twenty-five- or thirty-second answer.

The best reporter Q&As have a "here and now" feel to them. Whenever possible, a Q&A should not only give the news (e.g., "The jury convicted Mr. Rubble on all counts," "The plane was carrying twelve people, including two crew members"), but also add something that we could not get off the wires (e.g., "His wife burst into tears at the verdict," "I can see smoldering debris from the crash scattered all over the hillside"). To get that sort of description, a producer may need to ask the reporter the same question, or some version of it, a couple of times.

You may also want to ask series of questions, to get the reporter on the scene to look at an event from several angles. This is one way that a reporter—especially one where news is breaking fast—can provide newscasts with material for several cycles without being tied up for more than a few minutes on the phone. The following seven clips, for example, are all from the same reporter, provided over less than twenty-four hours. Together they present a detailed and powerful picture of life in a war zone:

> Things are still fairly tense out on the streets of Monrovia this morning. There hasn't been as much shelling today as there was yesterday. However, yesterday the shelling really was concentrated in the afternoon. People are out and moving about, but they're running from one place to another, sort of ducking from wall to wall. We are getting a little bit of mortar shelling at the moment, but nothing like what happened yesterday.
>
> It's very difficult to tell how many people have been killed over the last several days of fighting. Yesterday at one point in time, journalists just added up how many people they could confirm had died in a single hour, and it was about sixty, in one hour, during the shelling yesterday afternoon. So a number of 600 is definitely possible, but it's really hard to tell because people aren't moving about, it's not safe to go out in the streets, many parts of the city are inaccessible. So any complete total would be a guess.
>
> It's only foreign nationals that are being evacuated. They're being evacuated by the U-S government, they're being evacuated from the U-S embassy compound, which is a heavily fortified compound in the Mamba Point part of Monrovia—which, if the rebels did overrun the city, they'd have to come all the way through the city to that part of town. It's mainly aid workers, people from the International Committee of the Red Cross, from Doctors Without Borders, from other aid groups, who are scaling back their staffs here at the moment, because the security situation has become really bad. A Doctors Without Borders staff member was killed on Saturday by a mortar shell at his residence, so a lot of people are trying to get out.
>
> The fighting is still going on. Again, it's concentrated on the outskirts of the city. The rebels remain out of the very heart of Monrovia. But throughout the night there was gunfire in the streets. I assume most of that was people trying to protect themselves from looters. But mortar shells—we have been hearing some today, nothing like the barrage that we were getting yesterday. But clearly the battle for Monrovia is continuing here.
>
> Leaders of the Economic Community of West African States—which has pledged to send peacekeepers here—they're meeting in Senegal today and

> tomorrow. It's unclear when exactly they would send troops, if they're going to send troops. They've said that they will send troops. It's adding to the overall frustration of people on the streets that peacekeepers have been promised, [but] they haven't been delivered. And the fact that it's even a two-day meeting—some people feel that this is way too long. Giving the neighboring countries their credit, their due, however, they do need to work out logistics, they need to figure out who would actually lead a peacekeeping force, if it's an African peacekeeping force—all of these issues haven't been completely hammered out yet. And that's what they're trying to do in Senegal at the moment.
>
> The additional Marines that are being flown in are simply going to be for added security at the U-S Embassy compound. That's their mission. They're not going to be on the streets as peacekeepers. A military spokesman yesterday made it very clear that this does not signal that the U-S has decided that they're going to send in additional troops and have them out on the streets. These Marines who are coming in, in this deployment that got interrupted yesterday, are strictly for security at the embassy.
>
> The news came in that President Bush is going to be sending some U-S warships to the Liberian coast. The initial reaction to that was jubilation here—people were thrilled to hear it. And then they heard that it was going to take two weeks for them to get here, and that just simply turned to anger, because people's feeling here is that fourteen more days like the days they have just had—a lot of people don't think that they would survive that. They think that Monrovia would be in ruins if it takes fourteen more days of heavy shelling before there's actually some intervention in this conflict.

There is a variation on the Q&A called the "ROSR" ("radio on the scene report," pronounced "ROH-zer"), which is rarely heard on public radio, but is fairly common in commercial broadcasting. It's the A without the Q—an off-the-cuff report filed while an event is taking place—as the President is getting off Air Force One in Beijing, or as the crowd is chanting anti-American slogans at a protest, or as the funeral service begins. The sound is very much a part of this sort of report—we can hear "Hail to the Chief," or the chanting crowd, or the minister's prayer, as the reporter tells us what's going on. Getting reporters to file ROSRs is another way to make a newscast more than merely a collection of spots and stories. Both ROSRs and Q&As help to get the news from the scene on the air quickly, and can buy the reporter time to work on a more polished spot or a piece for one of the news magazines.

On the Air

Many people have a hand in putting together any one newscast—producers, editors, and anchors. But the listener only hears one of them—the anchor. We can and should expect a high level of professionalism from a newsreader, but there are also ways other newscast staff can make his or her job easier.

An anchor on public radio should present the news in an authoritative and engaged manner. As a newscaster, you need to have a real understanding of the stories you present—so that your understanding can come across in your delivery. You don't need to "sell" the news; just read in a natural, interested way. Your pace and delivery should vary—again, naturally—with the stories in the newscast. After all, if you were chatting in the hallway, you wouldn't tell us about evidence that North Korea has developed nuclear weapons in the same tone of voice as you would if you were describing a Fourth of July celebration in San Francisco; so don't become mechanical or monotonous when you read those stories on the air. "I just try to be affected by the story in a way that will reflect the content of the story," Craig Windham says. "I think about the content. What am I talking about here? If people are dying, I'm thinking about what it's like to be at a scene like that."

We also expect a certain "polish" from newsreaders. You should be able to pronounce the names of people and companies and places without stumbling; you should know how to keep the sound of rustling papers to a minimum; and in general, you should uphold public radio's high standards for on-air performances. The best way to do that is to take a moment and practice the newscast before you go into the studio. Even long-time newscasters rehearse. "I always read, and time, every story before I go on the air," Corey Flintoff says.

If you're producing or editing for newscasts, there are ways to help anchors out. Get in the habit of using pronouncers for any unfamiliar acronyms or name, including the names of new reporters (e.g., "Christine Peñaloza [pen-yuh-LOH-suh] has a report"). Question pronouncers you get from reporters or editors that don't look right or seem ambiguous—the pronouncer "Ouagadougou [wah-guh-doo-goo]," for instance, isn't as helpful as "Ouagadougou [WAH-guh DOO-goo]." And use pronouncers *wherever* the name occurs; just because we've indicated in one line that "Dunwich" is pronounced "DUH-nich" doesn't mean we can drop the pronouncer if the same place name comes up again.[5]

Anyone writing copy for a newscast should write out numbers so that they're unmistakable: Many anchors prefer "225-thousand" to either "two hundred twenty-five thousand" or "225,000." And use punctuation suited for radio. Many newscasters prefer "U-S" to "U.S.," for example, to avoid mistaking that last period for the end of the sentence if by chance "U.S." falls at the end of a line, as in:

> This is the first time there has ever been an all-women rifle team in the U.S.
> Army.

Similarly, it's easier for anchors to read station call letters if they're separated by hyphens: "W-A-V-W" is less confusing than "WAVW." Newscast scripts should also avoid abbreviations that might be misread on the air. The wires can get away with "Rep. Sen. John McCain," but in broadcast news copy it's safest to write "Republican Senator John McCain." "Didn't," "wasn't," "weren't," etc., all look too much like their opposites in news copy; it's generally a good idea to write "did not," "was not," and so on; otherwise a newsreader might mistakenly say, "A White House spokesman says he did mislead Congress," when he in fact he said he did *not.* (The newscaster can always substitute a contraction, once he or she knows what the sentence is about.) Some newsreaders use a separate paragraph wherever they want to pause or change tone. Anchors should try out different fonts and font sizes, to see which are most readable.[6]

Keep in mind that things can and will go wrong. Sound files will get corrupted, ISDN connections will break down, computers will crash,

5. See Appendix II, "Pronouncers," for tips on how to represent foreign names using the English alphabet.

6. MANY NEWSCASTERS WRITE IN ALL UPPERCASE LETTERS. Some anchors say they they've simply always written their newscasts this way; others say they don't want to waste the time it takes to capitalize the first words of sentences and proper nouns. But a great deal of research conducted by and for the publishing industry has shown that the most readable text is actually a serif font—one with small strokes at the end of the main strokes of each character—formatted in upper- and lowercase. Ironically, newscasters often use a sans serif font in all uppercase—perhaps a vestige of the era when news copy came from teletype machines.

and scheduled events will suddenly be postponed. So be prepared. Corey Flintoff and other anchors say they always bring extra stories into the newscast booth, just in case they need them in an emergency. "I bring in my previous newscast, for one thing," Flintoff says. "And I always try to write extra. It's so hard to maintain your composure when you know you don't have copy there." Craig Windham says the best way to be able to ad lib is to be thoroughly familiar with each of the stories in your newscast—including the spots. "You want to tell listeners more than 'We had a problem.' Obviously you had a problem! But what was that reporter going to say? If you're really writing news well, you should be able to sit in front of the mic and have someone take your copy away and still do that newscast."

Finally, if you're producing a newscast, you can help things go more smoothly by keeping the situation in the control room as calm as possible. News broadcasting is often a frantic business, with copy coming in at the last second, reporters phoning in their stories just minutes before air time, the news wires running updates and corrections—not to mention computers crashing or other technical glitches. As much as possible, producers, studio directors, and audio engineers should try to straighten things out on their own; you don't want any behind-the-scenes commotion to be evident to the audience. Newsreaders often *do* need to be kept abreast of what's happening—if there are computer problems, for instance, or if the White House spokesman has just made a statement that will change a story—but they can't get caught up in the pandemonium.

CHAPTER 11

Booking

Only one NPR news magazine is called *All Things Considered*, but in fact all things are considered on *Morning Edition, Day to Day, News & Notes, Talk of the Nation*, and other public radio programs as well. In a single show, a host might talk with the author of a new novel set in Madagascar, America's top dobro player, the president of an Asian country where a rebellion has just begun, a teenager who has just won a computer programming contest, a linguist who studies American slang, and a man who saw a sinkhole in Florida swallow up a McDonald's. But before any of those interviews can even begin, someone has to track the guests down and determine whether they can talk fluently and coherently on the radio.

This is the job of the booker—who has to be able to combine tact, persuasiveness, imagination, and persistence, sometimes in the face of indifference or outright rejection. When you're booking interviews, you may also have to get past stubborn press secretaries or agents, persuade the guests to be interviewed, reassure them about their speaking voices or depth of knowledge, and encourage them to come to a studio. The booker is one part reporter, one part sleuth, and one part salesman. And while the work is invisible to the public, it can be heard every day, in the voices of the people who often make public radio programming stand out from the rest of what you can find on radio, TV, and the Internet.

When you're booking, your duties depend in part on the show you're working on. Some bookers are expected to write introductions for the host to read on the air. Others routinely suggest questions to be included in the interview. Some collaborate closely with an editor. Some work almost independently. Even their job titles vary. On some shows, bookers are called editorial assistants or assistant editors; on others, everyone on staff is expected to book interviews, from the summer intern to the executive producer. The one constant is the need and desire to find the best possible person to talk about a particular topic—whether it's the assassination of a South American government official or the migration of golden eagles from Alaska to Arizona.

Booking Well-Known People

"Booking" actually describes a variety of tasks. The one most people associate with the job involves contacting prominent people and prevailing upon them to be interviewed. And that sort of booking has its special challenges and frustrations.

For one thing, many celebrities don't answer their own phones—so getting them on the air means working through their press secretaries, agents, or assistants. "Your fate will be in the hands of the handler," says Darcy Bacon, who has booked interviews for several shows, including *Weekend Edition Saturday*. She points out that the person screening media requests for a politician or other celebrity can facilitate a potential interview or kill it—depending, she says, on "whether it's presented as 'Gee, you probably don't want to do this, but I'm going to tell you about it anyway' or whether it's presented as 'This is something you might really want to do.'"

"It is important to understand the celebrity's motive when you try to book them," writes NPR executive Jay Kernis in a memo titled "The Zen of Booking." As a former producer for CBS's *60 Minutes*, Kernis is speaking from experience. "It is also important," he writes, "to give the impression to the publicist that you have done your homework and you are truly fascinated by the celebrity's accomplishment, performance, last published writing or entire body of work."

"Press secretaries for politicians are the most difficult," says Lizzie O'Leary, who worked at *All Things Considered* for several years before taking a job as a TV reporter. She says press secretaries are especially unhelpful during political campaigns, "when they know that their candidate is 'the guy'—the one you want to talk to. They can afford to be arrogant, and non-specific, and flaky." O'Leary says the proper response to a standoffish press secretary or any other stonewalling middleman is "extremely upbeat firmness"—in other words, calling back, repeatedly, to push (but always politely!) for the interview. After more than a dozen years booking at NPR, Carol Klinger has found it sometimes makes more sense to go *around* a handler, rather than work *through* him or her, particularly when you're booking someone in show business. "If the publicist is not returning your calls," Klinger says, "then you try to go through the manager, or—if it's someone in politics—you try to go through their chief of staff. You try to find any other ways to get through to the person." If you're striking out time and time again with an important guest, try to make time to meet with his or her representatives when you're *not* on deadline.

One way to help yourself land an interview is to make it as easy as possible for the guest. As a booker for *Morning Edition*, Emily Barocas found that if a handler insisted, "Sorry, we just don't have time to do that right now," the best response was "How can I make it more convenient for you? Can he get on the phone with me and just talk for five minutes." When Shannon Rhoades was booking interviews, she stressed "the PJ factor"—"if we're going to do a phoner," she'd point out, "they can do it in their pajamas!" That can be the key selling point when you're trying to arrange an interview for 5:20 a.m.

Bookers also rely on the host's names and reputations to make the sale. Ellen Silva—who worked on *Talk of the Nation* for ten years—says whenever she was booking a program on a military topic, she'd tell the prospective guest about the host's interests and experience: "Neal Conan was a foreign correspondent for NPR. He covered the Persian Gulf War, he was one of the people captured in the war—he knows this issue inside and out. Plus, military history is a particular interest of his." Shannon Rhoades says she would occasionally go a step further and ask a *Morning Edition* host or senior correspondent to write a short letter to the guest. "They can write a sparkling note, something that will stand out, and then I'd do all the follow-up calls."

Almost all the bookers say mentioning the size or composition of public radio's audience can help to nail down an interview. "You have to be a good salesperson," Ellen Silva says. "You're selling NPR. It's really important to know what our cume [i.e., weekly audience] is, for example." *Morning Edition's* Shannon Rhoades says she'd keep a document describing her program's audience size and demographics at hand, and quote from it liberally. Carol Klinger says she emphasizes the demographic information—the civic-mindedness of the public radio audience—and sometimes the fact that NPR programs can be heard around the world. "If I'm booking a head of state," she says, "I tell them we're carried on Armed Forces Radio and that it's listened to by the diplomatic corps, as well as military personnel. And to someone like the President of Pakistan, that's really important."

When you're booking a prominent person—say, a candidate for a top political office—you should keep in mind that yours is not the only news organization trying to reel him in. When you do get through to his media person, "a little empathy goes a long way," Shannon Rhoades says. She's found it can help, *before* you make your pitch, to say you know the press secretary is probably swamped with requests. Sometimes you may have to send a producer out to accost the potential guest in person, and try to persuade him or her to do the interview.

You may have a PR job on your hands if the guests are suspicious of the news media in general or of public radio in particular. Sometimes high-profile people assume that any journalist who approaches them only wants to put them on the spot. In his memo on booking, Jay Kernis outlines what you should try to get across when you're landing an interview—and some of the points apply whether you're working in public broadcasting or commercial radio:

> You have to transmit what you think is the best reason *in our minds* for having them on:
> - This is a very important story for others to hear.
> - This is a very interesting story.
> - No one will have the perspective that they have.
> - They can trust NPR to provide them a fair and balanced platform.
>
> You have to transmit what you think is the best reason *they* might want to tell their story.
> - Because it's the right thing to do.
> - Because it will help others in a similar situation.
> - Because their story will make a difference to others.
> - Because these calls will come from all over the place; once they do NPR, they can tell others that they gave the interview to NPR and that was enough.

Emily Barocas says she would try to win over such reluctant guests by emphasizing the commitment to presenting both sides of any controversial issue. "I explain to people, 'We're not trying to do an exposé. We're trying to understand your point of view.' And you hope that at some point they realize it's beneficial to them to come on the air." But Barocas says that appeal doesn't always work. "A lot of times politicians are so scripted and so polished, and their communication is so controlled, that you just can't get past that—they want to have such complete control over what gets written about them, or said about them, or what points of view they put out, that they're not willing to risk doing an interview." As in battle, success sometimes requires a strategic retreat. "Sometimes editors say, 'Keep calling! Keep calling!'" Shannon Rhoades says. "And I'm just straightforward about telling them, 'Look—on that last call, I was getting the sense they were getting a little annoyed. Let's just back off for a day, or half a day, and I'll touch base tomorrow morning.'"

Jay Kernis is adamant that bookers should consider any turndown as at best tentative. "Accepting a 'No' is not acceptable," he says. "Some bookings happen in minutes. Others take years. And if it does take years, and you're still the first one a newsmaker speaks with, then you've won." But while persistence is a virtue among all bookers, sometimes a show can't simply hold a slot open for the perfect guest. *Someone* has to be booked today, this afternoon, within the hour.

That's why they invented backup interviews.

While there are times when a program host wants to talk with a specific person—the author of a new biography of the President, a senator just back from North Korea, the first twenty-year-old to head a Fortune 500 company, etc.—there are many other occasions when it is not the individual so much as his or her expertise or experience that's in demand. In those situations, you should cast a wide net—hoping to catch the big fish, but willing to haul in the best of the rest.

Hundreds of universities, think tanks, and other organizations publish guides touting their resident experts. You should keep some of these on hand so that you can quickly find specialists—even if you just consult them to get background information. Your collection of these guides should reflect groups across the political spectrum, and include people from several parts of the country and various ethnic or racial backgrounds.

You can also find information through the online or printed *Encyclopedia of Associations*, which includes details on more than 100,000 nonprofit membership organizations in the United States and abroad.

Much of the time, your backup bookings will be with people you've been talking to in the course of trying to lock in your number-one guest. For example, imagine someone has created a computerized pinky ring that will wirelessly download and store a patient's entire medical history. You've decided that your ideal interviewee is the inventor of the ring. But just in case the she is unwilling or unavailable to be interviewed, you can consider booking a number of people who can tell the same story.

In this case, it could be a patient who tested the ring, or someone at the hospital involved in the trials, or the editor of a computer magazine that first mentioned the device, or a scientist who can describe the miniaturization process. Any time spent finding and talking to these people won't be wasted; even if the inventor does call back and agree to be interviewed, the information you glean from these conversations will pay off. Perhaps the patient says that downloading the data was difficult, or that the information was not always accurate, or expresses concern about what would happen if the ring were lost—all good points, and problems you can suggest the host raise with the inventor.

Since no host wants to do half a dozen "insurance" interviews in case the coveted guest doesn't materialize, don't ever commit yourself too early. Don't even tell prospective guests you're looking for an interview. Instead just talk about the *story* you're pursuing. Carol Klinger says, "If I'm not sure I want a specific guest before I call them, I say, 'We're trying to figure out if this *is* a story' or 'We're talking to a number of experts.'" The goal is to have a range of guests available on any given topic. Darcy Bacon says she's careful not to lead anyone on or make promises she can't keep. "Generally I'll say, 'We don't know quite what direction we're going to go in. Would you be available if we needed to come to you?' And then I keep looking to see if we can find somebody better."

And as part of the long-term planning for any program, Jay Kernis emphasizes what he calls "strategic booking"—targeting the people you want to get on the air *some* day, and keeping track of how well you're doing. The key to strategic booking, he says, is making lists:

> Who do we want to get in the next week or month or year? Who do we want to make sure that the show covers many different important subjects? Who do we want to make certain the program reflects diversity? Who do we want to reflect [the host's] interests and talents?
>
> It is essential that you make weekly and monthly and long-range target lists, discuss those lists with each other, continually edit those lists, and consult them to determine your booking success.

Booking Everyday People

Booking demands a special set of skills and strategies when the potential guest is *not* famous—not someone who talks on radio and television every day—but a "regular person" who for some reason is newsworthy.

That can range from the author of a new book on the history of central Africa to the winner of a competition to carve vegetables into musical instruments—and thousands of others in between. In those cases, the problem for bookers may be less that the guests are suspicious of your motives than that they have never heard of you, your station, or your network. Then, bookers agree, the "upbeat firmness" that works with professional middlemen has to be replaced by warmth and friendliness.

When she was at *All Things Considered*, Lizzie O'Leary says the first thing she'd do when she called a stranger to book an interview was introduce herself, brief the guest on NPR, and give a short description of her program. Then she'd make the pitch. "If it's some quirky little story that we picked up from a local paper, I'll say, 'Well, I read such-and-such about it, and we think it's really neat and we'd like to talk to you about it.' I think people who have had no exposure whatsoever to the press are startled to get a call—particularly if they've never heard of NPR, or have some sense that it's a national network and think, 'Oh my God, I'm going to be talking to millions of people!'"

That prospect can be frightening to someone whose only previous experience with broadcast journalism has been watching the evening news. So it frequently falls to bookers to reassure skittish guests. "I always go into the explanation about tape," says Shannon Rhoades. She tells nervous interviewees, "Whatever you say doesn't have to come out perfectly. You can start or stop as much as you want. If you're not comfortable with how a response came out, say, 'Hey, I just thought about that. Can we go back and let me answer that again?'"

Knowing that they don't have to speak with perfect fluency can calm people down. But it isn't always easy for a booker to know who's going to perform well during an interview. A professor who is at ease lecturing in front of a class of university students or a police officer who has faced dangerous criminals may sound self-confident talking to the booker on the phone, but get a case of the jitters as soon as the interview begins. Even so, it pays to get a sense of a prospective guest's ability to stay calm when a program host asks that first question. The last thing a booker wants to do is explain to a guest that his interview wasn't up to scratch. In fact, Carol Klinger says that's the hardest part of her job: "When I call them and say, 'I'm afraid it's not going to air,' they say, 'I'm so sorry—I knew I didn't do a good job.' And I say, 'You were nervous—and maybe next time you'll be less nervous.'" But she concedes that hardly ever happens. There rarely is a second time for interviewees who get stage fright (or in this case, "mic fright").

The Booker as Detective

There's another facet of the booker's job that at first glance seems more like detective work than anything else. One of the biggest challenges for any journalist is to get an interview with a particular person *without knowing in advance who that person will be*—for instance, someone who was involved in a seventy-car pileup outside of Cincinnati; a passenger who has just ridden on a new high-speed train connecting two cities in Japan; a soldier who will be heading out to Haiti as a member of a UN peacekeeping team; a teenager who is trying to adjust to having milk and fruit in his high school vending machine, instead of soda and candy. These sorts of conversations with people who have been directly affected by news events put the "public" in public radio. These same stories could be told through interviews, respectively, with a police captain, a railroad executive, an Army officer, or a school superintendent; but they would not have the same power, or give the listener the same impression of being on the scene.

Assignments like these almost always begin with, "Find someone who . . ." In other words, find someone who saw the accident, or who knew Mr. Jones when he was a child, or who can describe what it will mean to the town when they shut down the plant. One of the most exciting aspects of booking can be this sort of "hunt"—the "needle in the haystack" mission.

Fortunately, bookers have devised a number of strategies for tracking down such people.

The search often starts on the Web. Shannon Rhoades recalls booking interviews for a piece in which Susan Stamberg wanted to talk to someone who was living paycheck to paycheck. Rhoades's first step was to go to an Internet search engine, and type in "blue collar." "Sure enough," she says, "I found this 'Blue Collar Poetry' Web site. And that led me to this woman who is a fifty-one-year-old, working part-time at McDonald's, and working the rest of the week at a pizza joint, being paid under the table so she wouldn't jeopardize her husband's Medicaid. She was one of the best people I ever booked."

Rhoades offers another tip: Don't overlook the mail. Most radio news programs receive email from listeners, much of it describing people's first-hand experiences with organizations or situations in the news. "When I was doing a series of economic portraits," Rhoades says, "I got this great guy who worked for WorldCom, and he'd lost his entire retirement savings. And I found him just because of an email he sent."

If you know *where* but not *who*—that is, where your story is centered, but not who can talk about it—find someone near your mystery guest who can suggest the best person for the interview. "With the small towns, it's easy," Carol Klinger says. "I call the reference librarian. Or if there's no library—some small towns don't have them any more—they'll have a little newspaper, an 'advertiser.'" The editor of a local paper is likely to have lots of contacts in the town, and in the region; he or she may be able to tell you the name of, say, the nine-year-old who spotted the dinosaur bone, or at least the name of the school he attends.

The "letters to the editor" section of local newspapers can also provide names of people who either feel strongly about certain issues ("I am writing to protest the building of the new InstaMart at the site of an Indian graveyard") or who have personal stories that would bear repeating on the air ("My son was forced to leave the Boy Scouts because he admitted he was an atheist"). Once you've got a name and a location, it's easy to use a Web site like switchboard.com (or its human counterpart, telephone information) to get a phone number.

The important thing is to find someone in the right place who can help you with your search—as Darcy Bacon puts it, to "get yourself into the town and find a talkative person." She suggests phoning a local restaurant or bed-and-breakfast. You can even call city hall. "If the town is small enough," Carol Klinger points out, "everybody knows everybody."

Of course, many interviews focus on issues, rather than events. In other words, the booker may know the *what*—the general subject to be explored in the interview—but neither the *who* nor the *where*. These assignments consist of finding someone, for example, "who knows how regulations have changed since the dot-com bust" or "who can tell us about the cost of treating smoking-relating diseases" or "who is an expert on western dams." Here, too, there are some tried-and-true strategies.

You can start by checking whether a reporter or host has already interviewed someone on a similar subject; you can use electronic means—scripts archived online, or lists of experts and phone numbers kept on file, etc.—or call bookers on other programs. Web search engines, like Google, may also turn up the names of people who have written about a subject, or who have been featured in other journalists' stories. One Internet source that may not immediately spring to mind is the bookseller Amazon.com. Carol Klinger recalls using Amazon to find an expert on one of the lesser-known Russian republics. "We called all these former-USSR experts, and no one knew of anybody who knew all about Dagestan.

But through Amazon, we found this guy who had written a book on it, and he was really great." Typing "Dagestan" into Amazon's search engine gave Klinger a book title, which gave her an author's name—and more important, enough information on the writer to track him down. "Amazon.com will often have a bio on the author," she says, "that will say he's a professor at the University of Iowa, or whatever. Or 'She lives with her cat and her husband in Portland, Maine.'" Once you know where the author lives, it's usually easy to get a phone number.

Even after you've identified some likely prospects to talk about dot-coms or health care costs or dam ecology, it's always smart to check in with a reporter or editor who is familiar with the subject. "It's definitely worth a call," says Ellen Silva, "to say, 'Here's who I'm going after. Am I missing anybody? Am I missing any angles?'" Klinger says experienced reporters can also act as early warning signals if a booker is on the wrong path. "Sometimes they can tell you not to interview certain people because they're not respected or not good on the air for one reason or another."

The Pre-interview

Tracking down and getting commitments from guests is only part of a booker's job. The next part (and according to most experienced bookers, the fun part) often comes after the prospective interviewee is on the phone. That's when you conduct a pre-interview.

Depending on the nature of the story and the program, you may have several objectives during the pre-interview. Among them: to determine whether the story idea holds up (since it may be based on an impression, or hearsay, or an item published or broadcast elsewhere); to determine if the guest is knowledgeable; to gather information for the host to use in his questions; to find out if the potential guest is a good talker; and to determine when the interviewee will be available.

So even if the objective is to find the right person to interview, as a booker you are often also responsible for reporting the story, albeit off-air, to guarantee that the topic *merits* an interview. Ask some questions to make sure the show has the facts straight: "Did you see the accident? How long did it take the police to arrive? Did you see anything that made you think more people could have been saved?" Carol Klinger says the interviewee's responses can be surprising—and that may be good.

"I'll discover interesting things that might change an interview ever so slightly, or change the emphasis or focus of the interview." And on occasion, the answers will suggest that the story, as it's been reported elsewhere, is simply not accurate, or is being overblown. If a booker makes several calls that indicate the original assignment may be off-base, it's time to regroup, says Shannon Rhoades. "Sometimes you have to go back to the person who's asked you to book this, and say, 'Look—I think we need to reframe the story.' Sometimes editors are receptive to that and sometimes they're not. That can be one of the most difficult things—when you start to cry foul."

Keep track of the calls you make—even if they don't lead you to the perfect guest. Jay Kernis suggests keeping a "booking notebook," either on paper or in your computer:

> At the top of the page, write the name of the story and the date. At the left side, write the name of the person you're trying to book. Under the name, write the person's title or pertinent information. In the middle column, enter phone or email info. In the right-hand column, enter the date you last spoke with that person, and a few words about what happened during the encounter: *Phone busy. Left message. He said to call in a week.* Underneath the name of the booking target, write other names and include that person's relationship to the target guest: *Best friend, mother-in-law, attorney, next-door neighbor.*
>
> This note-taking is so simple to do, it saves so much time and it also is so helpful when you update a story months or years later...
>
> It is quite uncomfortable to have someone important call you back, and you don't quite remember how she is involved in the story and what she last told you to do.

Some assignments will involve many more calls than others. Since universities and think tanks are so eager to tout the expertise of their staffs, it's usually not hard to find someone to speak about, say, Russian history or city planning. But the fact that someone has an impressive title ("Dr. Charles Fillmore is director of Southern Agriculture Studies for the nonprofit Rockbottom Institute") doesn't mean he'll be a good interview—or even that he knows what he's talking about. Experienced bookers know they have to test whether the person they've found is authoritative—and to do that, they have to be well-informed themselves. "You can do your initial research on the Web, because you want to be up on the subject," says Ellen Silva. When Silva was booking for *Talk of the Nation*, she would then chat with the guest, making sure to follow up on answers that she thought seemed vague or that listeners might have trouble understanding. "In a pre-interview," she says, "it's better if questions are not really scripted, because you're basically feeling around—trying to find out where the guests are strong, and where they're not—and if you've read the material really well, you should be able to do that off the top of your head."

But expertise is not enough. Once a booker has located a guest who can talk about a specific topic, the next step is to find out whether he or she is a "good talker." And "good" has many dimensions.

"I want to make sure that he doesn't sound like he's lecturing," Shannon Rhoades says. "I want to know that he's comfortable, that he doesn't sound intimidated. I've found that some academics are great when they're in front of a classroom, but put them in a studio, and they do get nervous."

Darcy Bacon also listens to the guest's voice and manner of speaking. "I like people who use fresh language, who make people interested," she says. "Sometimes you hear voices on answering machines, and you think, 'I don't want to hear this voice for even five minutes in an interview.' The answering machine message will give you clues—if it's too cute, or too flat, or just a voice that you don't want to listen to."

Lizzie O'Leary says bookers should listen for good stories that will hook listeners. "The best thing to do in pre-interviewing someone is to say, 'Can you give me an example of that?' Because if they can tell a story—and sometimes you don't have to ask it, if they go already into creating examples for you—then you know that the interview's going to work."

The fluency of the guest is most important when the guest is going to be on live. Justine Kenin has booked guests for both recorded and live interviews—and she says the risks are always greater when the guest is on live. With recorded interviews, a skilled producer can edit the audio to make up for many of a speaker's deficiencies. "If the guest is even 75 percent coherent, then it's the producer's role, or the editor's role, to make him sound better," she says. Former *Talk of the Nation* producer Setsuko Sato agrees that on the news magazines, "you can do several takes, and cut out sentences, or 'ums' and 'ands.'" In contrast, "booking for a talk show takes a little more coaching—trying to get people clear about their thinking before they're even on the air."

Kenin has also found the news magazine producers more willing to take a chance on a marginal guest when they know the interview is being recorded. "You can say, 'Let's give it a shot. We have this commentary, we have this book review, we have this other stuff on the shelf that we can go to if it doesn't work out, so that there won't be dead air.'" On a talk show, where an interviewee may be on the air for twenty or forty minutes at a stretch, she says, the bar for booking a guest is higher. Dropping an inarticulate or uninformed guest from a news magazine, she says, means the producer has lost a "little snippet" of the two-hour show; on a show like *Talk of the Nation*, "there's a gaping hole."

A talk show is not only live, it can take twists and turns, depending on the questions callers ask—and Ellen Silva says that gives bookers one more thing to listen for in the pre-interview. "On a talk show, you need to book a guest who not only gets to their points, but is good with the callers, can interact with them really well. So you need a guest who is adroit at changing topics quickly. You need someone who is not only dynamic, articulate, and concise, but who can also be really flexible." For that reason, a pre-interview for a live talk program can take some time—as much as twenty or thirty minutes.

Keep in mind that as much as you may want to get a particular person on the air, there are some lines you can never cross. Don't offer to pay for interviews, unless the guest is a journalist who is acting in place of a staff reporter. And don't make deals with guests or their proxies in order to land an interview. That means you shouldn't agree to submit questions in advance, or say the interview will avoid certain topics, or reach any other bargain that compromises your, or your colleagues', journalistic integrity.

Final Steps

Once you have completed the pre-interview and are satisfied that the story is on solid ground—and that the guest is both knowledgeable and a good talker—it's time, finally, for the interview to be "booked." The first step is scheduling—finding a time when the guest, host, and studio are all available. Here again, a talk show differs from a news magazine. On *Morning Edition*, for example, a live interview is an option; on *Talk of the Nation*, it's a given. So a talk show producer is not only booking a guest; he or she is booking that guest for, say, 2:20 p.m. to 2:40 p.m. precisely. And showing up late is a cardinal sin. "People have to promise that they're going to be there when their on-air time comes," Setsuko Sato says. "You can't accept any 'maybes.' Not showing up is like breaking a promise."

And if you're booking a guest who may be in a different time zone, double-check that *your* 2:20 p.m. is *her* 2:20 p.m.—and if it isn't, go over the time precisely to ensure that there's no last-minute confusion: "It's 10:30 now where you are, right? We'll be calling you in a little more than two and a half hours—at 1:10 your time—and we'll expect to have you on the air from 1:20 to 1:40 your time. Okay?"

Once you've arranged the interview, you've got to make sure to communicate with everyone who needs to know about it—which, depending on the program, may include the host, the senior producer, the senior editor, the technical director, and the studio engineer. Some shows write up and print out all the information; some take care of business electronically; some do both, and then confirm arrangements orally.

If you're booking a guest for a special event—especially for live, continuing coverage—make sure to identify the guest for the host in language he or she can use on the air. Don't write, "Sen. Debbie Stabenow, D, MI"—that's too much for a host to try to decode when he or she is ad libbing. Write something like, "Debbie Stabenow is the Democratic Senator from Michigan." Neal Conan says that when he's anchoring special coverage, he also wants to know what the person will be talking *about*. Conan says he expects to see something that not only says "Retired General Joseph Hoar was formerly commander in chief of U-S forces in the Middle East," but also includes the fact that he served for a certain number of years in the Middle East, and has firsthand experience with whatever the subject of the interview is, or has written a book on the issue. "I want the *reason* we are talking to him." And in live coverage, he wants the booker to suggest one question. "If I'm reading something as an introduction to a guest, I have a hard time reading and performing and thinking at the same time. The first question gets me started."

This is often the time when you should make sure that the editor, producer, and host all agree on the guest and the focus of the interview. You never want misunderstandings or disagreements to surface after the conversation has begun on the air. Lizzie O'Leary says one of the most mortifying moments in the life of a booker is when the on-air interview and the pre-interview have little in common—when, as she puts it, "you're sitting in the control room and you're hearing these questions coming out of the host's mouth, and thinking, 'Oh my God, I didn't ask the guest this, and this general is going to be livid with me, because this is not the interview that I said we were going for!' " A discussion ahead of time also gets "buy-in" from the host—a prerequisite for a good interview. "If the host isn't on board with what the interview's about, it's going to fail. It *always* fails," Carol Klinger says. This is also a time when the booker can pass along anecdotes, insights, and other information gleaned from the pre-interview.

Justine Kenin says whenever she was putting together a *Talk of the Nation* segment, she tried to meet with the senior producer during the course of the day, "because it's not fun to have to undo your whole show at four o'clock [i.e., the night before the program]. It's a sickening feeling." And she says a briefing with all the interested parties is also the opportunity to tell them when things are not working out, or you're in a bind. "People sometimes don't want to tell the boss, 'I'm not finding anyone,' because it sounds like you're not doing your job. But that's not the point. It's not about your ego. It's about making sure you'll have a show."

Indeed, booking is not about ego; it is very much a backstage job. But it has many rewards, according to the people who do it daily. As Kenin notes, "You're actually reporting—and you're reporting sitting at your desk, so it's a rather leisurely way to report!" Bookers also like the opportunity to meet everyone from heads of state and cabinet secretaries to unknown people who have done something extraordinary. "To me, booking is establishing relationships with people," Darcy Bacon says, "whether it's people you're interviewing, whether it's their secretaries, and you're trying to entice them to find time and help you." (It's a good idea to call a guest after they've done an interview to thank them for taking part in the program. It's not only polite; it encourages them to work with you in the future. For extra points, try to persuade the host to send the thank-you!) Many bookers are also producers, and vice versa, so they may edit the interviews they arrange; but even when they don't, they are playing the same journalistic role shared by producers and reporters. As Emily Barocas puts it, "You get to see a story from conception—from the idea—through the process of finding the right contacts, and coming up with the questions. So you really exert editorial influence."

As a booker, you may sometimes wish your *colleagues* appreciated all the work that goes into getting an informed, well-spoken guest on the line or into a studio at precisely the right time; but don't expect kudos from the audience. "I'm just so satisfied to hear the interview on the air and to know I've done a good job," says Shannon Rhoades. "I don't want the listener to think, 'Man that must have taken a lot of work!'" Bookers don't expect listeners to wonder how their shows consistently come up with a procession of perfect guests; it should just happen. As Rhoades says, "I want it to sound seamless."

CHAPTER 12

Producing

The radio producer has no obvious counterpart in the print world. Even within public radio, the job title can mean different things, depending on whether the producer works with reporters, or with newscasters, or on a program. Some producers spend most of their time editing interviews and mixing pieces by reporters; others do research, conduct interviews themselves, and write scripts for hosts to read; some get to travel with reporters or hosts and audio engineers, helping to track down and record stories across the country or even around the world. Most producers do all of those jobs at one time or another.

Producers are idea people. Many of the reports and interviews that end up on the air began as suggestions from producers. "You're finding ingredients in the day's news that resonate with you in a way that would make *you* want to listen to the radio," says Art Silverman, who has been at NPR since 1978. He says producers have to be tuned in to everything from world events to the most frivolous fad. "What you're doing is zooming in and out, between the big picture and the little things you heard in college, friends, TV shows—it's the culmination of a lifetime of absorbing interesting and trivial things, and getting the sense that this one is right for people to hear." Then the producer needs to think of a way to turn those brainstorms into compelling radio. As veteran *Morning Edition* producer Neva Grant puts it, "You take the nugget of an idea from someone's lazy scrawl on a legal pad to what you might hear on the show the next day."

Good producers "own" their stories; they see their projects through from conception to birth. "We work in a very small unit," says Alice Winkler, talking about *Weekend All Things Considered*, "so I may come up with the idea, pre-interview the guest, do the research, write the introduction and questions, edit it, and if it's something with other sound elements—music, or clips—then I mix those as well."

When they work in the field, producers are collaborators with reporters and hosts. "It's sort of a visionary job," says Tracy Wahl. "You're trying to see the big picture; you're trying to figure out what the story is you want to tell. And you're doing that in concert with another journalist who's giving you input—sometimes more, sometimes less—on what they think the story should be."

Like the editors and reporters they team up with, the top producers are first-rate journalists. But as *Morning Edition's* Barry Gordemer points out, they are also audio artists. "As a producer, you're listening for content, you're listening for information, but you're also listening for other things. You're bringing to this job a skill set other people may not have. Your expertise comes from your own artistic or creative sense. That's why you're a producer!"

Finding and Pitching Stories

One of the great strengths of public radio is that anyone—an experienced host, a summer intern, and anyone in between—can come up with an idea that may find its way into a news magazine or talk show. Since *All Things Considered* went on the air in May of 1971, one of the defining qualities of NPR has been its eclectic approach to the news. For that reason, producers at all levels—from production assistants to senior producers—are responsible for finding potential stories, and expected to present or "pitch" them in a way that makes it obvious why they're worth the listener's time.

All it takes is curiosity—which in fact was identified as one of the "core values" of public radio programming by the Public Radio Program Directors Association. As the PRPD puts it, "It's the fuel that drives our learning and pushes us to ask why, to dig deeper." On a practical level, *active* curiosity involves raising subjects and asking questions that might not be considered by other media.[1]

Story ideas can come from anywhere—which can be both liberating and frustrating. "It's very confining being a producer sometimes, because

1. The PRPD core values are are "Qualities of the Mind" (including love of lifelong learning, curiosity, accuracy, credibility, and respect for the listener); "Qualities of the Heart and Spirit" (including idealism, humor, and civility); and "Qualities of Craft" (including "a uniquely human voice," attention to detail, and pacing).

you don't have a beat," Alice Winkler points out. "It's not like you can build up a repertoire of contacts on a particular subject or a particular region, and they call *you* when they've got something interesting. You're a generalist."

For that reason, successful producers read voraciously, and on all sorts of subjects. Some program executives assign staff producers to read local papers from different parts of the country—each producer is responsible for a specific region—and to search for stories that haven't yet gotten national exposure. However, some of the best news sources aren't newspapers. Many producers make a point of leafing through trade publications or specialty magazines; for instance, Winkler says she regularly reads the *Chronicle of Higher Education*. "What I try to do when I'm reading," she says, "is look for some nugget that's really interesting—a side issue, or a person they've talked to who was just one element in their story but you think, 'Wow, that would be a great interview.'" Barry Gordemer describes himself as a "techno-geek," so he reads magazines and e-zines that have to do with high tech. "I look at the Web sites of various computer makers to see what products they're offering or are in the pipeline, so often I'll get my ideas from those sources." Some producers make a point of reading journals catering to conservatives or liberals. In each case, they're searching for something they haven't read or heard or seen elsewhere.

Your personal interests can lead to good stories. Gordemer says that because he also has an outside business—namely, puppetry—he uncovers stories dealing with theater, especially the "process of theater and how things get put together." That led him, for example, to pitch a story on the state of the art of computer animation. Tracy Wahl says many of her ideas are prompted by things she does on the weekends. "I was out on the Eastern Shore [of Maryland], and I started thinking a lot about crab pickers. And I knew there were some immigration issues associated with crab pickers and businesses on the Eastern Shore. My best story ideas come from just driving around." Alice Winkler produced a story with science reporter Joe Palca on a Cambodian doctor's anti-smoking campaign targeting the country's Buddhist monks—a story she found out about by accident. "I have a friend who's an epidemiologist and she was doing anti-tobacco campaign work, and came across this project," she says. "She had nothing to do with the project, but she had heard about it and thought it was interesting." Winkler says one bit of advice for producers would be to make sure you have interesting friends!

Margaret Low Smith—who worked for many years as a producer at *All Things Considered*—describes the story-finding part of a producer's job as "keeping your antennas up" all the time. She tells new producers just to "pay attention to the world," in hopes of detecting trends, or issues, or people that could make interesting stories. Most of us make these observations all the time; we just have to be encouraged to think of them as newsworthy. When Smith was speaking to a group of producers a couple years ago, she asked them to take five minutes and write down something they had observed, or a question they had, that they had not seen reported anywhere else. Here are just a few of the potential stories they came up with—and how they thought of them:

- How the "democratization" of medicine is pushing decision making for medical treatment onto patients (based on a conversation with a staff member who had recently been diagnosed with breast cancer and another friend who had been diagnosed with a brain tumor);
- What it's like to decide *not* to adopt a child, when one's peers and colleagues have made the opposite decision (based on conversations with staff members who have adopted children);
- How pet owners decide what to do when their animals need expensive, high-tech treatments, like canine hip replacement or canine wheelchairs, that can cost thousands of dollars (based on a trip to the vet to get care for a sick cat and prescription cat food);
- How and why public schools have stopped teaching cursive writing (based on conversations with friends who are schoolteachers);
- How immigrants frequently live twelve or fourteen to an apartment in big cities like Washington (drawn from observations of the producer's neighborhood and a conversation with a commuter on a bus in Costa Rica);
- How community gardens spawn "garden police"—self-appointed monitors who make sure urban gardeners follow the rules (based on personal experience in Washington and chats with gardeners in New York);
- How silicosis affects professional potters, or a profile of ceramic artist Warren MacEnzie, who is afflicted with silicosis (based on conversations with members of a pottery class);
- How many immigrants who have spent decades in the U.S. still think of their native countries as "home" (based on an actuality in a previous NPR report);

- What Episcopal priests do when they are unable to find a ministry (based on an item in a church newsletter);
- How people who have suffered and recovered from mental illness have difficulty getting any health insurance (based on the experience of a relative);
- How female medical students who become pediatricians question whether they have made the "easy choice" (based on a visit with a friend from college and a party with med student friends).

Admittedly, an exercise like this is limited by the diversity—or lack of it—of the people taking part; some of the pitches above are the kinds you'll often get from urban, college-educated, artsy producers. It's also important to recognize that these are only the "seeds" of stories—insights or observations drawn from life, which may not have what it takes to make great radio. After producers come up with ideas like these, the next step is to verify that their subjective impressions are accurate—that their personal experiences are not rare or contradicted by the facts. (Many pitches fall into the category "important if true.") Alice Winkler says she always tries to do some research before pitching a story to make sure she's on the right track. "People who flounder around presenting their ideas annoy everybody else—you feel like you're wasting people's time. And it's also a waste of time if you pitch a story idea and it turns out there's nothing to it." (Producers did pursue a couple of the items above, and as of this writing two of them have resulted in broadcast stories. Can you guess which ones?).[2]

Producers should be prepared to defend their story ideas against skeptical questioning by editors, reporters, hosts, and other producers. How would you get national statistics about medical school students? How would we locate people who'd been turned down for insurance? Why would the dilemma facing Episcopal priests be interesting to people of other faiths? When he's pitching a story, Art Silverman says he welcomes the questions his colleagues pose; they uncover the weak spots in his pitch, or point him in a new direction. "That's the real reason we want diversity in this business—diversity of thought and diversity of backgrounds—to use it to shape stories in different ways."

Here are some other things you should keep in mind before pitching a story to a reporter, editor, host, or program producer:

2. *All Things Considered* ran stories on the demise of cursive writing and on potter Warren MacEnzie.

MAKE SURE YOU ACTUALLY HAVE A STORY IN MIND, AND NOT JUST A VAGUE IDEA. If you suggest we "do some sort of business story on old people retrofitting their houses so they don't need to go to nursing homes," you may have a clear sense of what the story is, but you haven't expressed it in a way that a reporter or editor could evaluate it. A better pitch would be, "I think we should do a story on the industry that has grown up around helping old people retrofit their houses so they can avoid moving into nursing homes. I've found surveys that show senior citizens overwhelmingly want to stay in their own homes—and articles that say some of the companies that are helping them do that are reaping big profits."

IF A NEWSPAPER ARTICLE PROMPTED YOUR STORY IDEA, MAKE SURE YOU CAN SUGGEST HOW TO ADVANCE THE STORY. "It's not an engaging lead to say, 'I read this interesting thing in the *New Yorker*,'" Alice Winkler says. After reading the story, do you still have questions you'd like answered? Who might be able to answer those questions? How might a reporter or host approach the subject in a new way? (And if the original item was the result of enterprise reporting—if it was a scoop—make sure the newspaper that published it gets credited on the air.)

IF YOUR PITCH WAS INSPIRED BY AN ACADEMIC STUDY OR GOVERNMENT REPORT, GO BACK TO THE SOURCE. Don't count on the wires or newspapers to tell you what's interesting or newsworthy; read the report yourself and make up your own mind. Also, make sure that the report actually contains news, or is based on new research.

CHECK TO SEE WHETHER ONE OF YOUR COLLEAGUES HAS DONE A SIMILAR STORY BEFORE. NPR and most individual stations maintain an electronic script archive. If you find that someone has already tackled the topic you're pitching, ask yourself whether there's enough news to justify a reporter or host returning to the subject again. Think about how the new version would be different from the old one.

FRAME YOUR PITCH SO THERE CAN BE NO DOUBT ABOUT THE FOCUS OF YOUR STORY. You don't need to put all the facts you have gathered into your pitch. But you do need to know where the tension or drama in the story is, what is happening now that justifies coverage, and why people outside your town, city, or state are going to care about it.

FIGURE OUT WHETHER YOUR STORY SHOULD BE TOLD BY A REPORTER OR THROUGH A HOST INTERVIEW. "A lot of times your degree of knowing that is in direct proportion to the knowledge you bring to the story," Neva Grant points out. "Do I want multiple voices? Does this scream out for a scene? Or is it something that could be done more economically if I just let this one voice tell the story? And a lot of that has to do with editorial decisions. Is this so controversial or so murky a story that it's my duty to bring in other voices and scenes to flesh it out?"

IF YOU'RE PITCHING AN INTERVIEW, SUGGEST QUESTIONS THAT WILL ELICIT ANSWERS THAT ARE SPONTANEOUS AND UNREHEARSED, OR EVEN SURPRISING. This is especially useful if you're pitching an interview with an author, government official, or other "professional talker." Think about ways to make the interview stand out from all the others the guest has done.

Successful producers know how to "sell" their story ideas by suggesting characters, places, or sounds that will make them come alive on the radio. Look for tension or conflict or conclusions that might run counter to conventional wisdom. A great pitch "makes you feel something," Art Silverman says. "It makes you feel sad, disappointed, emotional, or makes you laugh." Producers may never get on the air, but when they present their story ideas to their colleagues, they're on stage. As Silverman puts it, "We are the hosts for the moment we're pitching our idea."

Producing Interviews

Visitors to NPR are sometimes shocked to sit in on *Morning Edition* or another news magazine and discover that most of the interviews have been recorded (or "pre-recorded"—which is not quite as odd a term as it may seem, since the "live" show is itself recorded). *Morning Edition* is fed to stations at 5 a.m. Eastern Time (2 a.m. Pacific!), and it's difficult to find many potential guests other than reporters who are awake and lucid at that hour. And even the shows whose broadcast schedules are more in sync with people's biorhythms record most of their interviews. (The obvious exception is *Talk of the Nation*, all of whose interviews and calls air live.) Most public radio interviews will be edited, sometimes rather dramatically, before they go on the air. So producers need to develop the technical and editorial skills required to cut, say, a twenty-five-minute conversation down to four minutes—sometimes under intense deadline pressure.

At NPR, a senior editor will usually be on hand when an interview takes place, and will make suggestions about how the audio should be cut. But as a producer, you should understand that *you are an editor, too.* Your judgment is an indispensable part of the process of deciding which parts of the interview—which ideas—are essential, and your skills will be called on to package those ideas so that the final product is coherent, gripping, and even beautiful.

WHENEVER YOU CAN, YOU SHOULD PREPARE FOR THE INTERVIEW AHEAD OF TIME. On some shows, the producer will have come up with the story idea, found the person to interview, conducted a pre-interview with the guest, and written a suggested intro and questions for the host. In a situation like that, the producer will be fully informed about the topic. On other programs, the producer may be called on to cut an interview at the very last minute; she may know little more than the name of the guest, the subject of the conversation, and how big or small a hole it will have to fill in the program. However, most of the time, producers will have a few minutes, or a few hours, to inform themselves about the topic at hand. Use that time well. If you'll be producing a news interview, print out and read one or two relevant wire stories. If someone else has booked the guest, ask for the same newspaper clippings or background information the host is getting. You may need to rely on this material to make sure that the host raises all the critical questions.

UNDERSTAND WHAT THE GOAL OF THE INTERVIEW IS. Even if you pitched the story idea, make sure that you, the host, the editor, and the senior producer all agree about what you are hoping to get from the interview. Do you want an eyewitness account of events? An "explainer," where the guest is providing background information for a story covered elsewhere? Analysis? A profile of an important person? Everyone should agree on the focus of the interview before the host asks the first question.

LISTEN CAREFULLY WHILE THE INTERVIEW IS UNDER WAY. This can be harder than it seems. The editor or engineer may be talking to *you* at the same time the host is talking to the guest; often the phone rings, and either you or the engineer will need to take the call. "When you first start sitting in the control room, there are so many distractions that you aren't actually able to listen to the interview really closely," says Sarah

Beyer Kelly, who began producing on *Weekend Edition Saturday* in the mid-1990s. "But as you become more comfortable, you can listen more carefully. And as you listen, and you're able to digest it, then you'll know more when you leave the control room."

TAKE AS MANY OR AS FEW NOTES AS YOU NEED TO IN ORDER TO REMEMBER THE CONVERSATION. Some producers try to write down every question and answer; some write down only the questions—or nothing at all. See what works for you. Barry Gordemer describes his approach producing *Morning Edition* interviews: "I tend to just take rough notes on what the question was, and then I write down a key word from the answer—'No evidence,' 'Ridiculous,' 'It's been going on for fifty years'—key points, so I don't get myself bogged down in minutiae, but stay in touch with the *spirit* of the answer."

AS THE INTERVIEW IS TAKING PLACE, THINK ABOUT HOW YOU MAY WANT TO EDIT IT. You may want to make a note to remind yourself of each question-and-answer pair that seems like a "keeper," or of any one that doesn't. "I'm always cutting in my head as the interview is going on," says Matt Martinez, who has been a producer for various NPR programs. "I weigh things as we go along. If there's a weak question and a strong answer, I think, 'Okay. That might work. We'll see what comes after this.' And I'm always listening for an end"—an answer that will make the conversation conclude neatly. Also, think about the consequences of cutting out one of the less productive or interesting Q&As. Does the next question make reference to something you expect to delete? If so, remind yourself that the host may need to re-ask that question.

IDENTIFY ANY STATEMENTS THAT MAY NEED FACT CHECKING, OR A BALANCING STATEMENT OR RESPONSE. You don't want to put *any* falsehoods on the air, so listen for assertions that may need to be checked. And if the guest makes allegations about an individual or organization, make sure you solicit a response from the person or group being criticized—ideally a second interview, but at least a statement that the host can read on the air.

WHEN THE INTERVIEW IS OVER, SUGGEST QUESTIONS OR FOLLOW-UPS TO THE HOST. Did the guest make a statement that went unchallenged or unexplained? Has the interview covered all the bases that you and the host hoped it would cover? "I try to keep the focus on 'Why did we decide to do this interview?'" Gordemer says. "I keep asking myself, 'Are we getting to the *why?* Or have we now got so much information that we've *lost* the why?'"

Remember, once the host leaves the studio, you won't get another opportunity to suggest a question, or to get the host to say something in a way that will make it easier for you to edit the audio. Make sure you have what you need before the guest leaves (or hangs up).

Editing Interviews

Cutting interviews well is an art—and one that the audience will probably never appreciate. If you take a block of rough marble and chisel it down until you've made a polished bust of Plato, people will marvel at your creativity and skill. But if you take a wide-ranging forty-minute conversation on the federal government's financial priorities and cut it down to five coherent minutes neatly focused on tax cuts, no listener will be the wiser. It is one of the ironies of editing interviews that your work will be noticed more if you make even small mistakes than if do your job very, very well. But at least you'll have the admiration of your coworkers!

Here are some pointers from veteran producers on how to edit interviews skillfully, thoughtfully, and in the shortest possible time.

DISCUSS YOUR PROPOSED CUT OF THE INTERVIEW WITH AN EDITOR, AND PERHAPS ALSO THE HOST. Some hosts don't want to be involved at this point, but you should always tell an editor what parts of the interview you think are worth keeping, and see if your instincts jibe with his or hers. "The first thing I do is decide what *didn't* work, either because it was confusing or just didn't hold my interest," says Julia Buckley of her work on *All Things Considered*. "Then I ask myself, 'What was memorable?'" Matt Martinez says he bases his cut on what the interview was supposed to accomplish. "Let's say we wanted five things, and we *got* those five things—so they're definitely in. Then we might have room for one or two more ideas. But everything else? Be realistic. You can't have an extra twenty minutes."

LISTEN TO THE RECORDED INTERVIEW AND START BY MAKING A ROUGH CUT. Some of the editing strategies you proposed while you were standing in the hallway consulting with the senior editor and host may not pan out when you actually hear the audio. So start by cutting questions and answers that were dull, redundant, off-topic, or confusing. "If you have a twenty-minute interview and you only have room for four-and-a-half minutes, almost sixteen minutes is going to have to go," Sarah Beyer Kelly says, "so you have to be really ruthless at the beginning." A first pass at that twenty-minute interview may only get it down to ten, however—so keep cutting, but more slowly. "If it's a news interview, you have to get in the 'who, what, when, where, and why,'" Julia Buckley says. "But you can always rewrite an intro so you don't have to stick by the conversation that you had. You can always go back and put some fact you've cut from the interview into the intro. Your host will be much more economical with language than an interviewee is." If you're not on a tight deadline, you may want to take several passes at the interview, trimming a little more each time. "It would be sort of like getting a haircut," Alice Winkler says, "going from long hair to short hair: you cut off the first foot first, and then you can deal with a style."

TRUST YOUR OWN REACTIONS TO THE INTERVIEW; IF SOMETHING MOVES OR INTERESTS YOU, IT WILL PROBABLY AFFECT LISTENERS THE SAME WAY. Wherever possible, look for concrete examples to illustrate abstract or complex ideas. Personal anecdotes, firsthand observations, moments of humor, and any sorts of surprises all make interviews much more memorable.

PRACTICE "ETHICAL EDITING"; TAKE CARE MAKING INTERNAL EDITS IN QUESTIONS OR ANSWERS. Unlike newspapers, which use ellipses to show that quotes have been compressed, or TV interviews, which sometimes include visible video dissolves, radio interviews don't reveal their edits in any obvious way. (Or at least they shouldn't!) So be very careful that you don't change the meaning of what someone said when you trim an answer or question. As Sara Sarasohn puts it, the producer has to be faithful to the intentions of both the host *and* the guest. "You have to be able to think, 'What was the host getting at with this question and how can I make the answer be that?' And you also have to think, 'What did the guest really want to say in this answer, and how can I cut it to make it really reflect that?' If you keep those things in mind, it takes care of a lot of other ethical problems."

BE SENSITIVE TO CHANGES OF MOOD OR TONE AS YOU CUT. "A lot of times, when you're cutting down an answer, what you're looking for is where the person's mouth is vamping while their head is trying to think up the real response," Sarasohn says. "And often that's the first third of the answer." You may be tempted just to cut anything that doesn't directly answer the question—but Sarasohn says that's usually not a good idea; people speak differently in the middle of an answer than at the beginning, so when you cut the start of a response it can sound artificial. "I find a lot of the time that I take the first sentence, and *then* cut out a bunch, and then get to the real answer." Over the course of a long conversation, guests may sound cheerful, serious, introspective, defensive, or aloof, and hosts may sound curious, amused, skeptical, fascinated, or confused. If you cut five or ten minutes out of an interview, you may find that the tone shifts abruptly where you've made your edit. Sarah Beyer Kelly says small things can help mask those mood shifts. "Sometimes something as simple as putting in a little pause, or a big breath, or a particularly breathy 'um' just gives the listener a sense that there's going to be a transition. The idea is that when the listener listens, you don't want their mind to stop in a certain spot or they'll lose the narrative. If you have a jolting transition, they're going to do that."

MAKE SURE YOUR EDITS ARE LOGICAL AND TECHNICALLY PERFECT. Faced with a looming deadline, it's easy to accidentally shorten a question or answer so that it's ungrammatical, or to upcut or clip a syllable, or to leave two breaths between phrases or sentences. Listen with headphones to make sure you get the *whole* breath too, Alice Winkler says. She hates to hear "breaths where the end of the breath is cut off, or where someone starts to take a breath and there's normally little *teeny* pause between the breath and the beginning of their word, and that teeny pause is gone." Longtime *Morning Edition* producer Jim Wildman says he hates "blanks—where people will stop something and pick up [i.e., repeat what they said], and you can hear absolute silence instead of studio ambience." Don't end a sentence on an up inflection (with the guest's voice rising in pitch), especially if the original sentence didn't end there. "One of the things I hear is when producers cut an answer, and it sounds like the guy had more to say," says Matt Martinez. "Often I've gone back to the original file, and the producer has just cut out three words—keeping them in would only have added a few seconds to the interview! No show is that hard up for time."

IF YOU'RE HAVING TROUBLE GETTING THE INTERVIEW DOWN TO THE ALLOTTED TIME, SEEK A SECOND OPINION. If your deadline is approaching, and you're having difficulty making the interview short enough, don't try to go it alone. Sara Sarasohn has been cutting two-ways since 1991, and says, "Even as an experienced producer, I'd ask for help sometimes." She says both technical and editorial problems can crop up. "I'll tell someone, 'I can't decide what to do with this' or 'This is a really tricky edit—will you try to make it?'" Julia Buckley says when she gets stuck cutting interviews, she often goes back to the hosts. "They're on the air," she points out. "I think it's really important, when push comes to shove, to ask them, 'Which question and answer did you like better?' If they're both equal, let the person who *did* the interview decide." Alice Winkler says just bringing another person into the room where you're editing can help you to hear the interview in a new way. "They don't even have to open their mouths! All you have to do is play it with someone else standing there, and you will hear all sorts of edits. It works every time!"

CHECK THAT YOU STILL HAVE A CONVERSATION. Sometimes a producer gets so wrapped up in technical and editorial details—in making sure he preserves the essential elements of the interview, makes perfect edits, leaves the breaths intact, and so on—that he forgets to listen to the finished product to make sure it still sounds like a normal discussion. "One of the mistakes people make is thinking, 'Well, he said that one really good thing, and I'll just attach it over here,'" says Neva Grant. "But then it just sounds choppy, or as if it were put together by a jury. There has to be a seamlessness to it. Nothing should ever leap out at us." Or as Julia Buckley puts it, "Remember, you're having a conversation with somebody. It's not a computer, spilling out facts. Sometimes you just can't have a conversation about a complex topic in three-and-a-half minutes. Sometimes we end up cutting answers so short that we do damage to the subject." In situations like that, check with the show producer to see whether the interview could be a little longer, or even moved to a different segment of the show where time isn't as tight. And consider making a longer version of the interview available on the Web.

Editing interviews *can* be nerve-wracking, especially if you work on a daily program where you may have only an hour or so to do your work. But it can also be extremely satisfying. A good interview never goes on too long; it amuses us, or informs us, or touches us; and like any good story, it has a beginning, middle, and end. Neva Grant says that in a way, a successful two-way is like a great conversation in an elevator—one that ends just as one person steps out. "If it works well, there's a line, and a laugh, and then the elevator door closes. And sometimes that's a useful image to use when you're cutting an interview. There's a penultimate observation, and you know that elevator door's about to close. If you can just get in that final line and imagine the door closing, you'll have satisfied both people in the conversation—not to mention the listener."

Mixing Reporter Pieces

The "sound-rich" report that is emblematic of public radio usually begins its life as a bunch of discrete elements—the reporter's voice tracks, actualities, and bits of sound or music. They may be transmitted in one long audio feed—"acts, tracks, and ambi"—or as a series of separate digital files. In either case, it's often up to a producer to mix the piece—to put all the parts together in just the right way. Mixing radio pieces is a lot like mixing the ingredients when you prepare a meal; the script indicates what the reporter wanted, just as a recipe contains instructions for the cook. But the cooking metaphor only goes so far. "Let's hope it's not like baking, where if the recipe says it's half a teaspoon of baking powder, you've got to use exactly half a teaspoon," Neva Grant says. "Let's hope it's more like a western omelet, because then if you don't have red onions you can use green onions." Grant's colleague Jim Wildman seconds that. "The script represents what the reporter and editor have agreed needs to be said," he says. "But I also think the producer is part of that team. I don't ever feel any compunction taking what is requested in the script and making it sound better."

Learning how and when to assert yourself in the production process requires you to understand the fundamentals—and the nuances—of public radio's unique brand of multilayered audio journalism.

Ideally, reporters should be using sound to help tell their stories, not as a kind audio frill or embellishment. The best cut of sound *replaces* a track or actuality—we hear the funeral dirge for the elder statesman who has just died, or the bulldozers removing the rubble of the home destroyed in the hurricane, or the protests of college students against the return of ROTC to their campus. And because we hear these things, the reporter doesn't have to describe the scenes in the same way, or with the same details. Our ears will fill in the gaps in the narrative; the sound helps us *see* the funeral and the lost home and the college demonstration.

So, as a producer, you should apply sound in a way that's both logical and compelling. "Every single production decision that you make should support the content in some way," Sara Sarasohn says. "Every time you decide to fade something in, or post it [i.e., have it come up to be heard in the clear], or hit it hot [i.e., at full volume], or fade it out, or whatever—it shouldn't be just because it sounds good, or that's what the reporter said to do, but because it makes some sense in the context of the meaning of the piece." If a reporter calls for sound to be used in a way that conflicts with that meaning, it will be confusing, or at least distracting. "If your attention suddenly becomes focused on a production device," producer Barry Gordemer says, "you've pulled the listener away from the story, the editorial content. That's a problem."

Usually, the sound sends the message "we are in this place, hearing these things." For that reason, you don't want the sound to come in too soon—to suggest the reporter's in one place (at the student demonstration) when the actuality was clearly recorded elsewhere (at the Pentagon, for instance). You should often fade out sound of a location as a reporter leaves a scene—we can fade the dirge out, for instance, if he's stopped talking about the funeral and has begun to tell us some of the things the diplomat accomplished in his career. You want sound to "post" at a meaningful and interesting place—where the stubborn truck engine finally catches, or when the frustrated middle-school student spells the word correctly. You can sometimes fade sound to black—fade it out entirely, and follow it with a tiny moment of silence—to indicate that the scene or subject or time period is changing entirely. (A fade to black can be like starting a new chapter in a book.) Some reporters will suggest these and other artful uses of sound on their own. But if they don't, your job as a producer is to employ the audio you've got to make the reporter's piece as striking as possible.

"One rule of thumb is you should hear the sound before you begin to talk about the sound," Barry Gordemer says. "Life happens this way: if you're listening to a baseball announcer, he doesn't say, 'He swings, and it's a hit—a line drive' and *then* you hear the crowd. Instead, you hear the crack of the bat and the crowd, and then the announcer says, 'He swings—and it's a line drive!'" Getting the order wrong can sometimes sound contrived, or even comical. A reporter describes the scene in a Middle Eastern capital, concluding his track with "as the Muslim call to prayer fills the courtyard" and then we hear the call to prayer. Or a reporter says, "The Jacobsons wake up to the same sound their parents used to wake to... and their parents before them—a rooster crowing" and we then hear the rooster crow. This sort of heavy-handed production can give the impression the reporter is cuing a sound effects man. Gordemer says a producer can often move the sound to make things a little subtler. "Sometimes where there's a comma or a period, you can create a little space, so we can hear some of that sound—either in the clear or just beginning to sneak in—before we hear the reference." Sometimes a word or phrase—"a rooster crowing"—can be cut out to make the sound less redundant.

The same technique can be used to work around another production cliché, Sara Sarasohn points out. "Sometimes a reporter will say, 'Start with the sound,' but it isn't clear from listening to it what the sound is. So instead, you can may make a little 'window' in the middle of the first track and post it up there." Often just a few words will help to orient the listener. If a reporter suggests starting his story with ambiguous sound—something that could be a machine, or a person hammering in a nail, or someone knocking on a door—it may make sense for us to hear it *after* the reporter has started saying, "Cheryl Johnson gets up at 5 a.m. on Sunday mornings so she is ready when the first people arrive for her open house."

Some producers argue for dropping that first cut of sound altogether. "Opening up a piece with honking horns, or bells ringing, or cows mooing doesn't take me anywhere specific immediately," Matt Martinez says. "These are generic sounds." Martinez says reporters should be discouraged from beginning their pieces with sound unless it really does set a scene. "If you open a piece up with someone announcing that London will be hosting the Olympics—'I'm happy to announce the 2012 Olympics . . .' and you hear cheering which you take under—that's a perfect example of the kind of sound you should use at the beginning of a piece, rather than Big Ben chiming."

On the other hand, there are occasions when producers will want to give more time to evocative sound, even if the reporter didn't call for it. Sometimes reporters are so absorbed with including all the information they've gathered that they get stingy about using the sound they've recorded. That's when a producer can become an advocate for them—and their audio. Matt Martinez tells a story about mixing a piece by John Burnett, who had come upon a marimba player while on a reporting trip in Guatemala. "It was great music," Martinez recalls, "But he only called for ten seconds of music here, and five there. And I thought the music had to be twenty seconds longer in places. The music was the *subject* of the story, and the song examples needed to be longer. So it turned from a three-and-a-half-minute postcard to a four-and-a-half-minute piece, which I thought sounded much better." Producer Alice Winkler had a similar experience with another Burnett piece, this time about sea turtles in Costa Rica. "He managed to record what I think was one of the most amazing sounds ever—a female sea turtle vocalizing as she laid her eggs. And he wrote a very funny line into it, which was something like, 'It sounds almost like reptilian Lamaze.'" But Winkler says the reporter indicated on the script that the turtle sound should be in the clear for just one beat, then faded under the next track. "I heard the sound, and I thought, 'There is no way, after that great line and that amazing sound, that I'm just going to bring that up just once.' So I let this string of three grunts play." Winkler says Burnett wrote her a note afterwards to say he liked the way she had stretched out the egg-laying scene.

Occasionally—and only after consulting with the reporter and editor—a producer will actually add music or sound to a story that is lacking it altogether. Barry Gordemer's idea to do an update on computer animation ended up as an assignment for a reporter in Los Angeles. He says he liked the information she came up with, but was disappointed that she hadn't used excerpts from many films. So he added some. "I went back to the early part of the piece where they talked about the success of Pixar, which had done such animated films as *Toy Story* and *A Bug's Life*, or Dreamworks and *Shrek*. And what I did was put in a small clip of *Toy Story*—just a couple of words of Tom Hanks's voice—and just a little bit of *Shrek*." The result was a montage of film clips, punctuated by the reporter's voice. "Now the points were illustrated in a different way. It was one of the things that made the already good journalism even better."

Producers will also frequently have to exercise judgment—and caution—when they're working with actualities that have been translated. Typically, we'll hear the beginning of the foreign language, then the translation. Problems occur when there's no one around who speaks Arabic or Greek or Russian; you never want the translation to be at odds with the original actuality. With luck, the reporter will provide you some sort of production instructions and phonetic cues, so you know you're in the ballpark. Here's how one reporter in Moscow transliterated the Russian to help the producer out:

> . . . for the opening of the school year. <SNEAK IN SOUND OF ELENA KASABIYEVA TALKING> Alina had two daughters in Beslan School Number one . . . a first and fifth grader. Her five-year-old also went along, to see her big sisters off to school. All three girls survived the siege, but their grandmother says Alina died of shrapnel wounds.

<BRING UP BESLAN 3, SOUND OF ELENA KASABIYEVA IN RUSSIAN 15 SECONDS INTO THE FILE WHERE SHE SAYS "Pervy dayn', kogda ani uzheh . . .">

VOICEOVER: When the first grader came out of the hospital, she was in shock . . . She kept telling us what happened and kept saying, "I was trying to wake up mommy. . . . I was trying and trying, but she wouldn't wake up." The other hostages who survived say my daughter covered her children with her body and saved them.

When you are in doubt, however, always contact the reporter to make sure you're pairing up the right actuality with the right voiceover, and that you're starting the actuality at a logical place.[3] Also, don't hesitate to let people hear the native speakers, especially if their actualities evoke strong emotions. "If a woman has lost her child, you want to hear her sadness," Julia Buckley says. "So you really want to establish the sound." Of course, the more of the foreign language we hear, the longer the piece will become—and Buckley says many reporters and editors don't account for that extra twenty or thirty seconds when they time their stories. Think about whether you want to bring the foreign language back up at the end of the voiceover—something you'll often hear in television documentaries. "I always let the person [speaking the foreign language] finish first," Buckley says. "A translator would be speaking after the foreign language speaker, so it's logical." And whenever possible, try to match the age and speaking style of the person doing the voiceover with the one in the original actuality; it can sound incongruous, to say the least, to have a twenty-two-year-old playing the part of an eighty-five-year-old grandmother.

Buckley and other producers are also adamant about not "looping" ambience. Sometimes reporters will supply ten or fifteen seconds of background ambi that they want held under a minute-long scene. Digital audio makes it easy stretch that ambi bed by copying and pasting it three or four times. It usually doesn't work. "Sound has a natural rhythm," Buckley says, "and inevitably if you loop it you're going to hear a person cough again and again, or a pencil tapping." If you can't get the reporter to send more sound, do a little "sub-mix." "I will take a bed of ambience, copy it,

3. Former Moscow correspondent Lawrence Sheets makes an interesting observation about Russian. Because it has a highly inflected grammar, the words that begin an actuality in Russian may differ from the words that begin its English translation. For that reason, he says, Russian actualities should probably be held in the clear longer than those of other foreign language whose syntax is similar to that of English.

and mix it with itself. Instead of pasting the cuts of ambi, I'll crossfade them together."

Even the simplest "acts and tracks" piece calls on a producer's discernment and taste. "One of the mistakes many people make when they learn how to mix is they let tape cuts jump in very quickly," Barry Gordemer says. "Or other times there seems like a long gap before that tape starts. And they want to know, 'How much space should be in there?' Well, the answer is it's different depending on the reporter." Producers need to be sensitive to both the subject of the story—and to the cadence of the reporter's delivery. A "tight mix" that sounds good when a reporter is talking about gasoline prices might be inappropriate if the same reporter is describing a child who died after suffering from leukemia.

How quickly an actuality follows a voice track, how the producer stretches an ambience bed, how long a foreign language actuality runs, whether a report starts with sound in the clear—all of these decisions call for artistic, creative judgments. Your ability to make them on your own will grow with experience.

Producing Music Pieces

Two kinds of music pieces are featured in most public radio news magazines: interviews produced with music taken from CDs, and "perf chats" (performance chats), where the musician plays or sings in a studio. In either situation, the success of the interview will depend mostly on the questions that the host asks. But producers can play a big role in making both sorts of pieces memorable and esthetically pleasing.

Here again, experience will probably be your best guide to producing a stand-out music piece. Nevertheless, producers and hosts do have some suggestions—which may help you sharpen your ears, or keep you from making some common mistakes.

Many producers work with hosts ahead of time to sketch out the structure of an interview with a musician, whether the music will be played live in the studio or played from a recording. Often the "peg" for the conversation is the release of a new CD, but that doesn't mean the interview has to be tied closely to the album. Having hosted various programs over the years, Liane Hansen has done hundreds of interviews with singers, songwriters, and other artists. And she says nothing is duller than going through a laundry list of songs. "Ideally, you want to be able to take [the musician's] biography, take what you know about their previous work, and ask general questions where maybe something in the answer will lead you to ask about a certain song."

Producer Alice Winkler agrees that the host doesn't need to act like a DJ during a music interview. "You don't need to cue each song," she says. "Someone can be talking about something, and you illustrate it with music, and take the music under, and the host does not ask a new question—the person just continues to talk." Her colleague Tracy Wahl is even more emphatic. "I *hate* music cues," she says, "when hosts say 'Let's listen together to this piece of music,' because you're forced to hit the music in full and then fade it under if you use that construction."

Experienced producers stress that there are many ways to integrate music into an interview with an artist; Wahl calls it the "grammar of the music setup." "It's like the way a sentence can have subject-verb-object, or start with a clause. You want to mix things up as to how you deal with the music."

But there has to be a purpose behind to the way you "mix things up." Like the sound in a news report, the music added to an interview should be there for a reason—and the way it's introduced or faded should make a point. Sara Sarasohn describes the morphology of a music piece. "A hot hit means we're starting on something. A sneak-up means the music here is tightly connected to the thing before it. When the music comes up full and ends, and then someone starts talking, that's a change of direction—the thing the person starts talking about is completely different from the music that just ended. A warm hit [starting the music at low volume] in a pause means we're building up momentum on this same subject we're discussing. Sometimes you can have music come up and end, then you hot hit something else, and then that fades under some talking, and that's a really *big* change of direction."

In short: think about why you're starting or ending music in a particular place or at a particular volume. Don't use it merely as an adornment.

Just as an actuality can come in too quickly or too late after a voice track, so can music when it posts. "There's no such thing as just 'a post,'" Sarasohn says. "You're always making decisions about tight or loose. A tight post is when the thing you want comes right out of the talking, and that is energetic. And a looser post—with maybe a note of the instrumental before the vocal—is used with slower music."

Do your best to avoid mixing vocal music under the interview—or if you must, make sure the sung lyrics are mixed at a very low level. When someone is speaking at the same time someone else is singing, the song will almost always win, no matter how careful the mix. For some reason, the human ear favors the music; the listener can easily lose the sense of what the host or guest is saying when there's singing going on in the background.

As a producer, you'll make most of your decisions about how to use music after the interview is over. But at least one has to be made as the interview is taking place. If the host decides to ask a question about a specific piece of music, you need to decide whether to play that song in "mix minus." That setup allows the host and guest to hear the music—to talk about it and react to it—even though it is not being recorded; you can then go back and add the music after you've edited the interview. For Matt Martinez, the decision is easy: Always play the music for the host and guest so that the tone of the conversation will match the music after it's mixed in. "A person's voice changes—their tone changes, their pitch changes, they get louder, they get softer—it all changes when you play something for them. And it's as musical as the music they're hearing. They're talking in one key, and when the music begins, they're talking in another. So you have to play it to get it right."

You'll need to make a whole different set of decisions when you produce performance chats. Then you'll want to take advantage of the opportunity to get the artist to perform especially for your listeners. As Martinez points out, the fact that we've lured someone famous into the studio is not itself interesting to the audience! "An amazing amount of skill goes into producing performance chats, but it's often wasted. You know what people hear on the air? They hear, 'So-and-so is here in our studio 4-A, and they're going to perform live.' Well, nobody cares where you are! And then they perform live, but it sounds just like their CD."

To keep the performance chat from sounding like any other music interview, you have to think in advance about how you're going to use the music. After all, the conceit of the perf chat scenario is that the musician is making a special appearance, so you won't want to fade the music under a host's question or the guest's answer as if it were on a CD. Everyone involved should understand ahead of time that you'll want to hear excerpts of music, as opposed to whole songs, Alice Winkler says. "As a producer, it's helpful to remind the host—and sometimes I'll say this to the guest myself before we walk into the studio, or even when I'm arranging the perf chat—'We may just ask you to play the chorus or play a shortened version of the song.' Or I'll suggest to the host, instead of asking, 'Why don't you play that?' to say, 'Could you give us a verse?' Winkler also suggests getting a singer to perform a capella, or do anything that makes it clear to the audience that the music is not coming from a recording.

That's what a *Morning Edition* host did in a performance chat with the New Orleans musician Dr. John. "At one point," producer Barry Gordemer says, "Steve Inskeep says, 'Let me ask you about this. There's that rhythm: bump, bump, bump—the Bo Diddley rhythm. Where does that come from?' And Dr. John begins to explain that that rhythm comes from Africa—and then he plays it with his left hand. And then he starts to do this incredible stuff with his right hand. It seemed like he had ten fingers on each hand! Then he would stop, and you'd hear that rhythm—bum, bum, bum . . . And then he'd go back into it again," Gordemer says. "It was clear we were with the musician, and in a different situation than just having a conversation and playing music off a CD."

Liane Hansen says that as a host, she can tell people what it *looks* like when the musician is playing—or find other ways to remind the audience that she's sharing the same studio with the artist. "We had a performance chat with a young jazz pianist named Eldar Djangirov, who is nineteen years old now—he started when he was thirteen. And I asked him, 'Let me see your hands.' So he showed me his hands. And I wanted to know, 'Could you reach the octave?'" As a producer, you should look for similar ways to get the host to describe what he or she is seeing, and to interact with the guest.

In the end, a great performance chat is a combination of planning and flexibility. Consider working with the host ahead of time to focus the interview on one subject (the three 1940s songs that are all on the latest CD), or songs from the same genre or that share a theme (blues songs, or music about love), or the origin of an artist's music (as Steve Inskeep did with Dr. John). A performance chat or music interview, Matt Martinez says, "has to tell a story as much as any other two-way does. And the music has to further what they've just talked about." But be prepared to scrap your best-laid plans if the interview starts turning in another, unexpected direction. Liane Hansen says hosts learn this—and producers should, too. "You have to be ready to act on your feet," she says, "and to go with your gut instincts, get off the program—and just be willing to be open to wherever the conversation goes."

CHAPTER 13

Production Ethics

In the Soviet Union, photographs were often doctored for blatantly political purposes. A murdered commissar was removed from a photo where he had once appeared next to Stalin; a postcard showing a 1917 Bolshevik rally was altered to turn a sign advertising jewelry into a banner with a Communist slogan; Trotsky was airbrushed out of photos in a book about Lenin. And that was in the days when retouching a photograph could require hours in a darkroom, not a few seconds on a computer.

Responsible photo editors at American newspapers don't purposely change photographs. But they do occasionally print one without realizing it has been altered.[1] And they always have to decide how to crop each picture. Sometimes those decisions raise questions about the objectivity of the newspaper: Did the published photo make the demonstration look smaller than it was? Were the police purposely cropped out of the picture? Does the photo make the confrontation seem more violent than it appeared to people who were there?

As a radio reporter, editor, or producer, you will have to deal with analogous situations almost every day—to "crop" the audio, as it were. You will often be called on to decide which parts of a recorded interview will be broadcast, and which parts thrown away. You may want to

1. In 2007, the *Toledo Blade* discovered that one of its top photographers, Allan Detrich, had been digitally altering photographs. In a column titled "A basic rule: Newspaper photos must tell the truth," the paper's editor, Ron Royhab, described what had happened: "The changes... included erasing people, tree limbs, utility poles, electrical wires, electrical outlets, and other background elements from photographs. In other cases, he added elements such as tree branches and shrubbery. Mr. Detrich also submitted two sports photographs in which items were inserted. In one he added a hockey puck and in the other he added a basketball, each hanging in mid-air. Neither was published." Detrich subsequently resigned.

shorten some of the guest's answers, and leave others at their full length. You may decide to clean up some actualities—editing out ums and stumbles and hesitations—and not others. As a producer, you may be inclined to rearrange the order of questions in an interview, or even the order of sentences in an answer. Usually radio professionals take these steps for what they consider to be production (as opposed to editorial) reasons. That is, they want to make a news report or conversation easier to listen to—they're trying to improve the pace of the interview, tighten an answer to make it more coherent, or keep an actuality from ending abruptly.

But the fact that these things *can* be done—and digital audio editing makes it easier than ever to manipulate someone's words—doesn't always mean they *should* be. And like the decisions concerning newspaper photographs, some audio production practices have serious editorial consequences. While NPR doesn't have a comprehensive set of ethical rules for mixing pieces or editing interviews or cutting actualities, the "NPR Journalist's Code of Ethics and Practices" does set out some general principles:

> NPR journalists make sure actualities, quotes or paraphrases of those we interview are accurate and are used in the proper context. An actuality from an interviewee or speaker should reflect accurately what that person was asked or was responding to. If we use tape or material from an earlier story, we clearly identify it as such. We tell listeners about the circumstances of an interview if that information is pertinent (such as the time the interview took place, the fact that an interviewee was speaking to us while on the fly, etc.). Whenever it's not clear how an interview was obtained, we should make it clear. The audience deserves more information, not less. *The burden is on NPR journalists to ensure that our use of such material is true to the meaning the interviewee or speaker intended.* (emphasis added)

There is no easy way to ease that burden; in the end, there's no substitute for editorial judgment. But producers and others should take time to consider some ticklish production issues, in hopes that a hasty decision does not have undesired consequences.

The Unkindest Cuts

On live radio, a host's interview airs as it is being conducted (or, in the case of some call-in shows, within a few seconds). The interview is not—in fact, it cannot be—edited. Occasionally a line producer will suggest a question to the host through his headphones, or let the host know that he or she is running out of time, but that doesn't change the fact that the conversation taking place in the studio is exactly the same one that the listener hears.

However, with the exception of some talk shows, most public radio news programs rely much more heavily on recorded interviews than on live ones. And those interviews often have to be edited to a fraction of their original length. On one day selected at random, some of the raw interviews for *Morning Edition* and *All Things Considered* ran twenty, thirty, and even forty-five minutes; none was any longer than about seven minutes when it was broadcast, and most were considerably shorter.

If you're a producer assigned to edit an interview like that, where 75 percent of the questions and answers will have to be deleted, it will be difficult to stay true to the meaning and spirit of the original—no matter how much skill and experience you have. Also, you don't want guests to be shocked—or feel they were misled—when they hear themselves on the air and discover that most of what they said has been cut out. That's one reason hosts should routinely tell their guests that their remarks will be edited, and if possible give them a sense of the degree to which the interview will be shortened. This sort of fair warning gives the producer license to make the kinds of cuts required to get an interview on the air in a timely fashion.

But it doesn't absolve him of the need to exercise good judgment, or to consider the editorial ramifications of the editing process.

For example, if you're cutting an interview, it's understood that you may need to drop questions and answers, or shorten answers, or tighten up questions. But you may be tempted to go too far—collapsing two answers into one, rearranging the order of questions, and so on. When you make such extensive changes, the result may not reflect what actually happened in the studio.

Consider the highlighted sections of this interview on a proposal to cut federal funding for community development block grants—government programs designed to provide affordable housing, create jobs, and increase business opportunities. The host is speaking with a professor of urban affairs, who turned out to be a supporter of the grants. The version below aired in the first play of *All Things Considered*, before it was reedited for later feeds of the program.

HOST: Now what might be an example of a problem and a specific solution that was developed through the block grants?

PROFESSOR: Some of the problems were very prosaic. A crumbling water or sewer line in an old neighborhood would need to be replaced, and so these funds would be used to provide new infrastructure. Another example might be that the housing stock was decaying to the point of being unsafe for lower-income residents, so subsidized housing was built. In other circumstances, the moneys were used to provide community centers for recreation or to provide health-care services. The list covers the gamut of all sorts of goods and services and activities you could provide for needy neighborhoods.

HOST: The President's budget says that existing programs now are duplicative, that they're spread out over a number of agencies—he wants to put them all together—and, also, that this new approach would include what they're calling "rigorous accountability measures." Has there been a problem with oversight and accountability, making sure money's being put to good use?

PROFESSOR: I believe that the potential problem of waste or fraud is very overstated, because you have accountability at every local level with different branches of government. And I think that kind of checks and balance at the local level simply has to be trusted. And there's no reason to suspect that that kind of local check and balance would be any more or less effective as an oversight tool than stricter federal regulations and oversights in Washington.

HOST: You mentioned that the block grant program began in earnest in 1974. Who are its biggest supporters?

PROFESSOR: I think that virtually all the mayors in the country, regardless of their political stripes, like them because it's extra revenue that they have reasonable flexibility over spending.

The host here is trying to play devil's advocate by asking whether the money for these grants is put to good use. The guest responds that there are local checks and balances that are as effective as federal oversight.

But that's not how he actually answered the question. When the same interview was reedited to more closely approximate the conversation that took place in the studio, it became clear that the guest did not directly answer the host's question the first time around. What he said instead—that the use of federal money did not trigger private investment, as it was expected to—might be heard as an indirect admission that the program was *not* working as intended. And the host's follow-up question—which was cut out the first time the interview was broadcast—acknowledges that her first question had not been answered, or at least not answered completely. The host's tone in the second exchange is considerably tougher—and appropriately so. (The sentences in boldface were reinstated when the audio was recut.)

HOST: The President's budget says that existing programs now are duplicative, that they're spread out over a number of agencies—he wants to put them all together—and, also, that this new approach would include what they're calling "rigorous accountability measures." Has there been a problem with oversight and accountability, making sure money's being put to good use?

PROFESSOR: I think that the moneys have not necessarily always been put to the most strategic use. By strategic, I mean that cities have often spread their CDBG funds very broadly across lower-income neighborhoods—so broadly that there weren't significant concentrations of revitalization funds in any single neighborhood to trigger a change in confidence on the part of private investors there. And as a consequence, private investors didn't put their money in on top of the federal money.

HOST: What about oversight, though? This is federal money. Who makes sure that the housing project is, on paper at least, actually is built and is built without corruption and without kickbacks? There are all sorts of things that can go on when there's a big pot of money to be had.

PROFESSOR: I believe that the potential problem of waste or fraud is very overstated because you have accountability at every local level with different branches of government. And I think that kind of checks and balance at the local level simply has to be trusted.

HOST: You mentioned that the block grant program began in earnest in 1974...

The two versions also raise questions about the propriety of moving the answer to one question so it appears to be the answer to a *different* question. Producers sometimes do this when the host has to repeat herself to try to get a more complete response from the guest; the person editing the audio may take the first version of the question and tie it to the second, more coherent answer. In the case above, the producer justified the edit by saying the two questions were *effectively* the same. But the result gives the impression that the professor answered the question the first time it was asked—that is, it makes his answer seem more straightforward than it actually was.

This sort of unintentional editorializing can happen whenever an answer is shortened. In an effort to get time out of interview or actuality, producers or reporters frequently excise what appear to be irrelevant comments—asides, digressions, parenthetical phrases, and extra details. Most of the time, these sorts of "internal" cuts are not controversial. But look at what could happen by editing out the "digression" in the following fictional excerpt.

> HOST: Senator Jones thinks the administration's plans for the new weapons system will not make it through his committee.
>
> JONES: No—no way. The only possibility I can imagine is if the President cut funding for the D-X Missile program by fifty percent—and I can't believe that's going to happen. The administration has as good as said it isn't going to happen. This committee is not going to let the President run roughshod over us. We have the Constitutional responsibility—the obligation—to make decisions that are fiscally sound and responsible.

Take out the middle of the actuality, and the Senator sounds a lot more adamant:

> No—no way. This committee is not going to let the President run roughshod over us. We have the Constitutional responsibility—the obligation—to make decisions that are fiscally sound and responsible.

A reporter could argue that the excised portions don't need to remain in the actuality because the senator immediately dismisses the possibility he has just raised—that the administration will cut its budget proposal for the DX Missile. But in the original cut, one can almost hear the senator asserting himself, then backing off, and then reasserting himself. That moment of hesitation is lost in the second version. Through our editing, we seem to be making the senator take a stronger stand than he actually did.

Of course, some actualities *do* need to be cut down, and producers shouldn't become paralyzed out of fear of making an inadvertent journalistic blunder when they make internal edits. Many interviewees include more details than even the most patient listener could absorb. Here's another fictional actuality where a producer could safely cut any number of words and phrases without sacrificing the meaning or tone of the original:

HOST: Charles Johnson says it was actually a copier company that gave birth to the PC.

JOHNSON (48 seconds): Xerox invented, improved, or conceived of a lot of things we associate with modern computers. Look at a Macintosh computer today, or any machine running Windows—or Linux for that matter—any computer today will probably have a screen with folders, windows, menus, icons, a desktop, a trashcan, a cursor... blank "pages" that are white so they look like paper, fonts you can change, drawing tools... and the user will be selecting things with a mouse, and using a laser printer, and sending information over ethernet cable, which is used to link workstations together in a network. Even the idea of a stand-alone personal computer rather than a mainframe that people shared using dumb terminals—all of these things came out of Xerox's Palo Alto Research Center, and nearly all of them were on display as early as 1974. That's ten years before Apple came out with the Macintosh.

While the details of Xerox's achievement might be interesting, dropping a few specific examples (the whiteness of the pages, the drawing tools, etc.) or making the speaker less verbose (by trimming the mentions of Windows, Linux, etc.) would make the actuality more concise and still allow him to make his point. An edited version might look something like this:

HOST: Charles Johnson says it was actually a copier company that gave birth to the PC.

JOHNSON (25 seconds): Xerox invented, improved, or conceived of a lot of things we associate with modern computers. Any computer today will probably have a screen with folders, windows, menus, icons, a desktop, a trashcan, a cursor... and the user will be selecting things with a mouse, and using a laser printer, and sending information over Ethernet cable. All of these things came out of Xerox's Palo Alto Research Center, and nearly all of them were on display as early as 1974.

The following actuality—again, this is a fictitious example—would pose a bigger editing problem, both because of its subject and its structure.

HOST: James Johnson is a deputy assistant secretary of defense.

JOHNSON (53 seconds): I think we can see the results of the administration's Mideast policy in a number of ways. First we have had elections in Afghanistan, and we have had elections in Iraq, and in both cases, things went much more smoothly than our critics predicted, and the turnout was bigger and more enthusiastic than many people predicted. We have seen the first municipal elections in Saudi Arabia in forty years, which amount to a small but important step toward democracy in a country that hasn't known it. In Egypt, President Mubarak has called on parliament to amend the constitution to allow for open and competitive presidential elections, potentially presenting him with the first real challenge since he came to power in 1981. And in Lebanon, we're seeing demonstrations calling for Syria to remove its troops from that country. It is not a coincidence that these are the same countries the President singled out in his State of the Union speech as being in need of democratic reform.

Here our imaginary deputy assistant secretary is making an argument—that the President's State of the Union address somehow precipitated political change in the Middle East. He supports that assertion with several examples. Since the State of the Union speech mentioned specific countries, we can't simply cut one of them out of the actuality. And trimming the dependent clauses ("which amount to a small but important step toward democracy in a country that hasn't known it," "potentially presenting him with the first real challenge since he came to power in 1981") undermines the case the speaker is making. Whether we think the speaker's argument is cogent or specious, we owe him the opportunity to make it intelligently; if this actuality has to be edited, it should be done judiciously and with forethought. (One possibility might be for the reporter to paraphrase the beginning of the cut dealing with Iraq and Afghanistan, and begin the actuality with, "We have seen the first municipal elections...")

You can also cross onto shaky editorial ground if you keep all the sentences of an actuality intact, but change their order. This is a trick many reporters or producers use to keep an audio clip from ending on an "up" or "rising" inflection—the result of ending an actuality before the speaker took a breath or completed a thought. (For instance, imagine that the first sentence of this paragraph were part of an interview, and you wanted—for whatever reason—to end your actuality with the words "...actuality intact." Chances are that the pitch of the word "intact" would be rising just as the cut ended.) Generally producers do whatever they can to avoid ending a phrase on a rising inflection. There's no consensus in public radio on whether this kind of audio editing is unethical. Some producers and editors object to it in principle because they feel that what is broadcast should accurately reflect the actual conversation that took place. Others insist that it's their prerogative to rearrange sentences if the speaker was forewarned that his or her comments would be edited. Still others say it depends on the particular actuality that's being edited.

Compare the two variations of the following actuality. (Again, this is an imagined audio cut.) To most producers, reordering the sentences to improve the sound of the clip would clearly alter the excerpt's mood, if not its meaning:

> I have never been able to make friends with anyone in the United States. When people find out I am an Arab, they avoid me. If they did not need my skills as a computer programmer, I would probably not even be able to find work. And I have been in this country ten years! My children speak Arabic and English—so maybe things will be better in a generation. They think of themselves as American, and I think that's how most people think of them. But my experience has been difficult. [Up-inflected]

> I have never been able to make friends with anyone in the United States. My children speak Arabic and English—so maybe things will be better in a generation. They think of themselves as American, and I think that's how most people think of them. But my experience has been difficult. When people find out I am an Arab, they avoid me. If they did not need my skills as a computer programmer, I would probably not even be able to find work. And I have been in this country ten years!

The first cut ends on a rather optimistic note; the second buries that sentiment between two gloomy descriptions of life in the United States. And yet they both have exactly the same words. Here's a case where a producer should hesitate to "fix" the rising intonation by changing the order of the sentences—or at the very least, seek a second opinion from an editor.

Other Production Issues

Some other common production devices raise editorial questions. For instance, reporters frequently like to "butt-cut" two actualities to show contrasting points of view. This can be very effective and ear-catching, as in this fictional example:

REPORTER: Janet Franklin is president of the National Committee for Women and Children. She says America's children are the ones who have benefited most from changes in welfare policy.

JANET FRANKLIN: I have been to New York, I have been to Chicago, I have been to Anacostia in Washington, and I can tell you welfare reform is working. Women who used to be dependent on welfare checks now have more self-respect. Children who could only look forward to living on welfare themselves now have a positive role model at home. And families have more money overall, which is hardly insignificant.

SUSAN ROGERS: That is complete nonsense. Nothing could be further from the truth.

REPORTER: Susan Rogers heads the group, Single Mothers of America. She says her research shows many families are still struggling to get by.

SUSAN ROGERS: I think welfare reform has been an unqualified disaster, no matter what the statistics say. Essentially the government has pushed large numbers of people off welfare into low-paying and stultifying jobs. Single mothers especially are now forced to work long hours at minimum wage, and put their children in day care, because they've exhausted their allotted time on welfare. It's a disgrace.

This sort of point/counterpoint production gets a lot of information into a small amount of time, and as long as the voices of the speakers are not too similar, it can highlight the crux of a controversy very powerfully. It can also be misleading.

In the example above, it sure *sounds* like Susan Rogers is responding directly to what Janet Franklin is saying. But it's not clear who was interviewed first, or what "that" refers to in the sentence "That is complete nonsense." For example, the reporter might have elicited that response by saying to her, "Your critics say you're using the welfare issue as a political ploy because you're planning on running for state office. What do you say about that?" If that's what provoked the "nonsense" answer, it would be unethical to butt-cut the two actualities this way. It's incumbent on the reporter and editor to make sure that they don't sacrifice editorial integrity for the sake of a production gimmick. In situations like this, the reporter would always be on safe ground if he or she played the recording of one interviewee to the other to get a reaction—or at least, quoted the first speaker verbatim to the second.

No one in public radio argues that it's ethical to deceive the listener. What people are constantly trying to define is when deception occurs. After all, the production process necessarily involves a certain amount of manipulation of audio, whether it's simply picking the actualities out of a raw interview or fading the sound of a farmer's combine under a reporter's voice track. Our art depends on a certain amount of artifice. So how much is too much? Does every ambience bed suggest that the reporter is really on site, and not in the studio? Should a host always make clear to the audience when an interview has been recorded? If a live interview is rebroadcast on a "rollover" of a program, should it be preceded by an announcement that it was previously recorded? Should the entire show start with such an announcement?[2]

After a while, these questions can feel like the debate over how many angels can dance on the head of a pin. But whether you are a producer, reporter, editor, or host, it's worthwhile at least to discuss these issues, and to try to come to some agreement with your colleagues about which production techniques might be off-limits.

Some issues seem straightforward. For instance, public radio reporters and producers do not "manufacture" scenes for news programs. If you arrive at an office fifteen minutes after the employees finish holding a prayer vigil for their kidnapped boss, you cannot ask them to reconvene so you can record a *simulation* of the event. By the same token, you shouldn't ask people to pretend they are answering the phone, or typing a letter, or fixing breakfast, so that you can get sound of those activities. You should never use sound effects that could be mistaken for actualities or for ambience that has been recorded on site. For example, if you interview an environmentalist who is trying to determine whether pesticides are causing frogs to be born with abnormalities, you cannot add

2. These and other issues were raised in John Solomon's report "Pulling Back the Curtain," broadcast in December 2004 on the NPR program *On the Media*. Solomon says, "By making everyone sound better and increasing the amount of content in the broadcasts, it would seem to be a win-win-win for the network, its sources, and most importantly, its listeners. Yet is there a small sin of omission? NPR may not be actively misleading listeners, but we all know that they don't know how we create the cleaner and more articulate reality."

the generic sound of frogs croaking when you mix your piece.[3] The only sounds that are appropriate are those you record yourself, presumably in the field with the researcher. That rule applies even if you are producing a feature, as opposed to a hard news story. One independent producer submitted a documentary about a shipwreck that included what sounded like scratchy archival audio from the 1940s:

> Marine forecast for Newfoundland: For south coast, storm warnings. Winds northerly, fifteen to twenty knots—increasing to northwest, thirty, this evening, and to northerly gales...

When asked how he happened to have a recording of a radio broadcast for the very day of the shipwreck, the producer confessed he had written it himself and put it together it in his own studio. He was forced to remix his entire piece to remove this and other re-creations.

Some situations are trickier. One editor was concerned that she had allowed a reporter to write this description in her story on the economic consequences of prison labor:

> REPORTER: Another state, another noisy factory—but this one in Zeeland, Michigan, is not behind bars. Men line stations churning out desktops and office cubicles. This factory belongs to furniture maker Herman Miller. The company was hit hard by the recession. It laid off more than four thousand workers, almost half its work force. During those same years, the federal prison industry saw its furniture manufacturing grow 30 percent.
>
> MARK SHERMAN: They pay no taxes, and they pay their employees a buck or less an hour. It's an outrage.
>
> REPORTER: Standing just above the factory floor, Mark Sherman, head of public affairs for Herman Miller, looks around at the employees below. He says there's more at stake than company profits.

3. In the late 1990s, birdwatchers were exchanging notes on the Internet about what they were hearing during the CBS TV broadcast of the Masters golf tournament in Augusta, Georgia—namely, the sounds of birds that shouldn't have been so far south in early spring. For example, one writer said he heard the distinctive call of a red-tailed hawk on the ninth green—and then the identical call, recorded from the exact same distance, on the tenth. "A couple of things I heard made me wonder whether CBS wasn't playing a soundtrack or something," he wrote. It turned out he was right. CBS confessed it had been adding taped birdsongs for its broadcast, and in 2001 made good on its promise to air only the sounds that were actually coming from the golf course.

Only the reporter and editor knew that the reporter never visited the factory. Her description was accurate, but it wasn't drawn from her own observation, as her writing implied. In retrospect, the reporter felt the passage probably *was* misleading; the audio had been recorded on site by a producer, so the quality of the sound would have led people to think she was there, and she describes where Sherman was standing as if she had seen him herself. This is hardly a serious crime, but the editor thought it crossed a line, and discouraged the same reporter from implying that she'd been on the scene in a later report that included an interview that had actually been conducted by phone.

It's natural for listeners to assume that a detailed description comes from a reporter or host's firsthand observation—and almost invariably, it does. But there are exceptions, and they often surface when producers write "tape and copy" for hosts. Most of the time this involves "public" audio—a presidential news conference, an announcement from the New York attorney general, or the National Transportation Safety Board's description of an airline accident. In these cases, the host often provides continuity and context—paraphrasing parts of the speech or briefing, repeating reporters' questions that were not recorded, or giving historical information to help the listener make sense of the day's news.

But some tape-and-copy segments—like this one from a series on immigrants' first Christmas in the U.S.—come very close to sounding like regular news reports:

HOST: Earlier this year an airplane landed in Burlington, Vermont. It was carrying a refugee from the Republic of Congo. He fled his country in 1998, when male members of his tribe were threatened by militia forces. He was sent to Vermont through the international resettlement work of the United Nations and the United States State Department... [W]e hear today from Joshua Dimina. In one way, he's like many new refugees this time of year. He's feeling alone.

[Music]

JOSHUA DIMINA: Good. You sing in the Congolese language. It helps me to remember my family. And Christmas is something for a family, to recognize that family is committed. But you see, I'm alone. I think Christmas will be good when I see again my son.

[Music]

HOST: When he fled the Republic of Congo, Joshua Dimina left his three-year-old son behind. They haven't seen each other since, though Dimina has been told that his son is safe, living with Dimina's mother.

> Pictures of Dimina's family and keepsakes, like his father's Bible, fill his one-room garage apartment in Burlington, Vermont, which is where he likes to practice his guitar. The thirty-eight-year-old says he's grateful for the resettlement staff that helped bring him to Vermont...

The same report goes on to include a clip from Dimina's case manager, ambience from a hospital and actualities from a doctor who talks about getting Dimina involved in medicine in the U.S. But in this case, the host not only didn't go to Vermont; he didn't conduct the interviews. The piece was reported and written by a producer.

Does it really matter? Again, this was hardly a controversial topic, so few listeners would be outraged to find out that the host served here primarily as a narrator, as opposed to a reporter. They may assume that's the way broadcast news works. Indeed, many TV producers *expect* to write for their hosts or correspondents, and are surprised to discover that public radio reporters come up with their own questions, do their own interviews, usually gather sound by themselves, and always write their own scripts.[4] But it's precisely because radio reporters and hosts are working journalists and not just announcers that a piece like this poses problems; it blurs the line between reporting and performing. "When I read an intro to a reporter's piece, I'm performing a fairly recognizable role of continuity reader," says *All Things Considered* host Robert Siegel. "But when I go to the Gaza Strip for NPR and do a report, I hope it's widely assumed that I have written what I have written—it's been edited, but I wrote it. And then there's this strange area in between, where somebody does a couple of phoners, writes copy to go around it, brings it in the studio, and I read it. I would think, if I were a listener and heard that, that the host had done this reporting. I don't want to be in that situation." Once listeners know the host wasn't the author of the report, they have a right to wonder how much of what else they hear on the air is what it appears to be.

The simple solution to this, and to many similar cases, is to reveal enough of the production process to make it clear that the host was not on site. Explaining at the top of the story that "We sent producer Joe Burgess to Vermont"—or even inserting a phrase like, "As Dimina told an

4. In 1998, CNN was forced to retract an investigative report alleging that U.S. forces used nerve gas during the Vietnam War. The producer responsible for the report was fired, but the "on-camera reporter" was not. In fact, he was able to defend himself by saying he had "contributed not one comma" to the piece.

NPR producer" before the first actuality—lets listeners know how the audio was obtained without undermining the effectiveness of the report.

For the same reason, it's important to be explicit about how *any* audio is acquired if the circumstances are unusual or could raise questions. If the only available cuts of a politician under investigation come from a two-year-old interview, you should say when and where the interview was originally conducted. (After all, a politician might well answer questions one way when he's just been elected and a completely different way when he's been accused of a crime and his popularity is sagging.) If your report includes excerpts of an author's works, you should include a track explaining whether they're read by the author or by an actor. If a foreign language speaker's actualities are heard only in translation, you should explicitly say so in the script. And if your piece relies on man-on-the-street comments, it's a lot more honest to tell the audience, "We asked tourists outside Union Station in Washington what they thought" than to say, "Here's a random sample of what people think."[5]

Similar issues arise when a news story includes music. "At NPR, certain feature stories and mini-documentaries—on rare occasions—use music when it is clear to the audience that the music is being used to enhance the listening experience," Jay Kernis says. "But in most cases—especially for news stories—the source of music should be authentic and identified." For instance, if you're interviewing people in a diner and the jukebox is on in the background, you could include music from the jukebox; if you profile a baseball player and you ask him to play jazz guitar for you, you should make it clear that you coaxed him to do the performance; if you interview a doctor who says he likes to play Mahler during surgery, you could ask him to take out his favorite CD and play it for you, but you shouldn't mix it in as if it were part of a scene, if in fact it wasn't. In short, Kernis says, "the audience should not be led to think music was present where it was not, unless it is told why."

For better or worse, public radio has relatively few hard rules about how to mix and edit audio; these and most other tricky cases are judgment calls. It may be a sign that we're doing things right that very few

5. Pieces that rely heavily on "vox"—a montage of actualities without a reporter's voice tracks in between—often pose editorial problems. One NPR music feature did not disclose the fact that its sample of opinions came from people at the same radio station where the reporter worked. In another piece, a commentator included actualities of her friends and even family members, again without identifying them as such.

guests complain about how their interviews are edited. Or it may simply mean that most people don't remember precisely what they said, or don't hear the interviews or pieces that feature them. In any case, the same ethical principles apply to producers as to reporters: Be fair and responsible, never mislead or deceive people—and always aim for the truth.

CHAPTER 14

Program Producing

Producing a radio news magazine is a little bit like putting together a jigsaw puzzle that has been bewitched.

You start with just a handful of pieces, but discover some will only go into specific places on the board—one always seems to place itself in the middle, for instance, and another keeps snapping into a corner. As you acquire more pieces, you find ways to connect them to the ones you've already laid down, but sometimes their shapes and sizes unexpectedly change—a piece may grow or shrink without warning, and occasionally one disappears altogether! Just when you've found a way to turn the seventeen or eighteen discrete puzzle elements into some sort of coherent picture, the pieces from the top move to the bottom—or you discover that you have to make an entirely different picture.

To top it off, you have a limited time to finish, and millions of people are waiting to find out how you'll do.

That may give you a feeling of what it's like to be a producer of one of the NPR news programs. The pieces of the puzzle consist of news reports, interviews, commentaries, book and film reviews, and a variety of other items, from audio postcards, to tape-and-copy blocks, to poems. You know that the most newsworthy elements, whatever they turn out to be, will usually begin the hour. But you'll have to arrange—and frequently rearrange—all the others to fit the time allotted to each segment. (And there may be as many as five segments of different lengths each hour—an A segment at the "top," through a D or E segment at the end.) Stories that look like sure bets at 10 a.m. can evaporate by mid-afternoon; late-breaking news can force you to completely reconstruct the show just minutes before it goes on the air. And however complicated and frantic the process of putting the show together, the producer has to make sure the broadcast sounds coherent, polished, and authoritative.

The Art of the News Program

If producing a news program is like assembling a puzzle, ideally the end result—the "picture" it forms—should be a work of art. It should have contrasts of light and dark, it should be balanced but not symmetrical, it should have a certain rhythm that is suited to its material. In other words, a good radio show should always be greater than the sum of its parts—especially now that people can download the parts individually (and in any order) from the Web.

One of the ways a program does that is to organize the pieces and interviews in some sort of logical way that allows one story to complement and amplify another. "What a show does is allow you to put information in context," says Chris Turpin, who has been both a program producer and Executive Producer for *All Things Considered*. "It allows you to show its relative importance; it allows you to suggest how it relates to other material; it allows you to work out where it stacks up in the universe of ideas." Or as Jeff Rogers describes *Day to Day*, "There's a sense of connectivity and cohesion if you listen to it the way we put the show together, as opposed to listening to it in pieces. The information is the same, but the way the pieces are put together enhances the listening experience over the course of an hour." A news program, like each of the individual items it contains, should have a coherent structure; it should move—it should progress—from the beginning to the end. The host serves as a guide through all this information—taking listeners from one topic to the next.

It's an article of faith at NPR and throughout public radio that a news magazine should encompass many fields of human endeavor. "Each program should have a little bit of everything in it," Rogers says. "National news, political news, international news, health, and science—but there should also be opportunities to introduce characters, to meet people; there should be places for music and humor, and opinion and analysis and commentary." After all, you're producing a show that should appeal to "the whole person," says Sue Goodwin of *Talk of the Nation*, not just the news junkie. She and her staff look for the angles of top news stories that really warrant discussion—that fit the format of a talk show—"but we also look for something arising that day that sparks curiosity—something amusing or creative about the human spirit. We like to find a topic that is about how we live our lives—health care, family relationships, how we deal with new technology." The goal of any program producer is to get that mix just right, and to present the different subjects in such a way that the audience remains immersed in the material—whether it's a report on a car bombing in a foreign capital, an interview about a new atlas of American slang, or a visit to a French restaurant that is teaching high school dropouts how to prepare haute cuisine.

Here are some other tips that will help you make sure your news program is more than a collection of discrete elements.

BUILD SEGMENTS LOGICALLY. Whether you're planning a talk show or a news magazine, it isn't enough just to put several related stories together. Whenever possible, you want the second item to answer questions raised by the first, and the third element to take you somewhere you didn't get to in the first two. One edition of *All Things Considered* started with a report on sectarian violence in Iraq that broke out after the bombing of a Shiite shrine; that was followed by a two-way with a reporter about the mood in Najaf, the seat of the Shiite clerical establishment; then listeners heard a commentary saying the conflicts between Sunnis and Shiites in Iraq are just decades—not centuries—old; and finally the host interviewed Iraq's national security adviser, Moaffak al-Rubaie. Each report or interview filled in some aspect of a story left unresolved by the preceding one. That's not always easy to pull off. "You can think simply by sticking three things together that have to do with the same subject you're somehow adding context, and then you listen to it, and it's deadly," says Chris Turpin. "That's because the different stories, the different threads, are not organically tied together. They take you in different directions."

MAKE SURE THE SHOW HAS THE RIGHT RHYTHM OR PACE. The most important story may not necessarily be the longest one; a hard news report may be short and intense, and you may want to follow it with a more reflective, in-depth interview. "Hopefully, if it's a little bit slower and a little more drawn out, the content is compelling and powerful enough that the audience will want to pause and listen more actively," says Jeff Rogers. A single program may be composed of pieces or interviews or tape-and-copy blocks that range from just a minute long to six or eight minutes. The producer needs to consider how quickly to change topics, or whether a given subject deserves the time budgeted for it. "If we've done our job right, over the course of an hour the show has a rhythm and a pacing so that it never drags, and it's never so rapid-fire that there's too much information, and it goes by too fast," Rogers says. "If it's too slow, they'll tune out, and if it's too fast, they'll tune out." On a talk show or call-in program, the pace is often adjusted on the fly, while the program is on the air. At *Talk of the Nation*, the Senior Producer is always prepared to cut off a topic sooner than planned, or extend one if the listeners' calls are adding a dimension to the subject no one had anticipated.

LAY OUT THE "GEOGRAPHY" OF THE SHOW FOR THE AUDIENCE. Producers may understand the logic of their programs, but they also need to convey it to the audience. A strong billboard can point the way and pique the listener's interest, as can an "umbrella lead," where the host tells people what they're going to hear over the next ten or fifteen minutes. "The best radio feels planned and organic," Turpin says. "Tell people there are several aspects to this story, and that we're going to tell you about all of them—but first we're starting here, with the news of the day." And use promos throughout the program to let people know what's ahead.

PLAN SOME SURPRISES FOR THE LISTENER. When someone signs up for a podcast or clicks on a Web link, he's choosing material he's already interested in. But one of the great strengths of public radio is that it frequently presents people with a fascinating interview on a subject they've never considered, or one that they thought they didn't care about. They may have tuned in to learn the latest about fighting in the Middle East, but get caught up in the interview about a woman in Florida who was knocked unconscious by a flying sturgeon. Sue Goodwin calls it "the browse factor"—the radio listener's tendency to stick with a story that suddenly captures his or her imagination. "It's serendipity," she says. Leave room for it when you plan your program.

CHECK THAT THE SHOW IS BALANCED. Almost any day of the week, you'll find NPR program producers staring thoughtfully at "the board"—the wipe-off (or electronic) board where they plan out each show. They may move items from one segment to another, or even from one day to the next. Even when all the pieces fit perfectly, a news program may be unbalanced in a number of ways. "It might have lots of pieces, and not enough interviews, or more foreign news than national news, or maybe more news about death and destruction than news about something hopeful, or funny, or beautiful," says Jeff Rogers. "There may be too much talk, and not enough sound, too much idea and opinion and not enough characters or people to meet. Achieving the right balance is all part of programming a news magazine."

Creating the Lineup

It's one thing to know you want your program to have attributes like balance and good pacing; it's quite another to achieve those goals under the pressure of a looming deadline and as the news is changing. Even the best-laid plans may need to be jettisoned. "We may have a story on the board for Tuesday," Carline Watson says of *Talk of the Nation*, "but when we come in Tuesday morning we'll ask, 'Is this still the story we want to be doing today? Or should we be doing something else?' The news changes."

One of the first challenges a program producer has to face is selecting a top news story to be at the beginning of the show. (Or in the case of some two-hour public radio shows, one to start each hour.)[1] Show producers continue to talk about the "lead story" even though they know listeners tune in at various times during the hour, and even though the "first" item in a show actually follows a minute-long billboard and a five-minute newscast. But you have to start somewhere—and, as Chris Turpin points out, the convention of leading a show this way serves the listeners who are eager to hear the biggest news of the day. "We've become used to newscasts—either commercial newscasts or public radio newscasts—that put the information that's considered newest, freshest, most timely, at the top. Probably town criers started with the most important piece of information!"

Often the lead is self-evident: the President proposes sweeping changes in U.S. immigration laws, a wildfire approaches a national landmark, a politician resigns in the middle of a scandal—in other words, news that any curious adult American should want to hear about. At other times, the big story of the day doesn't have a single event to hang it on. At the end of May 2007, for instance, *Morning Edition* wanted to do a story on rising U.S. casualties in the Iraq war; that May had been one of the bloodiest months in the four years of the war. "We built an entire lead segment ourselves," says editor Bruce Auster. The hour began with an interview with a retired Army general, who talked about why the violence had increased and what might lie ahead for the summer. "We

1. Most NPR stations broadcast both hours of the two-hour news magazines, but not necessarily in the same order that they're fed from Washington. In other words, a station on the East Coast may decide to begin its broadcast of *Morning Edition* at 6 a.m., even though the show starts an hour earlier, in effect switching the positions of hour 1 and hour 2. Or it may air three or more hours of the program, in which case an hour may be broadcast twice. Because stations have this freedom, a two-hour show fed from Washington needs to have two beginnings.

followed that with a very ambitious tape-and-copy segment. One of our producers went back and found the families of soldiers who had died in the same incident during this month, and he then arranged for the father of one to speak with the mother of another." By combining the general's analysis with the family members' personal anecdotes, the show produced a vivid and powerful lead segment on an important story, even though it never called on a reporter.

There are some days when there isn't an obvious lead because no single story seems more prominent than the others. "The news days are easy! It's the days when there isn't obvious news that are hard," says Madhulika Sikka, who began producing *Morning Edition* in 2006. On those news-poor days, Sikka says, she gets creative—perhaps using a less-than-fascinating event as a platform for a topic that might be more engrossing. She gives the example of the international forum of the Group of Eight, or G8, which takes place every year. "We knew that nothing much new would be coming out of it in 2007. We definitely didn't want to do three or four minutes every day on who said what, so we worked with science editors to try to build something climate-related in the top segments for the days during the G8. That way we could play off any news there might be, but then make the turn to a story that NPR as a news organization was committed to, and that would be a little different sounding and not obvious." By focusing more on what she saw as the real story (the climate change discussion, in this example), rather than event of the day (the G8 meeting), Sikka was able to craft a lead to her show that she knew would be meaningful to listeners.

Once you decide on the lead story, all the other factors—context and pacing and balance—come into play. It would be easy to look at a program clock and assume that the relative importance of the stories will determine their placement in the show—hard news will go at the top, a little softer news in the next segment, until we wrap up each hour with a "kicker" feature. But experienced producers avoid this sort of paint-by-numbers approach.

For example, *Weekend Edition Saturday* often begins one of its two hours with foreign news.[2] And many foreign stories—with their unfamiliar places or people, their demand for background information and

2. Weekend shows—especially weekend morning shows—rely on foreign reporters for much of their hard news. At 8:00 a.m. in the eastern United States, when the programs go on the air, more news will have been generated in Europe, the Mideast, and Asia than here in the U.S.

analysis— take longer to tell and are relatively hard for listeners to digest. So when a show requires listeners to do some heavy mental lifting at the top of the hour, the show's producer, Peter Breslow, tries to "reward" them with something fun in the next segment.

Producing a news magazine should be an active, not a passive, process. "It's possible to plan for news—that's not an oxymoron," Madhulika Sikka says. Or as her colleague Bruce Auster puts it, "Our ambition is to look ahead, and figure out what is important for that day. So you don't start your day by looking at what's on offer and what you can use to fill what slot in the show." Whenever you're producing a news magazine, find out what reporters are working on, and then ask yourself whether there are additional angles that can be covered by a host in an interview, or by an analyst through a commentary. "That's the beauty of National Public Radio, as opposed to TV—that there's time to bring different perspectives to the same subject," says Robin Gradison, a longtime television journalist who produced *Morning Edition* in 2004 and 2005. She gives the example of a segment that the show put together dealing with college drinking. "There was a commentary from a college kid about what she experienced when she first went to school, there was a three-way that Steve Inskeep did with somebody who is the head of a fraternity and somebody who led an abstinence program, and then we had an interview on the worst 'party schools' in the U.S.—and you learned everything that you needed to know about what college kids face when they deal with binge drinking," Gradison says. "You should always bring different angles to a story."

But be careful about piling too many similar stories on top of one another. Imagine a program that starts with a story on a plane crash in Europe, which is followed by an interview on a ferry disaster in the Philippines, which segues into a report on an outbreak of a mysterious infectious disease in China, which leads into an essay from a doctor about fighting AIDS in Africa. While each item is related by geography or subject to its neighbor, the cumulative effect of such a string of disaster stories would itself be disastrous. You can almost hear the listeners crying out, "Enough!" There are much subtler and more palatable ways to tie stories together, according to Bob Malesky, who produced *Weekend Edition Sunday* for twenty years. "Completely disparate pieces that have no relationship to each other but have a similar *mood* will feel right to listeners back-to-back."

And sometimes, variety is more important than flow. In fact, Chris Turpin says having too *much* news is as bad as having too little. "That's the day when you have to tell reporters their stories are not going to make it on the air, and justify putting a funny story about an organist who got thrown out of the ballpark for playing 'Three Blind Mice' in place of the story on the latest Saudi crackdown on terror." Of course, a producer can mix things up *too* much; transitions can't be jarring or inappropriate or offensive. Peter Breslow says he'll call for a long musical button to ease the transition if he wants to go from something serious to, say, "adult diapers for chimpanzees."[3]

One thing all the top program producers strive for is *texture*—and you can achieve it in lots of ways. For Robin Gradison, texture meant taking the listener somewhere. "When you close your eyes, you're transported to a place. You can smell it, and you can feel what the air feels like, and you're somewhere else, rather than in a hermetically sealed studio, with two people talking." Texture also comes from including pieces and interviews of different lengths. Chris Turpin notes that segments on *All Things Considered* are roughly multiples of four minutes, so producers on that show need to resist the temptation "to march in four-four time"—three four-minute items for one segment, two four-minute pieces for another. *Morning Edition* actually will make a special effort not to fill a segment with a single story, says Bruce Auster. "We will sometimes purposely program a seven-and-a-half-minute piece in an eight-and-a-half-minute slot, just to open up that minute and force ourselves to do something different in it." Commentaries can help put a little syncopation into a program whose rhythm has gotten too predictable. Other shows rely on short audio "postcards" to change pace and mood. Most weeks, *Weekend Edition Saturday* is composed almost entirely of host interviews, as opposed to reported pieces. And it's the one news magazine that doesn't use commentaries. So Peter Breslow says he maintains texture in the program by looking for ways to add sound to the host's interviews—a clip of archival audio from an old radio show, an excerpt from a recent presidential speech, and so on. When Scott Simon spoke to a Broadway percussionist who had to perform from his dressing room because there was no room in the orchestra pit, he asked the musician to end the interview with a drum roll. "Just little moments are enough," Breslow says.

Just as there should be those "little moments" inside each show that catch the listener's ear, there ought to be individual programs that stand

3. Breslow assures us that—at least at the time this is being written—*Weekend Edition Saturday* has not actually done a feature on adult diapers for chimpanzees.

out; today's show shouldn't sound like yesterday's, in either content or form. Many news magazines have recurring slots for politics, or sports, or business, or other topics—specific times on specific days when they can program recorded features. This can make it easier to build the show—but it should never be too easy. "We are all comforted by predictability," says Bob Malesky—and that applies both to the audience and to program producers. "There's a balance you need to strike between making your show as comfortable as an old shoe and as boring as an old shoe!" Chris Turpin says producers need to "play around" with the forms of radio more often. "We need to make ourselves less predictable and try to come at things in a different way"—to vary the tried-and-true formulas of radio news coverage. For example, one common way *All Things Considered* gets at a subject in depth is to start with a piece by a reporter, follow it with a sidebar or background interview, and conclude with a commentary from an analyst. The challenge for program producers is to keep that piece-interview-commentary sequence from becoming humdrum and repetitious. When you're producing a show, on public radio or anywhere else, try to meet that challenge in different ways—for instance, by having the host conduct a series of short interviews on a topic, or by running several commentaries from people with different experiences of an event, or by devoting an entire half-hour or hour to one subject, or by breaking format—and allowing the material to cross over what are usually sacrosanct time posts.

Most of the time, if you're producing a news show that has defined segments and breaks—either for commercials, or underwriting announcements, or local inserts—you'll need to work around, or with, the program's strict clock. *Morning Edition*, for instance, has five segments an hour, varying in length from nine minutes to four. Each one has to begin and end at exactly the right time; in the thirty seconds or minute between segments, local stations add information on their own programming (or on local traffic, or weather, or anything else). In addition, there are NPR-produced newscasts at the top and bottom of each hour. The slices of the program "pie" may be all different sizes—but their sizes remain the same every day, regardless of what's happening in the world.

That means you not only have to think about which stories are most important, and which items go well with each other, and whether the film review that was scheduled for broadcast today can be postponed to accommodate breaking news; you also have to consider whether the pieces of the program will fit into the time budgeted for them. The lead segment on *Morning Edition*—the "A segment"—is nine minutes long when there's a corporate scandal in Europe, a tornado in the Midwest, turmoil on Capitol Hill, and a World Series upset; and it's nine minutes long in mid-August, when all of Europe is shut down for summer vacation, it's sunny and warm in the Midwest, Congress is in recess, and the Yankees have a ten-game lead in the American League East. Material has to expand or contract to fill out each segment, every day, on every program.

Let's say you are producing *Morning Edition* (or any other news magazine), and you plan to pair an investigative piece from a reporter in Germany with a host interview on a related topic. Imagine the reporter was told she could have five minutes for her piece, but the interview ended up being six minutes long, even after it was edited down from the original twenty. With eleven minutes of material to cram into a nine-minute segment, you now face some tough decisions. Should you separate the two items and lose whatever synergy their combination provided? Drop the interview? Trim one or both of the elements? The solution involves much more than just making the pieces fit (although that's an essential consideration); you need to take into account the relative merits of each item. Perhaps the reporter's original story trumps the interview—as it would if she had uncovered something that had not been reported elsewhere. In that case, you might ask the editor of the story from Germany to cut it by thirty seconds, and ask a producer to trim the interview by another minute and a half.

Or consider the situation where that same interview is slated to air in a seven-minute segment. As the program producer, you can shorten the interview to make room for another interview, or a short piece, or a commentary; you can add copy for the host to read (often called "fill copy," since it is written at least in part to fill out a segment); you can move the interview to a different segment; or you can ask the editor or producer to restore one of the sections of the interview that was cut. There is no single right way to block out a segment—except that you can't ever risk boring the audience. If four minutes is the optimal length for an interview, it's almost always a mistake to stretch it to five, says Bruce Auster. "Often people will say, 'It's not quite a full B segment, but if I add this . . . ' And I'd say, 'No! Just give us what you think it needs to be."

If items must be trimmed to fit the program's segments, you may need to bargain with producers, editors, reporters, and hosts. And it's easy for relations to get strained when the request for cuts seems arbitrary or editorially unsound. You have to be tactful when you're trying to persuade a reporter to cut her story at the last minute, or get a producer to take an extra minute out of an interview. Chris Turpin says the show producer has to be ready to give, as well as take, in negotiations with reporters and their editors: "If you cut their piece down one day, you may let it run a little longer the next." And the task is no easier if the cuts need to come out of a host's interview. "My degree is in psychology, not in journalism," Bob Malesky says. "And frankly, it's been much more useful in this business than a journalism degree would have been!"

Producing Live Programs

In the eighteen months that Robin Gradison produced *Morning Edition*, she watched her job evolve "from rarely doing live coverage, and leaving the nice textured pieces alone, to doing live coverage all the time." In a world where cable television is ubiquitous, and where people have access to the Internet at their homes and offices (and bookstores and coffeehouses), program producers have to be ready at any point to replace their carefully planned programs with live, unscripted coverage if news breaks on their watch.[4]

But "going live" can mean many different things. Sometimes the best approach is simply to have many short news bulletins throughout the program. In other words, you may decide to leave some or all of the prepared pieces in place, but to insert updates into the show frequently. This is often the best strategy when the importance of an event is not yet obvious or when few details are available. For example, imagine you see on the news wires that the power has gone out in Los Angeles. At the very least, you know it will mean that traffic will be snarled up—but at the worst it could lead to a regional blackout (as another power outage did in the east), or could be the result of sabotage. Since you can't determine what caused the loss of power or how long it will last, the best approach is probably to tell people what you do know, what you *don't* know, and how you're trying to find out more. You might ask the host to say something

4. Radio producers should be careful to make the decision to go live on its merits, and not to let their judgment be clouded by cable TV coverage. TV networks can present a story just by pointing their cameras at a burning warehouse, or a street demonstration, or a lectern outside a courthouse. That doesn't mean you should necessarily go live to cover the same stories.

like, "We're getting reports of a power outage that has blacked out a large area of Los Angeles. It appears that businesses and homes from downtown L.A. to the coast and north into the San Fernando Valley have lost electricity. At this point, that's just about all we know. Our reporters are contacting the L.A. Department of Water and Power to find out if they have tracked down the cause of the problem. Stay tuned for more information over the next few hours."[5] Inserting even a thirty-second bulletin of this sort forces you to cut thirty seconds from a piece or interview; but it's often the easiest and most appropriate way to accommodate breaking news.

Sometimes going live requires a more radical change. On October 27, 2005, *Morning Edition* had planned to start its first hour with the lineup below:

(1.) {A-1} ARMY CORPS—The Army Corps of Engineers has been at work on the nation's waterways for nearly one hundred years. Its priorities are to support navigation and flood control. But in New Orleans, rivers cut through the city, making it difficult for the Corps to simultaneously manage waterways and control floods. N-P-R's Laura Sullivan reports.

~~~~~~

ATC Promo (0:29) Music Bed (1:29)

~~~~~~

(2.) {B-1} CAPITOL HILL—Conservatives in Congress want to cut the budget to pay for the mounting bills attributed to Hurricanes Katrina and Rita. But G-O-P moderates shrink at many of the proposed cuts to programs for students, children, and the poor. Republican leaders are trying to cut a deal between their party's factions. N-P-R's Andrea Seabrook reports.

(3.) WISCONSIN TWINS—It's been a year and a half since the Witmer family lost their daughter, Michelle, in a roadside ambush in Iraq. She is the first woman in the history of the Wisconsin National Guard to die in combat. Michelle's two other sisters are struggling with whether they should stay in Iraq or come home as their parents wished. We visit the family to talk about what's changed for them since Michelle's death. Brian Bull of Wisconsin Public Radio reports.

5. This "excerpt" is a re-creation, though the event is real. In the end, the outage was blamed on a mistake by utility workers, and it was resolved before the evening rush hour.

This sequence of pieces had run at 5 a.m. Eastern Time, and again at 7. But just before 9 o'clock, news broke. Around 9:05, Station Relations Manager Gemma Hooley got on the "squawk" channel—a phone number that allows her to give oral warnings of format changes to member stations—and at 9:07 she sent out this alert over the DACS system:

> NPR NEWS WILL BE COVERING THE NEWS OF HARRIET MIERS' WITHDRAWAL FROM THE NOMINATION [*SIC*] IN MORNING EDITION. WE WILL BE FOLLOWING THE REGULAR MORNING EDITION CLOCK. MORE INFORMATION ABOUT COVERAGE WHEN IT IS AVAILABLE.

Three minutes later, this is what listeners heard:

> MONTAGNE: This is Morning Edition from N-P-R News. I'm Renee Montagne.
>
> INSKEEP: And I'm Steve Inskeep.
>
> Harriet Miers has withdrawn as a candidate for the Supreme Court. This announcement came amid increasing criticism over her nomination, and questions about her qualifications. We're joined now by N-P-R White House correspondent Don Gonyea. And Don, good morning.
>
> GONYEA: Hi, Steve.
>
> INSKEEP: So—um—there were questions about her qualifications, but this letter that Harriet Miers sent to President Bush doesn't mention her qualifications?
>
> GONYEA: It does not. Let me read from the letter. It was just released to us at about five minutes before nine this morning. There was one of those announcements in the White House: "There's paper coming to the bin," is what they say. That means there's some sort of a press release. And this one is a blockbuster.

Every segment from then on incorporated additional interviews relating to the breaking news story, or at least short updates. By 11 a.m., the last time that first hour of *Morning Edition* was fed to stations, the lineup had completely changed:

> (1.){A-1} MIERS WITHDRAWAL/GONYEA—Steve speaks with Don Gonyea about the withdrawal of Harriet Miers.
>
> (2.) MIERS WITHDRAWAL/NEAS—Steve speaks with Ralph Neas, president of People for the American Way, about the Miers withdrawal.

~~~~~~

ATC Promo (0:29) Music Bed (1:29)

~~~~~~

(3.){B-1} MIERS WITHDRAWAL/BROWNBACK—Steve Inskeep speaks with U.S. Senator Sam Brownback about the withdrawal of Harriet Miers from her U.S. Supreme Court nomination

(4). MIERS WITHDRAWAL/GREENE—Renee speaks with NPR White House correspondent David Greene about the Miers withdrawal.

~~~~~~

Funding Credits :39—Cutaway :29—Return :29—News IV 2:59—News V 1:59

~~~~~~

(5.){C-1} MIERS WITHDRAWAL/SENATOR DURBIN—Renee speaks with Senator Dick Durbin about the Miers withdrawal

~~~~~~

TOTN Promo (0:29) Music Bed (1:29)

~~~~~~

(6.){D-2} ARMY CORPS—The Army Corps of Engineers has been at work on the nation's waterways for nearly one hundred years. Its priorities are to support navigation and flood control. But in New Orleans, rivers cut through the city, making it difficult for the Corps to simultaneously manage waterways and control floods. N-P-R's Laura Sullivan reports.

~~~~~~

Tapes & Transcripts (0:15) Music in the clear (1:44)

~~~~~~

(7.) {E-1} MIERS WITHDRAWAL/FRUM—Renee speaks with author and former Bush speechwriter David Frum about the Miers withdrawal.

(8.) MIERS WITHDRAWAL/LEAHY—Steve speaks with Senator Leahy about the announcement from Harriet Miers earlier this morning that she is withdrawing from her nomination to the Supreme Court.

Over the course of one week in November 2005, *Morning Edition* scrapped its planned lineup to go live three times—when Miers took herself out of consideration for a seat on the Supreme Court; when White House aide Lewis Libby was indicted for his role in a CIA leak scandal; and when the

President nominated Samuel Alito as Miers's replacement. "The entire show was jettisoned in the cases of Miers and Alito," Gradison says, and "much of the show was changed when we learned that [presidential advisor Karl] Rove would not be indicted, but that Libby would be."

When major news breaks, NPR and other broadcast networks may decide to label the programming "special coverage" and make it available to all stations. (In NPR's case, that can even include stations that don't subscribe to the program broadcast at that hour.) As Managing Editor, Barbara Rehm had to decide numerous times whether to go live for just a segment, or much longer. It was always a tough call. "At the beginning of the hour, I'd think, 'Is this sustainable? And for how long?' Hour-long chunks are better for the network, particularly for small stations that don't have the staff to come in and out [of the live programming] with a lot of grace."

In the Miers, Libby, and Alito cases, the *Morning Edition* producer was able to "color inside the lines"—switch to live programming, while maintaining the normal program clock. That's not always possible. Sometimes a news conference will be in full swing just as a scheduled station break is about to begin. In those cases, someone has to alert stations so they know about the change.

When the news is of national or even worldwide importance, NPR and other networks will ramp up the level of coverage, by broadcasting a special report in place of, or in addition to, its scheduled programming:

NEARY: This is a special report from NPR News. I'm Lynn Neary.

Pope John Paul II died today. The announcement of the pontiff's death came just minutes ago from the Vatican, where thousands have gathered in St. Peter's Square to pray for the pope in his last hours. The pope was 84 years old and had been struggling with serious health problems, including Parkinson's disease. When he became pope 26 years ago, he was a robust, active man. At 58, he was the youngest pope in the 20th century. John Paul II traveled more than any other pope and is credited with helping to bring about an end to communism in his native Poland.

Coming into power on the tide of change brought about by Vatican II, he held firmly to a conservative course. Joining us now from Rome is NPR correspondent Sylvia Poggioli.

Sylvia, what are we hearing now about the pope's final hours?

POGGIOLI: Well, you know, his health suddenly took a turn for the worse on Thursday evening. His blood pressure plunged, and he developed a high fever, but he was still lucid. And on Friday, he approved the appointment of several new bishops all over the world. But by Saturday morning, he was beginning to lose consciousness.

NPR was able to mount a special report on the death of the pope so quickly because he had been very sick in the weeks and months leading up to his death. As Barbara Rehm explains, this was the sort of special event that she could plan for; she had mobilized producers, reporters, editors, and bookers long before the pope died. Rehm says that the goal of this sort of special coverage is to "have a narrative arc—you have a story to tell, and these are all the chapters that you want to get told in this period of time. For instance, this author has done a wonderful biography of the new pope, so he can speak to us, or this one is his boyhood friend, or this is the critic that he silenced. You try to build the whole story of the event." When the subject of a special report can be foreseen—the death of an ailing leader, a presidential election, even the first salvos in a war—a program producer can often start working days or weeks before the show airs.

Often, however, the news comes at us without warning. Scott Simon was nearing the end of a two-hour *Weekend Edition Saturday* on February 1, 2003, when the news wires reported a problem with the space shuttle. The first thought of his producer, Peter Breslow, was to call Pat Duggins—a Florida-based reporter who had covered the space program for years.

SIMON: This is *Weekend Edition* from NPR News. I'm Scott Simon.

And we are breaking in with sad news this morning. The space shuttle Columbia has been seen apparently breaking up in the skies over Texas as it returned to Earth shortly after 9 a.m. Eastern time this morning. Search and rescue teams are reportedly scouring the grounds below in central Texas. NASA says there was no indication of trouble at the time of reentry. We're going to go now to Pat Duggins of member station WMFE, who I believe is in the Cape area. Pat, can you hear me?

DUGGINS: Yes, Scott, I'm right here.

SIMON: Pat, the situation there—what do we know?

DUGGINS: Right now, a lot of confusion, Scott. When the shuttle begins its descent, there's a very set number of things that have to happen, a number of signposts as the space shuttle comes down. The spacecraft was supposed to cut a path across California, Nevada, New Mexico, and it was over Texas that apparently something catastrophic went wrong with the space shuttle. As you mentioned earlier, we have

> unconfirmed reports of the shuttle breaking up. There's talk of possible problems with one of the hydraulic units on the spacecraft. But whatever it is, for those of us here at the Kennedy Space Center, the double sonic booms that usually herald the shuttle's arrival did not occur. The countdown clock ticked to zero. There was no sign of Columbia, and then, obviously, something had gone tragically wrong.

What had started out as a conventional news magazine—with news stories on a looming war in Iraq and containing North Korea's militarism, and feature interviews on the National Endowment for the Arts and a scientist who had taught children to recognize different chimpanzee calls—suddenly became a special report on one of America's worst space disasters. Breslow and Simon and their staff stayed on the air live for six hours. The programming integrated interviews with press conferences and eyewitness accounts. This was nothing the host or producer could prepare for; the success of this sort of program depends on your dexterity in booking guests, in going to whatever news conference or statement is happening live, and in taking calls from listeners. "People in rural places had seen pieces of the shuttle come down and they called in to the show," Barbara Rehm recalls. "So we become kind of like the campfire people gather around."

From the producer's point of view, "Live is totally different," Robin Gradison says. "You have to be a step ahead: Where are we going next? Does the host know where we're going next? Does that person on the line know he's next? Is it a clear line? Has anybody welcomed him—said, 'You're in the board, thank you so much?'" Gradison says a live program, by its nature, is changing and fluid—and that makes great demands on the show's producer. "You have to have somebody reading the wires to make sure that whatever you're reporting right at the moment is the latest news, and that it's been confirmed, and the editor's okay with going with it. It's very different from producing a daily show, which to me is kind of a static, finished product."

During live events, it is essential for you, as the program producer, to establish clear lines of command. It's hard enough just to figure out what the facts are when news is coming at you thick and fast; that certainly isn't the time to be deciding who will be choosing guests, creating the lineups, talking to the director, assigning producers, and so on.

When you are producing a live show on short notice, you will need to communicate even more effectively than usual with the studio director and host. You may even want to station a second producer in the control room to relay messages. Keep a phone line open so you can tell him or her about changes in the lineup, as guests are booked or you learn about news conferences or other events. It's important to have some sort of roadmap, so the director and host know what you're planning. When Neal Conan is anchoring live coverage, he likes the producer to put the tentative lineup on a white board that he can easily see from the studio. "They can erase it, and change it, and use it to give me some idea of where we're going next—including if they don't really know *where* they're going next."

As you produce live coverage—as you make sure that guests are booked and brought into the studio, and that the host and director know what's ahead—you also have to monitor what you're putting on the air. It's easy to get so wrapped up in constructing a *potential* program on the board that you forget to listen to the *actual* show that's being broadcast. And what looks good on paper or on a computer or a wipe-off board doesn't always sound good on the radio. For instance, if the questions at the live press conference are not clearly audible, the listener won't hear much more than clicking cameras and room noise for fifteen or twenty seconds before each reply. If that happens, you may want the host to paraphrase questions—or you may decide to drop the news conference altogether. If a guest is dull, you may not want to keep him on the air for the full five minutes you budgeted. If you are too busy to keep an ear on the program yourself, find someone to be the "designated listener."

This sort of live, unscripted programming is inherently stressful—for the host, the studio director, the bookers, and the program producer. The more frantic things become, the calmer you need to be. "The worst thing that can happen in live broadcasts is to have somebody screaming," Robin Gradison says, "because it makes everybody else shut down." You may not know how you'll respond until the call comes to "go live." But few producing jobs command as much respect as keeping a program on the air for hours without a script during a national emergency. More than one NPR producer has had greatness thrust upon him or her by breaking news.

CHAPTER 15

Program Editing

It's twenty minutes past four on a weekday afternoon, and *All Things Considered* is on the air.[1] A host calls the editor to ask how to pronounce the name of a Polish leader mentioned in the next report. The editor puts the host on hold while he gets the information. At that same moment, a reporter calls to say that the introduction for her piece that airs in four minutes needs to be changed; there was an error in the first version, she says, and a rewritten one will be sent shortly. The program editor sends a one-line message to the host who's scheduled to read that intro: "NEW TRIAL INTRO COMING—SCRAP OLD ONE." Then he gets the Polish pronunciation from a producer (who had called the Polish embassy), and passes it on to the other host who is still on the phone—seconds before he has to read the name on the air.

Now a producer walks up to the show editor and announces, "We're in trouble. I made all the cuts you suggested and the interview's still more than two minutes long. And it's on in less than fifteen minutes." As the editor is absorbing this bad news, the director calls from the studio and asks, "How serious is this trial piece? If I play music after it, can it be upbeat?" The editor replies, "No one dies—it's about money," which is all the director needs to know. The revised introduction for the trial story appears on the editor's computer; he reads it over quickly, shortens one of the sentences, removes a phrase that he has seen repeated verbatim in the first paragraph of the report itself, and changes "lie" to "lay" to fix the grammar. He prints the intro script and yells for someone to get it into the studio immediately. He then turns to the producer—who has been hovering anxiously, waiting for suggestions on how to trim the interview—and

1. As they say on TV police dramas: Although inspired by true incidents, the following story is fictional and does not depict any actual person or event. But it's mostly true.

asks, "Can we cut the part about the history of the program? Or funding? Let me hear the last couple minutes." Even as the producer is getting the interview up on the computer, another one-line message appears on the editor's machine, this time from the co-host, asking, "YOU SURE ABOUT LAY NOT LIE??" To which he replies, "80 PER CENT."

All this can take place in three or four minutes. And that's on a day when the flow of news is manageable. Whatever is happening in the outside world, the show editor is at the nerve center of the program. He or she is a vital node on an editorial network that comprises the control room, studio, newsroom, production team—and the places where the news is actually being made.

Setting the News Agenda

Much of the news that's on the air every day is dictated by events, so to a certain extent the news agenda is not under anyone's control. On weekdays especially, when businesses are open and governments are in session, a show's roadmap may be dominated by pieces from network correspondents and reporters around the country. But important stories can nonetheless be overlooked, and reporters are not always deployed where the news is being made; so as a program editor, you need to be thinking about what may be missing from the planned lineup of news reporters.

To accomplish that, you need to have at least a passing knowledge of a wide variety of news stories. Martha Wexler worked on the NPR foreign desk before she started editing *Weekend All Things Considered*, and says she had to learn to read the newspaper differently when she changed jobs. "In the past, I only had to pay close attention to foreign affairs—Europe, the former USSR. Now I need to keep up with national news. I'm reading the *Los Angeles Times* online, and papers from Minneapolis, or Milwaukee, or St. Louis. I've gotten more aware of and interested in local issues around the country." Or as former *Weekend Edition Saturday* editor Gwen Thompkins puts it, "It's like being a general assignment reporter, except that you're general assignment for the rest of your life when you edit a show!"

While newspapers and magazines can give you some ideas for stories, you don't want to be influenced too much by what other news organizations are doing. Try to avoid the sort of "follow-the-leader" news reporting that exposes people to the same angles of the same topics on nearly all news media. For example, a story will often appear first on the front page of a national newspaper; it will be picked up by bloggers almost immediately; the next day it will show up on cable TV; a day later it will be on the evening network news; and perhaps a few days after that it will appear in the weekly news magazines. That's *not* a story you need to add to your show's lineup! It's sometimes hard to resist doing the radio version of a compelling feature that you or other members of the staff have read or seen elsewhere. But as a rule, when you're thinking about lines of coverage for your program, you should strive to break news, or at least to explore sides of stories that may have been underreported by other media.

Your news agenda may look quite different if you're working on a morning show than if you're editing an afternoon or evening news program, as Bruce Auster discovered when he began editing *Morning Edition* in 2004. "On this show, if we wait for the events, we're already late. The news cycle has so accelerated that we can't afford to do that." Auster says he generally doesn't even want his audience to hear the same stories on *Morning Edition* that they may have read in that day's newspaper. "There are certain things that warrant coverage the next day, but if that's what we build everything on, we're going to be a late show all the time. So we're always trying to anticipate, even on an ongoing story. Pick any Washington scandal and you can anticipate where it goes next. You can ask questions about who's going to respond, what's the likely outcome? You have to essentially be thinking ahead twenty-four hours all the time." Interviews with newsmakers can help advance a story, he says, if the host asks tough and probing questions. And if you think a given topic is important, you can get it on the air right away by checking in with reporters who are in the process of covering the story; you don't necessarily have to wait until they've done all their interviews and written a polished script.

On the weekend, an editor may play a key role in deciding what the program will encompass, simply because fewer events are scheduled for Saturday and Sunday, and there's usually a smaller supply of pieces from reporters. Thompkins always saw her main responsibility as creating "the skeleton of a show"—choosing the foreign, domestic, and cultural news stories that she thought listeners needed to hear on Saturday morning. This meant working hand in hand with the show producer "to come up with a fresh angle on a story that has a lot of different iterations." She also tried to discover undercovered stories; like a good producer, she says, a show editor should be drawing on personal experiences, conversations, and pursuits. "It helps to walk in with a lot of interests. Over time, you're going to exhaust all of them."

As you consider the stories to include in your program, think about ways to tie them to other pieces and interviews. "With so many outlets for news, if all you're doing is putting out more information, without showing how it connects to anything else, you're just contributing to people's sense of being overwhelmed with information," says Bruce Auster. So he tries to connect the dots whenever he can. Sometimes that's done just by clustering related items in a show—letting people hear how two subjects or issues are linked. But you can also decide to cover a topic over several days, or even on different programs—relating an interview that airs one morning to a story from the night before. Here's how *Morning Edition* did just that during the Iraq war:

INSKEEP: Join up with an American night patrol through a Baghdad neighborhood, and this is what you may hear. It is, in any case, what N-P-R's Anne Garrels recorded—gunfire and grenades directed at U-S troops.

UNIDENTIFIED MAN #1: Hey, 3 o'clock, 3 o'clock.

UNIDENTIFIED MAN #2: Three o'clock.

UNIDENTIFIED MAN #3: Where'd that come from (unintelligible)?

UNIDENTIFIED MAN #1: Over here to our left, down this road.

[Sound of gunfire]

UNIDENTIFIED MAN #1: Here it comes.

GARRELS: Is this a typical night?

UNIDENTIFIED MAN #1: For the most part, yes. On a pretty normal night for us here . . .

INSKEEP: That's N-P-R's Anne Garrels, from N-P-R's *All Things Considered* this week. She's reporting on the effort to bring security to the capital, and we're following up now with General Joseph Fil, the commander of U-S forces in Baghdad. General, welcome to the program.

FIL: Well, thank you very much.

INSKEEP: We should stress that we just heard one Baghdad area, what's normal there. Is that normal in a lot of Baghdad, though?

FIL: Well, it's mixed in an area that we're still clearing. That sounded like—Baya'a there, I think, is the area that you were in . . .

Bruce Auster says by placing an interview in the context of previous reporting, "you create patterns in people's minds that help them understand it—and hopefully make some of the stories more memorable than they would have been if they just stood alone."

If all of this sounds like the job of the program producer, it's not surprising—a news show's editor and its producer play overlapping roles. At some stations and networks, the jobs may even be filled by a single person. The important thing is that the staff should never be in doubt about whose decision is final.

Error Checking

Broadcast news is an intense, high-energy, rapid-fire business. Stories may be reported, written, and produced in only a few hours. If, as it's often said, journalism is a first rough draft of history,[2] then broadcast news might be regarded as a first rough draft of journalism. "The editor is our protector, who is trying to keep us from making errors," *All Things Considered* host Robert Siegel says. "It's a very important role of the editor because there will be a certain degree of error on any given day. I know I will make a certain number of mistakes every week—mistakes of reading and performance and some mistakes of content—and I've got to figure out how to recover from them, correct them, but in most cases avoid as many as I possibly can. We count on the editor for that."

After all, the show editor is often the host's connection with the rest of the program; the editor focuses on the parts of the show that feature the host—specifically, billboards, interviews, and the intros to stories. Unlike a story editor, who often is an expert on a specific subject or a certain part of the world, a program editor has to be a generalist—to know a little bit about a lot of different subjects, and to know where to get more information quickly, when necessary.

When you're editing a news magazine, you're also responsible for catching errors or omissions. Alas, there are no limits to the ways you can get the facts wrong, but a few kinds of errors do crop up again and again.

SOME MISTAKES RESULT FROM NEGLIGENCE. The program editor should always scrutinize names, places, dates and numbers. Be prepared to ask anyone who writes copy for the show, "Are you sure?"—and then check up on them anyway. Here are a few of the problems editors face all the time:

2. The quote is usually attributed to *Washington Post* publisher Philip L. Graham.

- Big numbers can easily be off by a factor of a ten, a hundred, or even a thousand. "Sometimes you look at it and say, 'That can't possibly be right,'" says John Buckley, who was an editor for several NPR programs, including *Morning Edition* and *Day to Day*. "For instance, if someone refers to Britain's 'sixty billion' population, you know they must have meant 'million.'" More often, though, you'll need to verify big or unfamiliar figures.
- Dates can be off by a year, a decade, or almost any other interval; for instance, a commentator who covered the Reagan administration might set an event in 1985 when he means 2005, just because he always associates the 1980s and the White House.
- People or organizations may be misidentified. A reporter might refer to the "Veterans Administration," instead of the "Department of Veterans Affairs," of call someone a "former representative" even though she's still in Congress.
- Superlatives almost always deserve a hard look—"and that includes 'firsts' and 'onlys,'" Buckley says. "For example, 'He is the only English bullfighter in Spain' or 'This is the first time a rocket has hit northern Israel.'" Statements like these are almost always hard to nail down—and when they *can* be checked, they frequently turn out to wrong.
- A producer writing in haste may repeat a mistake made by the news wires—referring, for example, to a "growing gap" between long- and short-term interest rates, when in fact the news writer at the wire service meant "narrowing gap."

SOME ARE ERRORS OF PERSPECTIVE. A reporter on the scene—in Moscow, or Chicago, or Flagstaff—sometimes forgets that "here" to him is "there" to the program host or listener. Intros sometimes include phrases like, "Shop owners here in the Midwest," and it's often up to the show editor to cut the extraneous "here." More frequently, reporters include references to people or places that make perfect sense to local listeners, but not to people in other parts of the country. One intro, for instance, relied on people's stereotypes of Connecticut—"Highest per capita income, in the richest country in the world. . . . celebrities, white collar executives"—when most Americans may have no impression of Connecticut one way or another. As the program editor, you have to make sure the geography of an intro is clear to your listeners.

By the same token, you should avoid giving the host a script that includes an observation or conclusion more appropriately made by a reporter. Someone who has been covering a story for weeks may feel confident making a point that would be presumptuous coming out of the mouth of a host. For instance, if a reporter has been on the campaign trail with a candidate, and then hears her change her stump speech dramatically when she speaks to a crowd of unionized workers, it's fair for him to draw a connection between the changes and the audience; a host who did the same thing would sound like he was spinning the story. As a program editor, you should strip out adjectives that suggest the host has seen something that couldn't have been seen from Washington, or New York or Los Angeles, or wherever he or she is sitting. The intro to one obituary for a fallen U.S. soldier included the line, "To those who knew him, he was someone who made you laugh, and seemed more likely to have a video game than a gun in his hand." To the editor, this suggested a familiarity with the slain soldier that seemed inappropriate for the host to be feigning. Another intro said, "As any visitor to Havana notes, there is one feature of the city that eclipses all others—a miles-long sea wall called the Malecón." But since most hosts and listeners have never been to Cuba, the editor changed it to, "In Havana, one feature of the city eclipses all others: a five-mile-long sea wall called the Malecón."

SOME ERRORS ARE TYPOGRAPHICAL, OR THE RESULT OF USING A WORD PROCESSOR HASTILY. Words can be misspelled, sometimes calamitously. Some people insist spelling doesn't matter in a radio script, but if you're a host reading an intro cold, you don't want to see "pace" instead of "place" in a phrase like "The police investigation moves to a different pace this evening." As a script gets revised by a reporter or editor, sometimes bits of an old phrase poke through the new intro; it's not uncommon to see sentences like, "The President plans to meet with Senator Warner Republican members of Congress," when the reporter meant to *replace* the Senator's name with the generic description. Sometimes a reporter or editor decides at the last minute to turn the first paragraph of a report into the intro for the piece; often this results in an intro that is missing the reporter's name—e.g., "NPR's Josephine Jones has a report." A producer on *Morning Edition* accidentally copied the DACS line—the brief description of the story that is sent out to NPR member stations—instead of the intro. Fortunately, editor Doreen McCallister noticed something was wrong before the script got to the host. "This went through several people," McCallister says. "One was on loan from another show, and she didn't catch it. Then it went to the segment producer, who didn't catch it." So it was up to the editor. Reading an intro out loud—and thinking about what you're saying—is the best way to detect this sort of mistake, and many others.

Massaging the Script

Many of the changes you will make when you're editing a show won't be to fix errors per se, but to give the copy a consistent style. News magazines at NPR and elsewhere draw on material from lots of different sources. Hosts may write billboards and the introductions for their interviews; producers may be responsible for tape-and-copy segments and the intros to commentaries; reporters, producers, and editors send intros to the show; and the program editor may write copy based on wire stories, or information provided by a reporter or interviewee. In a two-hour NPR program with eighteen or twenty elements, a dozen people may have a hand in writing the various parts of the script that the host, or hosts, will read. As a show editor, you may be the only person to see all of the pieces of copy before the program goes on the air, so you're the only one who can make them harmonize.

Here are some of the ways you can turn a collection of discrete intros and tape-and-copy blocks into a coherent program script.

ELIMINATE REDUNDANCIES. As a program editor, you have to ensure that the same information isn't repeated in two different items. Let's say *Morning Edition* is putting together a business segment on the underground economy. It will start with a report on cash-only workers—nannies, yard workers, shade tree mechanics, and so on—and will be followed by an interview the host has done with someone at the IRS. It would not be surprising if the editors of both the piece and the interview felt it was important to include the fact that the government is losing more than $200 billion each year because people fail to report their cash income. As the show editor, you would need to cut one of the sentences—or at least change the second one into a second reference, indirectly acknowledging that the same fact was presented earlier: "The hundreds of billions the government loses each year . . ." This sort of unintentional redundancy often crops up after a very big news event, when each reporter and editor feels it's important to remind listeners, for instance, how many weeks have gone by since the ferocious hurricane struck, or the current death toll in a war, or how many days a new president has been in office.

GIVE THE SCRIPT A CONSISTENT STYLE. A program editor also acts as a sort of homogenizer of all of the different pieces of copy that will become the script for the show. That requires, at the most basic level, making the scripts *look* similar from piece to piece, so that a host isn't reading one intro in single-spaced type, and another double-spaced and uppercase, and a third with short, sixty-character lines. It may also involve ironing out some of the stylistic differences in the various bits of copy. Hosts are used to writing in their own voices, as it were; they know which words and phrases they would never use in speech. But other writers don't know that, so the show editor has to process all the different intros to make them "host-friendly." "Different hosts are comfortable saying different things," former *All Things Considered* editor Mary Louise Kelly says. "So you find out the limits and strengths of who you're writing for." Sometimes that means expunging specific words or phrases: one host stumbles over the word "legislature," another avoids "rural," a third always prefers simple words of Anglo-Saxon origin to longer Latinate synonyms ("roads and bridges," instead of "infrastructure"). Some hosts prefer to sound more formal than folksy. "After you do it long enough, you start hearing their voices in your head," Kelly says.

That *doesn't* mean you should try to ape the writing style of the host. "This is the downside of having a host who's known for humor," says Gwen Thompkins, speaking of *Weekend Edition Saturday* host Scott Simon. "People think they know how Scott would read something. They'll try to write humorous scripts, or scripts they think 'sound like Scott,' but they really don't." Thompkins says when she was editing the program, she would end up rewriting "faux Scott" copy, since she knew Simon would put the material into a style that suited him. "We just want simple declarative sentences."

CHECK PRONUNCIATIONS. Reporters and editors who really know about art, or Afghanistan, or sub-atomic particles may not think twice about how to pronounce Magdalena Abakanowicz's name, or which syllables to stress when you talk about the Wakhan Corridor and the Pamir Knot, or whether hadrons and mesons take short or long vowels. But the rest of us need help. As the last person to see a script before it goes into the studio, you are responsible for making sure that unusual names, places, and things all get reliable and intelligible pronouncers.[3] Some-

3. See Appendix II, "Pronouncers," for a key to phonetic spellings.

times just listening to a piece or interview can resolve a question; often you can turn to an in-house expert; if there's time, a librarian will always track down the right pronunciation for you. Once you know how to say the word, if you have any doubt about whether the host will understand your pronouncer, read it out loud for him or her.

TIE RELATED ITEMS TOGETHER. On big stories, the program editor can write an umbrella lead to a segment—something that gives people the news, and also sets the stage for the next piece or two. By telegraphing what's coming up, this sort of introduction lets people know that the coverage of this particular topic will be extraordinary. It can be the radio substitute for a banner headline and a big front page spread:

SIEGEL: From NPR News, this is *All Things Considered.* I'm Robert Siegel.

BLOCK: And I'm Melissa Block.

The tension in Iraq between Shiites and Sunnis is the highest it's been since the U-S invasion three years ago. More than one hundred Iraqis have died in violence that has flared since yesterday's bombing of a Shiite shrine north of Baghdad.

SIEGEL: Shiites have attacked Sunni mosques and neighborhoods, and throughout the country, people have taken to the streets. Here are the sounds of some of those protests.

[Sound of protesters in Iraq]

BLOCK: In Najaf [NAH-jahf], protesters marched through the Shiite holy city, carrying flags and banners.

[Sound of shiite cleric addressing crowd]

SIEGEL: A cleric in Kut [koot] addressed a crowd through a loudspeaker, asking "How long must the faithful put up with this kind of attack?"

[Sound of shiite rally]

BLOCK: And in Samawah [Sah-MAH-wah], in the south, Shiites rallied against the mosque bombing. Similar scenes took place throughout much of Iraq.

The Iraqi government has announced it will enforce a daytime curfew tomorrow to try to stop the violence.

SIEGEL: In this half hour of the program, we're going to take an in-depth look at what's happening in Iraq. We'll speak with our own Anne Garrels, who was in Najaf yesterday. We'll hear about the divide between Shiite and Sunni Muslims, and we'll talk with Iraq's national security advisor.

BLOCK: First, to N-P-R's Jamie Tarabay in Baghdad. And Jamie, give us a sense of the scope of today's violence . . .

Sometimes you can relate two stories just by referring to the previous item: "As John Burnett just mentioned, police are now on the scene . . ." or "As we heard in that report from Debra Amos, Iraqi lawmakers will be meeting tomorrow . . ." This works especially well if the fact that knits the stories together is in the middle of the first report. Here's an intro where a slight tweak made *two* connections with the previous piece (about President George W. Bush's nomination of Harriet Miers to be a Supreme Court Justice):

> Julie Myers is NOT related to Harriet Miers—in fact, the two women spell their last names differently.
>
> Julie Myers is the President's pick to head the Immigration and Customs Enforcement agency at the Homeland Security Department.
>
> Today she was approved for that job by a Senate committee. She still needs confirmation by the full Senate.
>
> As N-P-R's Pam Fessler reports, there are questions about Myers' qualifications . . . spurred by Michael Brown's performance at FEMA.

This sort of editing is not merely cosmetic; it's a large part of what keeps listeners engaged in the program, according to Mary Louise Kelly. "If they feel the next thing that's coming is logically answering questions that they may have had raised by the piece before it, it keeps them hooked," she says. "It helps to give the listener a sense they're being guided through the show. That's really what the hosts are there to do—to guide listeners through what they're hearing and help them make sense of the material. It's very difficult to do that if you have disjointed intros, and it doesn't sound like there's a logic to the presentation of the program."

Just remember: Some stories are *not* related, and it can sound contrived—and make the host appear downright silly—to link items that are really mismatched. If you have one story on scientists in Chicago sequencing the cat genome and another piece on the Kentucky Derby, you shouldn't try to connect them with a facile observation like, "From housecats now... to horses." "That's why God made musical buttons," says Stu Seidel. "When you're drawing links between one piece and the following piece, you have to be sensitive to the possibility that you're pushing it too far."

FRESHEN THE LEADS. As a program editor, you will have many opportunities to "punch up" an introduction or other script written by another editor, producer, or reporter. Sometimes that involves moving the news toward the beginning of the intro or copy block. Here are the beginning of a script written by a producer, and the version rearranged by the program editor to make the first sentence more timely:

> The first hand transplant performed in the United States took place last weekend at Louisville Jewish Hospital in Kentucky. The donor hand was attached two inches above the wrist. The recipient was a 37-year-old New Jersey man who lost his hand in a fireworks accident . . .

> The recipient of America's first hand transplant has already wiggled one of his new fingers. The operation was performed last weekend at Louisville Jewish Hospital in Kentucky. The patient was 37-year-old Matthew David Scott of New Jersey, who lost his hand in a fireworks accident . . .

There are times when you can improve an intro by adding a line to orient the listener—to set the scene, especially if the previous piece or interview is on a different subject or about a completely different place. If the original intro says, "At a union meeting in Chicago today, Teamsters president James P. Hoffa called on the Administration to reverse its stand on re-regulation of the trucking industry," you may want to add a short headline-style sentence—something like "In Chicago, the head of the Teamsters Union took on the Administration today"—so that listeners have a moment to absorb where the next story takes place and generally what it's about. Or if the intro says, "A federal appeals court in San Francisco has blocked the government's plan to allow logging of backwoods areas of the Tongass National Forest in southeastern Alaska," you might start off just by saying, "An appeals court dealt the government—and the timber industry—a setback today."

You can also sharpen an intro by updating it with news that has occurred since the reporter recorded his or her story. Some reporters will actually include a note on their intro as a reminder to the show editor—for example, "Update with latest figures on number of acres destroyed by the wildfires." It's generally a good idea when you're handling any developing story to check in with the reporter or his editor—or at least scan the news wires—to see whether the information in the intro has changed, or whether there is news that should be added. "Quite often, foreign correspondents are filing hours before an intro's going to be read on air," Mary Louise Kelly points out, "and it's always worth checking the death toll, or the budget number they're reporting."

When you modify an intro for a story you haven't reported yourself, you always run the risk of making it worse—so revise with care. Always make sure you don't obscure or eliminate an important fact or turn of phrase. If you are in *any* doubt that you might have gotten things messed up, ask the reporter whose intro it is to look over any changes you've made. And be careful that you don't in effect bury the lead in the process of freshening it. Let's say a reporter has done a story about new treatment for spinal cord injuries using human stem cells that has been tested successfully in New York over the past year. As a show editor, you see there's a meeting of stem cell researchers in Los Angeles taking place today. It's tempting to tie them together somehow—in the process giving the science story a "today" peg. But unless the New York physicians were, say, presenting their findings at that conference, you'd be conflating two unrelated items, and probably confusing listeners about what the story was about and where it took place.

In fact, there are times when adding news—even breaking news—to an intro does a disservice to the listener, unless you follow up with details on the new development. Read this intro and try to determine what the piece is going to be about:

> In Baghdad today, hundreds of Shiite pilgrims were killed when the rails of a bridge they were passing over collapsed. Hundreds of thousands of Shiites had gathered at a mosque in the north of the city for a religious festival when rumors circulated of a suicide bomber. That led to a stampede across the bridge. Many of those killed were women and children. The incident comes amid a deep Sunni-Shiite divide over Iraq's draft constitution. Sunni leaders have called for rejection of the document in a referendum in October. N-P-R's——reports.

Here the program editor already had a story about Iraq's constitution and—no doubt under deadline pressure—tried to work in the news of the bridge collapse as part of the intro. But the result is a sort of journalistic bait-and-switch, where we think we're about to learn more about the terrible accident, and instead hear about the Sunni voter registration drive. Assuming it was not possible to get a reporter on the line to provide an update on the bridge failure, the show editor should have separated the items—and written something for the host like, ". . . Many of those killed were women and children. We'll have more on this story as soon as we can get a reporter on the scene. [Pause] Iraqis are just two months away from voting on a new constitution..." Just because two events take place in the same region (Iraq, in this case, but it could Chicago or New Hampshire or anywhere else) doesn't mean they can be combined into one story—or one intro. The last thing you want to do when you're editing a show is make it *harder* for a listener to sort out what's happened, and to whom.

ADD SOME FACTS AND REMOVE OTHERS. The show editor always needs to weigh how much listeners may know about a story, and how much they need to be told to make sense of what they're about to hear. "On certain big stories, I'd assume the listener had a certain familiarity with the news," Mary Louise Kelly says, "I'd still remind them who the key players are, and the key events, and what's happened since they last listened—and that could have been yesterday or a week ago, so it has to make sense for both." You don't want to talk down to listeners, but you also don't want to leave them in the dark. John Buckley says, "An old BBC editor once told me, 'Assume ignorance—you won't be far wrong.' I think that's patronizing at one level, but I always keep it somewhere at the back of my head." Buckley says that when an editor is working with any script, he or she should have the "courtesy" to introduce ideas to the listener. "If I were explaining something to someone across a bar at a restaurant, I'd have the courtesy to explain myself without, in a didactic way, spelling things out." For instance, now that decades have passed since researchers discovered the human immunodeficiency virus, it is no longer necessary to say, "HIV—the virus that causes AIDS." But you may need to expand other acronyms or provide other guideposts—explain where Mali is, for instance, or remind people whatever happened to former Haitian dictator Jean-Claude "Baby Doc" Duvalier.

At the same time, you want to remember that intros are not little encyclopedia entries. A news program is not a college course, and listeners are not students; they don't *have* to pay attention, and they won't be quizzed on the facts. If you provide so much background information or so many definitions in an intro that it begins to sound like a glossary, you'll lose people.

VARY THE INTRO FORMULA. One way you can perk up the script of a show is use different approaches to the stories in the introductions. Most reporters (and their editors) usually just end their intros with a line like, "NPR's Janet Johnson has this report" or "From member station KXWZ, Doug Davidson has the story." But there are lots of permutations that will allow you to vary the menu. *All Things Considered* host Robert

Siegel has worked with lots of editors over the years, and suggests these alternatives:

- Put that information at the start of the sentence leading into the piece: "John Doe reports from Chicago that the strikers are unlikely to return to work anytime soon."
- Place the reporter at the site of the story and state what is happening there: "John Doe is in Chicago, where the strikers say they are not about to return to work."
- Insert the report's attribution in the middle of the sentence leading into report: "The strike has been going on for two weeks and—as John Doe reports from Chicago—there's no end in sight."
- Separate the location from the reporter. "The strike is the biggest Chicago has seen in years. Plants have been closed for two weeks, and John Doe reports there is no end in sight."

Another way to get out of the intro rut when you have two related pieces is to scrap the intro altogether. Just let the audience know what they're going to hear:

> We have two reports. First, here's N-P-R's Jack Speer with more on the situation at Ford.

Then you can allow the second piece to start with the reporter identifying himself:

> SPEER: . . . Jack Speer, N-P-R News, Washington.
> LANGFITT: I'm Frank Langfitt in Detroit. In better days, unions used to win benefits for their members. . . .

Obviously, this sort of hand-off between reporters has to be planned in advance—or at least in time for the second reporter to add his self-identification to his opening.

CHECK BILLBOARDS (OPENS) AND RETURNS. At NPR, hosts often write billboards based on "lines"—summaries of pieces—provided hours or even days before stories were produced and edited. On top of that, they may not get to the billboards until the last minute. As the show editor, you're often the only person who has heard the bulk of the program, so you should take a moment and make sure any billboards are accurate and engaging—that they give listeners a reason to stay tuned. Are the references cryptic? Misleading? Do they sacrifice clarity to be cute?

As the program editor, you have to make sure the billboard or show open comes as close as possible to providing audio headlines that will promote the pieces and interviews that are coming up. Don't waste that airtime.

PROVIDE GRACE NOTES. Whether rightly or wrongly, there is a similarity to much of the writing in news broadcasts. Part of that is a result of the small amount of time devoted to each story in most newscasts; if an intro runs only ten seconds (as is typical for an NPR news spot), there's only so much you can do to make it sing. But in a half-hour- or hour-long news magazine, a show editor has an opportunity to make the program sound qualitatively different from the kind of news served up at the top of the hour on most radio stations. "It's going to move into something more spoken, maybe more individualistic, reflecting the people who are hosting the program—and more literate, I think," *All Things Considered* host Robert Siegel says. "When I edited the program many, many years ago, I liked to use a music metaphor. I had the ability to set the bass line for the show and add some grace notes. The editor can provide us with an interesting turn of phrase, something we hadn't thought of before. I think of the editor's role of both keeping us steady, and then having moments that make us feel that we're good and different."

Sitting In on Interviews

At NPR, program editors usually listen in as interviews are conducted and recorded, and work with a producer and host to ensure that the conversation is focused, informative, and interesting. So if that's part of your job, make sure you understand the subject and the guest's area of expertise. Before sitting in on an interview for *Weekend Edition Saturday*, Gwen Thompkins would get relevant articles from the library, or talk with a local reporter about the story, or check the Internet. "I can't stand walking into an interview without knowing anything about the topic," she says.

Many editors help craft the intro and questions for the interview—or at least review the planned line of questioning with the host and producer. "In order to have the interview flow smoothly, you want to have the two-way in mind almost before you do it," editor Martha Wexler says. "What you're looking at is for your host—with the input from the guest—to tell a story. So it has to have a narrative line. The way the questions are phrased, and the order of the questions, can help you structure that narration."

Your most important job when you're in the control room is to listen intently—to make sure the conversation has substance, that the host challenges the guest if that is called for, that references are explained or clarified, and so on. (You may want to take notes to remind you of particularly good parts of the interview, or you may leave that to the producer.) "I see the editor's role as providing another set of ears," Stu Seidel says. "When the host is doing a political interview—particularly if it's an adversarial one, where the subject might not want to be as candid as we would hope—the editor's job is to help catch the nuance. The host has to focus so much on the flow of the conversation and the next question, and having another set of ears is very helpful."

As you listen, you may want to alert the producer to a question and answer that you particularly like or one you think can be cut—but don't get into a long conversation or you'll miss the rest of the interview. Host Scott Simon says he can sometimes see a vigorous discussion going on in the control room while he's in the middle of an interview—and it's disconcerting. "They're busy talking about what they're hearing, in a way that would be unacceptable if they were watching a movie. It's not good for the concentration of the person on the other side of the glass, to say the least. We're *working* in there, and you like the sense that everybody is working toward the same result."

For the most part, you should suggest questions or follow-ups only at the end of the conversation; hosts hate to be interrupted (and sometimes get flummoxed) by an editor getting on the intercom to suggest a follow-up while they're in the middle of the interview. "I just can't listen to somebody talking to me, and listen to an editor, and think about a question at the same time," *All Things Considered* host Melissa Block says. And a lot of times what the editor is about to tell me to ask is what I was going to ask anyway." Scott Simon says from host's perspective, the experience comes out something like this:

INTERVIEW SUBJECT: Well, I think there are several ways of addressing the problem, and we should have a solution in the fall.

EDITOR IN HOST'S EAR, TALKING OVER WHATEVER INTERVIEW SUBJECT IS SAYING: *What* ways? Get him to specify the way of addressing the problem!

INTERVIEW SUBJECT: —if we did that, the situation should be corrected.

The one time you may want to talk to the host while an interview is under way is when the guest or host says something that you know to be factually incorrect. If the interview is live, you can prompt the host to correct the misstatement right away; if the interview is being recorded, you can ask the host to correct and repeat the question or statement so the guest won't echo the same mistake. (Otherwise, the correction can usually be made later—after the interview, but while the guest is still available to hear the change.)

Once a recorded interview is over, you should be ready to suggest how it might be edited. After you've listened to enough interviews, you'll get a sense of how much you can squeeze into three minutes, or four, or five, and you'll be able to help the producer make sure that the key points are all represented in the budgeted time. Some hosts always want to be part of this conversation, and some don't.

And after the interview has been cut, listen to it to check for both editorial and production integrity. Is it informative? Interesting? Is the pacing right? Pay particular attention to places where the producer had to work hard to condense a guest's response. Often you'll hear that the grammar of a long sentence has been accidentally mangled, or you'll catch a surviving reference to a part of the interview that has been excised (e.g., "As I mentioned" when the "mentioned" item has been removed, or a "she" when you no longer have the antecedent).

Check the time of the interview to make sure it's the length it's supposed to be. But don't try to shoehorn a five-minute interview into a four-minute hole (or vice versa—expand four minutes of information to fill out a five-minute segment). Some conversations just take longer than the amount of time they have been allotted—and as the editor, you need to make sure that the listener is not shortchanged just because the guest spoke more slowly than you expected, or her answers were especially complicated. The story may end up to be more gripping than you imagined it would be when you originally carved out the time for it. Back in 1996, *All Things Considered* host Noah Adams interviewed Jon Krakauer, who had just come down from the summit of Mount Everest on an expedition where eight people died. Krakauer called the show from Katmandu; the phone connection was barely airworthy. The interview was only budgeted for four minutes, but when the host, producer, and editor left the control room, it was clear that it needed—and deserved—much more time. It aired at twelve minutes—long even for an in-studio interview, and almost unheard of for a scratchy phoner. The response from listeners was overwhelmingly positive. As an editor, you have to recognize a great story when you hear one.

The opposite is true as well. If despite the booker's best efforts, the guest turns out to be uninformed, or determinedly on-message (i.e., doesn't answer the host's questions but keeps delivering the same talking points), or excruciatingly dull, you may need to kill an interview. That's always a hard call, because it means scrapping the work of the booker, producer, and host. And it will generally be a joint decision, Mary Louise Kelly says. "Usually, whether it's verbal or not, there's a dialog going on back and forth through the glass between the editor and the host, where you're looking at each other and raising eyebrows, and smiling—you both have a sense of how the interview is going." So you and the host may agree long before the interview ends that it didn't go well. Frequently the interview can be redone with a backup guest. And even if it can't, the alternative—wasting the listener's time—is unacceptable.

The Editor as Manager

At NPR, the program editors are supervisors, but the editor of any program plays an important managerial role, whatever title he or she holds. You represent your staff at editorial meetings, and in frequent transactions with other senior editors and producers. If you want to give an extra minute to an interview, for instance, you may have to ask another editor or producer to cut a reporter's story by thirty seconds; if you intend to follow a piece with an interview, you may need to ask a reporter to leave out a section that will be covered in the two-way—or to structure her story so it ends where the interview begins. These sorts of negotiations require diplomacy. Gwen Thompkins says you need to learn how to approach other editors and bureau chiefs so they work *with* you and not against you. "It's almost like going into a big dancehall, and everybody's got one dance they really know well, and you want to dance with everybody. So you have to figure out with this person you've got to rumba, with this person you've got to waltz, with this person you've got to Lindy Hop. This one isn't good on the phone—you've got to talk to him personally. This one only wants to hear from you on email."

You also have a responsibility to maintain the journalistic standards of your program and of your news organization. As the editor, you monitor and shape the news content of the show, edit all the writing done for the hosts, sit in on the interviews, and approve the way they are cut. Fundamentally, you are in charge of quality control for most aspects of the program. Over the course of a day, or a week, you will have many opportunities to share your thinking about your decisions—and in the process, train younger or less-experienced members of your staff.

Take copyediting, for instance. *Morning Edition* editor Doreen McCallister says that on her show, producers often write tape-and-copy blocks—scripts for the host to read incorporating actualities and sound (sometimes excerpted from interviews producers have conducted, sometimes from publicly available audio). "I always ask people to provide their source copy. I repeat what my math teacher used to say—'Show your work!'" If in the process of editing a script you see that a producer has gotten some of the facts wrong, or written around the actualities poorly, you should describe the changes you're making—and explain why you're making them. A program editor should use his or her position to teach the basics of news writing, fairness, and production, a little at a time.

You can do the same thing when you're listening to an interview. "It's ideal to have the producer sitting with you and listening," says Mary Louise Kelly. "You can say to them, 'Listen to this bit: you see, this edit didn't quite work. Listen to this bit: you've got the host following up on a question that's been cut out. We need to find a way to address that. Listen to this ending: it sounds really abrupt the way it ended.'" If you simply make the changes yourself, the producer may commit the same errors in the future. You want to send the message that we are *all* journalists, regardless of our job titles. "This is why I hate to hear producers referred to as 'cutters,'" John Buckley says. "To call them 'cutters' implies that they are divorced from the whole editorial process, and that they exist in this artistic audio world, and I don't think that's their role." Since so many editorial decisions are made so quickly in broadcast news, everyone—from the newest production assistant to the most experienced host—has to have an eye open or ear peeled for possible errors. "If everyone shrugs and says, 'Leave it to the editor,' enough nonsense will be shipped down that canal that it will eventually overwhelm the locks," Buckley says.

Indeed, Bruce Auster says one of his roles at *Morning Edition* is to foster "a culture of reporting" among all staff members—not just his hosts and field producers. "What fascinates me about NPR is you can get people on the phone here that other news organizations could never get—and sometimes all we do is ask them, 'Are you available tomorrow morning at six?'" So Auster tells producers and bookers to think of each contact as a way to gather intelligence on news developments. "When we book an interview, we should be asking 'What else are you watching? What else is happening? Did you hear anything about this?' There's all this information you could get if you have a culture of reporting even among people who are not strictly reporters."

And make sure to show your staff how their work is making a difference—for instance, if a story or interview got a big email response from listeners, or if it motivated a person or organization to take action. "I find that young people sometimes don't understand the mission of service that we do," Gwen Thompkins says. "They're thinking of this as if we were accountants, as if this were one of the professions. But the whole point is to feel, at the end of the day, like you have helped somebody, you have told someone's story, you've been a voice for someone. That's what we have to work toward."

When things do go wrong—when a script has mistakes in it or an interview is cut poorly—the editor can be an example of how to provide *constructive* criticism. You want to stress the importance of getting facts right and of striving for excellence, but you never want to demoralize the people you work with. A good rule, for this job and many others: give praise publicly and criticism privately.

The program editor—and this is true of all senior news managers—can also show others how to stay calm and work efficiently under pressure, when the news is changing rapidly, or when technical problems occur. Panic is contagious—in a newsroom, as much as in a crowded theater! By demonstrating grace under fire, as it were, you can show real leadership.

CHAPTER 16

Commentaries

Everybody seems to have an opinion on everything these days—or so you'd think by examining the news media. Newspapers often supplement their op ed and letters pages with features in which readers get to express their personal views on anything from national politics to local zoning changes. Cable TV anchors frequently blur the line between news and commentary; one cable newsman uses his nightly program as a platform for attacking U.S. policy on illegal immigration. Many Internet-only sites that purport to deliver the news actually run it through the filter of their own biases, so that what emerges on the Web contains more spin, speculation, and half-truths than verifiable facts. Highly opinionated men and women host call in-shows on talk radio around the country—preaching to the converted, in most cases, but also attacking those heathens who are bold enough to phone in. And on public radio, many people get to express their personal views through commentaries—and they always seem to touch a nerve.

People may *say* they listen to NPR and other public networks for the up-to-date news, for textured features, or for thoughtful and informative interviews. But actions speak louder than words: commentaries often top the list of the most-requested transcripts and most-emailed items, even though they take up a relatively small percentage of the airtime. They also regularly provoke the strongest comments from listeners. "There is probably no part of our programming that stirs as much reaction as commentary," writes programming chief Jay Kernis, "with the possible exception of Middle East coverage."[1]

1. This and other quotes are taken from a 2004 memo, "Raising the Commentary Bar," written by Kernis and former news vice president Bruce Drake, and issued to program Executive Producers and other NPR managers.

Commentaries, like all the other elements of a radio news magazine, should provoke thought. "That means airing commentaries that are provocative and original, even adventurous—opinion and analysis that you will hear nowhere else," Jay Kernis says. The standards he ticks off for commentaries would apply to news coverage as well:

- compelling subjects or themes
- subjects that matter; or, if on the lighter side, that truly entertain
- intelligence
- superior writing
- first-class delivery
- accuracy
- language and subject matter within acceptable bounds of public discourse
- assurance that the listener knows s/he will hear more than one side of controversial issues
- proper labeling: if something is not an act of reporting or an informational interview, all commentaries, essays, two-ways and conversations offering analysis, opinion or criticism must be labeled as such, *before and after.*

Meeting those standards can take work, in part because the people writing and reading commentaries generally are not journalists. They may have never worked with an editor before. Their two or three minutes on a public radio program may be their first and only time on the air. So if you're going to be working with commentators, you'll want to be able to find writers who can respond quickly to news events. You'll need to identify people who can write thoughtfully and meaningfully about their own experiences and insights. You'll have to encourage some people's ideas—and discourage others. You'll need to know when a commentary requires an opposing view. And you may have to learn how to coach people who've never been behind a microphone before.

Finding Commentators

Commentaries on the radio and on the Web bear on so many different subjects—from Notre Dame's chances in the next football season to what it's like to live with Lou Gehrig's disease to the decline of organized labor—that almost anyone can be a commentator, at least in theory. "The one thing I look for in a commentary is a new idea," says Sara Sarasohn, who spent several years soliciting and editing commentaries for *All Things Considered.* After all, many people can make a cogent argument for or against estate taxes, a Supreme Court nominee, or the designated hitter in major league baseball. The issue for Sarasohn is always whether they can tell us anything we have not heard before. "Imagine I walked into a restaurant, and there was a group of people there, and I gave them the general topic of the commentary, and they just sat around and talked about it for fifteen minutes. If what they could come up with is what is in your commentary, it's not good enough."

Finding people who *have* those new ideas is the big challenge. After years as the *Morning Edition* commentary editor, Maeve McGoran suggests that you start by reading a wide variety of magazines, and by surfing the Web. Don't look for columnists or commentators per se, she says, but for writers or analysts who seem to have original thoughts and fresh ways to express them.

NPR also solicits commentaries from listeners with this blurb on the Web.

> We are looking for commentaries or essays that tell a tale, reveal a personal reflection, or add an informed perspective to events in the news. We want pieces that express an original idea with clear, creative writing. While many essays are from regularly scheduled commentators, we also want to hear from people who can comment on a once-only basis.

And it works; the commentaries come pouring in. McGoran says she receives "at last a hundred a week"—and reads them all. "This is where you're looking for diamonds in the rough."

Like all journalists, commentary editors have to keep their eyes and ears open. If you meet people who have perspectives you find startling or revealing, you should want to get them on the air or on the Web.

You also need to act quickly to solicit topical commentaries—opinion pieces that can be paired with stories from reporters or with host interviews. Here radio has a distinct advantage over newspapers or magazines; you can commission a commentary in the morning and have it on the air in the afternoon.

With that in mind, Jay Kernis suggests some creative booking ideas to help you focus your search. When news breaks, he says, ask yourself these questions:

Who would be the most remarkable or special person to hear from on this subject?
Who has had a similar experience and can write or talk about it?
Who has fought against this idea for years and has succeeded in changing public opinion?
Who is the person who has dedicated her or his life to this cause?
Whose life was changed by this occurrence?
Who is the writer who can offer the funniest take on an event or idea?
Who might change your opinion, if given the chance?

Public radio offers people a soapbox that is unrivalled in American broadcasting. Those few minutes on the air can inform, persuade, provoke, or amuse millions of people. Writers generally don't get paid very much for their commentaries—perhaps a few hundred dollars—but that has rarely been an issue. They are compensated in other ways. "We have lots of commentators who have gotten book deals on the strength of a single commentary," Maeve McGoran points out. "That's just a testament to the size of our audience, and who our audience is."

Topical Commentaries

When you're trying to find someone to comment on a news event, the Rolodex is your best friend. In that respect, as Kernis points out, one of the first and most important steps is to keep an up-to-date list of "the top thinkers in various fields who have something new to say."

Most reporters and bookers maintain extensive lists of experts and analysts who have been sources for stories or guests for interviews. So if you are searching for commentators and a name does not leap to mind, look to your colleagues for recommendations; they're likely to have interviewed someone who is informed and opinionated about a given topic. But remember: you don't need to solicit a commentary from that person as soon as you make contact; the first step is just to put out some feelers. That's what Ellen Silva did when she was looking for an *All Things Considered* commentary to pair with a news interview about the sacking of a four-star Army general; the Defense Department had relieved him of duty after allegations surfaced that he had had an affair with a civilian. "I called all the generals who were in our records that morning, just to talk to them, off the record, about what they were thinking about it," she says. One of them, a two-star general, had been on the air frequently in the first weeks of the war in Iraq. As it turned out, he knew the officer who had been relieved of command, and was interested in writing about the case. Silva could have turned to one of Washington's think tanks to find a policy analyst to weigh in on the story. But "to have a two-star general writing about it is a lot more interesting. You want someone who's been in the trenches."

Many subjects demand more than one perspective. Former NPR Vice President for News Bruce Drake set out the network's policy in a memo a few years ago: "When we air commentaries that take a stand on a controversial subject, we must make it our practice to *ensure we will air other points of view* [his emphasis] on that subject in a timely way, and do our utmost to let the listener know that this will happen. This must be a conscious effort." That puts an extra burden on you if you're looking for, or receive, a commentary on an especially hot topic.

How do you know what's too hot for a single opinion? "The line on that moves," Sara Sarasohn says. "There are certain topics that raise flags and where I'd always consult on them—abortion, or the Middle East, for instance." And some cases are simply obvious to anyone who is keeping up with the news. "Intelligent design, and evolution and creationism—I can't think of anything we'd do on that subject where we wouldn't need more than one commentary," Sarasohn says. Such controversial issues require you to commission and edit *two* commentaries, to run either back to back, or on consecutive days of the week. In either case, it's double the work. But, Maeve McGoran says, "You just can't think like that. It's like self-censorship. If I think it's an idea worth doing, I just grit my teeth and know that I'll have to find two spaces on the show for it, not one."

That highlights one way broadcast commentaries differ from newspaper op eds (and from columns on the Web, for that matter), where the writer is identified only at the end of the piece. Whenever you bring in a radio commentary on a contentious issue, it's a good idea to give the commentator's credentials (though not necessarily his or her title) in the introduction, as well as in the backannounce. If you simply identify someone as "commentator Ellen Ritchie," people may wonder, "Why should I care what this woman has to say about parental notification laws?" It's better to say something like, "Commentator Ellen Ritchie has been counseling pregnant teenage girls for twenty years."

Frequently you'll be able to land a well-known writer, academic, or politician to write one of these op ed pieces. That can actually make your job harder. Whenever you work with celebrities, Ellen Silva says, "you're going through layers of people—agents, and publicists and assistants." And it may be difficult to ensure the piece was actually written by the person with the famous name—not by some speechwriter or aide. "I've had to kill a commentary by a really well-known politician because I found out that a staffer wrote it," Maeve McGoran says. Finally, even when everything else works out, some celebrity essayists are not used to being edited. But you can't be intimidated by someone's reputation. "I've edited people who've won the Booker prize, and *they* needed to go through five rewrites," McGoran says.

Personal Commentaries

When you're working with commentators who are writing about their own lives, you need to be strict about making sure they've really got something to say that merits time on a national news program or Web site. "We get a lot of commentaries about dogs and babies," Maeve McGoran says. "Generally, if something's on *Morning Edition*, it's because it illustrates a larger point—it may be a beautiful story, but usually there's a little punch to it at the end—more of a payoff, journalistically speaking." So be selective. Even regular contributors to the news magazines sometimes pitch several ideas before one is accepted. One *All Things Considered* commentator, Paul Ford, even wrote a little dialog describing his experience of pitching a commentary to NPR:

PAUL, in a touching voice: ". . . and my girlfriend and I were standing there in New Mexico, negotiating at that moment what our life would be like, trying to plan for our future together."
SARA: "That's important to *you*, not to the audience. Next pitch!"

Encourage potential commentators to send in their ideas by email. If something looks promising, you can always follow up with a phone call. "It's usually best if people share their ideas before writing a commentary, if only so I can say, 'We've already done this' or 'That sounds like a great piece,'" Maeve McGoran says. "Usually we'll have a conversation and figure out which one of the ideas I'm most interested in, and then they'll send a draft and we take it from there."

Often a chat about one idea leads to a much better one. McGoran got a pitch from a Chicago woman, Beth Finke, who wanted to write about what it was like to be working as a "life model" in art classes—posing in the nude, when, as a blind person, she couldn't see the students who were drawing her. McGoran says that she wasn't keen on that topic, but she realized that Finke could write well—so she kept in touch. "One day, in an offhand way," McGoran recalls, "she said, 'When you lose your sight, there's a blind school you go to learn everything you need to know. The thing that worried me the most—that I fixated on—was thinking that I don't want to spend the rest of my life dressing like it's 1980, because that's the year I became blind and I don't know what people wear any more.' I told her, 'See what you can come up with about that.' And she wrote a really funny, touching piece."

Whether you accept or reject a commentary, when you work with people who are not professional writers, make sure they understand the rules of the game. You need to make it clear that even if you express interest in someone's idea, they are still sending you the piece on spec. Sara Sarasohn says she told all her commentators, "Just because we are having this conversation about your idea does not mean you're going to have a commentary on *All Things Considered.* There are many points along the way where I may say, 'Thank you very much, but this is not working for me.'" *Morning Edition* editor Maeve McGoran also kills commentaries at various stages of the editing process, and says you have to be courteous but firm when you turn someone down. "Sometimes I'm *too* polite," she says. "Someone sent me a note back and asked, 'Was that a no?'"

Editing Commentaries

Whether a commentator is an experienced journalist or has never sold a story before, he or she may be surprised at the kind and amount of editing involved in bringing a piece to air. "We will routinely go through six or seven edits," Maeve McGoran says. "Our commentaries are about 450 words—two-and-a-half minutes—which is much shorter than the average newspaper column, which is two or three times that long! So a lot of it is trimming it down." In addition, she says, many professional analysts are used to working in print. One responsibility of the editor is to make sure their pieces don't come out like newspaper columns read aloud. You have to get them to write the way they would speak.[2]

2. See Chapter 3, "Writing for Broadcast," for some basic rules.

They also need to make their points as persuasively as possible. "It's different from editing reporters, where you want them to keep their opinions out," McGoran says. "A common problem with commentaries is that people don't have *enough* opinion." If necessary, ask questions to push the commentator to express his point of view more strongly, or to argue his case more coherently: "Why do the child's rights trump those of the parents?" "You seem to be suggesting that all ideas about the origin of the human species are equal—is that what you believe?" "Why don't you accept the argument that pulling troops out now would end up costing lives?" You want to encourage the commentator to fill in the gaps in his argument, to express himself effectively and forcefully. But remember, you are looking for the commentator's view, not your own; you don't want to tell him *what* to say. "I would never, ever rewrite," Sara Sarasohn says. "I would never type something and send it to a commentator, and say, 'Do this.' Not ever."

Although every commentary poses its own editing challenges, Sarasohn says she found herself giving similar comments to many different people during the edits. Among them:

START STRONG. "Write a really good first sentence. A lot of times people will just start writing, but you need a great first sentence—for anything, and especially for a commentary."

DON'T LET THE WRITING BE TOO DENSE. Experts are experts because they know a lot, but you can only impart so much information on the radio. Sarasohn says that she frequently had to tell commentators, "This sentence is too baroque for my ear to follow it; we need to restructure it and break it up into two or three sentences."

EMPHASIZE TRANSITIONS. Sometimes it's easy to miss the opinion amid the rest of a commentary, but underscoring it often just takes a few words, Sarasohn says. She'd advise commentators, "You're really turning a corner here. This is the heart of your argument, and you need to emphasize it."

END STRONG: "The last two or three sentences are extremely important in a commentary," Sarasohn says. "That's the last thing people hear; it's the last thing they're going to remember."

Commentary editors—whether they're working on the radio or on the Web—also need to be especially rigorous about fact checking. While reporters realize that their reputations depend on presenting details accurately and without embellishment, commentators may not be so conscientious. Sara Sarasohn tells a story about editing a commentary on the fortieth anniversary of the Watts riots in Los Angeles. "During the editing process, the commentator had a story saying that a certain incident where a man was shot by the police led up to the Watts riots. I checked, and it turned out that that incident happened in 1966, and the Watts riots were in '65. The writer insisted on the phone to me that this incident was one of the things that led up to the riots. But after I found it, he said, 'Oh my God—you're so right. That's just the way I had misremembered it.'" If you're in any doubt about the facts in a commentary, ask a reporter or librarian to look the script over.

Your fact-checking job even extends to verifying the writer's identity, especially if the commentary comes in unsolicited. After all, you may never meet the writer face-to-face. So check that the commentator really is a professor at Dartmouth College, or has worked at the Nassau County Department of Public Works for thirty years, or does have a mailing address in Anchorage. "I live in fear of being fooled," Maeve McGoran says. One tip-off, she says, is when a commentary is too good to be true. "We got this one commentary over the transom from a soldier who'd returned from Iraq. He was back in his old job—he was a teacher—and he described how he'd be washing the dishes, and he would suddenly remember when this guy died in front of him. And I thought, 'Wow!'" McGoran emailed the writer to say that she planned to run the commentary, and that she just needed some information about where he was posted in Iraq. "We never heard from him again."

Coaching Commentators

Even people who are used to public speaking might need coaching when they get behind a microphone. The professor who is poised and fluent in front of his freshman history students may be scared silly by the thought of presenting his commentary to millions of public radio listeners. A CEO has a captive audience when she speaks in a boardroom; on the air, she has to *earn* our attention. The best-written, most cogent analysis will fall on deaf ears if the reader drones on and on, or stresses the wrong words, or comes across as pompous.

And for many people, it's a new and unsettling experience to hear their own voice coming back at them in their headphones. (When we talk the rest of the time, we hear our voices resonate through our skulls.) As a result, they may speak more softly than usual, and in more of a monotone. As a voice coach, you may need to prompt commentators to put more life in their delivery. After all, when we're standing right next to someone we use facial expressions, hand gestures, and other nonverbal cues to help us communicate, but when we're on the radio, we have to make up for the absence of body language by speaking with greater expression. "Most people feel like they're overemoting in the studio when they're getting it just right," Sarasohn says. So she would tell commentators, "I will never let you make a fool of yourself on the radio. If I feel you are overdoing it, I'll pull you back. I don't want you to feel you have to check yourself."

In addition to trying to energize commentators, make sure they emphasize their key ideas while they are reading, perhaps by slowing down a bit; changing the pace is often the best way to draw a listener's attention to a sentence or phrase. Look for moments of transition in the text—going from a specific incident, say, into a more general observation, or from an abstract idea into a personal story—and encourage your commentators to use their voices to indicate that change. Finally, don't let a commentator read the last line as if it were any other sentence in the piece; get him or her to slow down and think about what the words mean, so that the listeners will think about them, too. "When you're reading a magazine article, your eye knows the end is near," Sarasohn says. "But in a commentary, no one can see the end coming, so you have to signal it with your voice."

If you don't feel you have the coaching skills required to get the best delivery out of a commentator, ask for help.[3]

Commentator Paul Ford's description of reading his script for NPR also includes a few practical tips for first-time contributors:

Bring a clean, double-spaced copy of the latest version of your piece, and a pen. If the piece is longer than two pages, try to make the page breaks at a natural pause in the piece, so you can shuffle them easily.

Don't wear clothes that rustle, i.e., a suit jacket. You'll only be meeting an intern and an engineer, and they don't care that you dressed up.

Be comfortable. Ask for water or tea.

Ford has one last piece of advice for anyone who's reading his or her first commentary for NPR: "Bring a camera so you can take pictures of the studio, since God knows this stuff doesn't happen to you every day."

3. See Chapter 8, "Reading on the Air," for tips on delivery.

Commentary Intros

A good intro for a news report often begins with the latest developments. But a topical commentary will frequently be paired with a story that *itself* includes all the news. Editors are sometimes at a loss to come up with something for the host to say to introduce the commentator.

One way out of this dilemma is not to have an introduction at all—i.e., to start the commentary with a sentence or two, then give the commentator's name and credentials, and then continue with the rest of the piece. (At NPR, this is usually called a "drop-in.") For example, if you've been asked to bring in a commentary on a new Supreme Court nominee named Jane Jamison, and you know it will be broadcast after an interview on Jamison's record as a lawyer and Circuit Court judge, you might edit the commentary so it begins something like this:

COMMENTATOR: Jane Jamison has all of the obvious qualifications to be on the Supreme Court—except one. She lacks humility.
HOST: Commentator Joshua Mayer is a Washington lawyer who opposed Jamison when she argued before the Supreme Court in *State of Virginia versus Smithfield.*
COMMENTATOR: Jamison thinks so much of herself that...

Another solution is simply to start with the commentator's ID, and give us the thrust of his or her point of view:

HOST: Commentator Joshua Mayer is a Washington lawyer who opposed Jamison when she argued before the Supreme Court in *State of Virginia versus Smithfield.* He says Jamison has all of the obvious qualifications to be on the Supreme Court—except one. Mayer says she lacks humility.

The real intro challenge comes with first-person and evergreen commentaries. Look through the NPR transcripts, and you'll see they often fall into the pattern: glittering generality, followed by commentator's personal observation. Here's a typical (and fictitious) example:

Many people look back on the 1960s with a mixture of nostalgia and embarrassment. For commentator Randy Miller, the decade never seems to have ended.

You never want your program host to say things that are dull, obvious, or platitudinous. So here are a few tips for improving commentary intros:

REMEMBER THAT FACTS ARE YOUR FRIENDS. When you're writing under deadline pressure, you may not think you have time to find out which 1960s shows Nickelodeon is currently broadcasting, or whether *Hair* is still being staged somewhere, or how much people are paying for a 1964 Mustang. But getting that information to strengthen a commentary intro may actually take less time than wracking your brain for another way to get into the subject.

DON'T USE FAMOUS QUOTATIONS, RHETORICAL QUESTIONS, DICTIONARY DEFINITIONS, OR OTHER DEVICES. These are simply tired gimmicks for avoiding the hard work of writing a real intro.

CONSIDER MAKING SOME SORT OF TRANSITION FROM THE PREVIOUS PIECE. Remember, listeners often don't know where the program is going; they only know where it's coming from. So one of your jobs is to find some sort of link with what they've been listening to. The goal is to make the commentary sound like an integral part of the show and not just filler. However, also observe the following rule.

DON'T MAKE A FORCED TRANSITION. Sometimes there really *isn't* a connection between the previous piece and the commentary. That's why many news magazines use musical bridges or "buttons" between items.

WRITE IN THE RIGHT MOOD. If the piece is topical—for example, a news analysis—you want your writing to be straightforward and newsy; if the commentary is light and frothy, you've got to try to bridge the mood from the rest of the show to the two minutes of fun that's about to follow. Suit your style to the commentary.

IF ALL ELSE FAILS, SEE IF THE INTRO IS ALREADY IN THE COMMENTARY. Sometimes the commentator's first paragraph really *is* the intro. You may be able to lop it off and give it to the host to read.

To Comment or Not to Comment?

Commentaries serve both production and editorial purposes.

They add texture to a program and change its pace. You can use commentaries to add the voices of women, or young people, or westerners to a show that otherwise lacks them. Maeve McGoran was hired as a commentary editor to get those perspectives on the air. "They really wanted to expand the pool of people doing commentaries for *Morning Edition*," she says. "They wanted it to be more diverse—by age, geography, ethnically, and so on."

Regular commentators can also be magnets for listeners. They tune in hoping to hear Frank Deford's take on sports, or Bailey White's stories, or the essays of Andre Codrescu. That's one reason the programs try to have their own commentators. "All of our programming sounds so much the same," Maeve McGoran says. "Commentaries are one small way we can differentiate between *Morning Edition*, and *All Things Considered* and *Day to Day*."

Commentaries provide a variety of perspectives on news events. They let us hear from people with expertise, or with firsthand experience.

Finally, commentaries make shows easier to produce; commentaries of various lengths can replace pieces or interviews that at the last minute fail to materialize, and they can fill up those pesky two-minute holes at the end of segments.

But there are limits to what you can get into a commentary. You may decide that an idea pitched as a possible opinion piece requires a variety of voices and perspectives—that it would be better if it were pursued by a reporter. When that happens, remember that the commentator "owns" the original idea. You can't just steal it, Maeve McGoran says. "Sometimes I'll ask, 'Would you object if we asked a reporter to do a story about this?' And almost always they don't mind, because it's an issue they care about." Sometimes the commentator's idea is simply too complex to be delivered in four hundred words, and would work better as a magazine article—or even a book. "Some ideas are just out of scale to a commentary," Sara Sarasohn says.

Public radio can't produce news magazines without reports or interviews, but it can, in the end, put a show together without any commentaries at all. So every time you solicit, edit, or program a commentary, whether it's for broadcast or for the Web, you should be trying to add something of real value—to bring in a new point of view, add a different voice, change the pace, draw a laugh, or evoke a feeling. The first rule of commentaries, as of so many things, should be "First, do no harm."

CHAPTER 17

Studio Directing

Almost every day, something goes wrong during the production of an NPR program. Sometimes lots of things go wrong.

A guest who's supposed to take part in a live interview may not answer the phone when she's called. A computer error may keep a sound file from being saved to a hard disk. Engineers might discover they can't establish the audio line with the station across the country where a reporter is located. Scripts sometimes have typos in them. CD players fail.

Many of these sorts of problems get diagnosed and resolved in the control room, and at the last minute, by the studio director. In addition to serving as one of the last gatekeepers before material is broadcast, the director times program segments, selects music, talks to hosts, gives cues to engineers, communicates with newscasters, gets updates from the program producer or technical director, and greets guests—sometimes all at the same time.

And multi-tasking is only one of the job requirements!

Most commercial radio stations, and many public ones, do without directors; audio engineers and producers communicate directly with news anchors and hosts. But at some stations, and at NPR, the director is viewed as a linchpin in the program production process. In addition to acting as a sort of air traffic controller, the director chooses the musical buttons between pieces or interviews, as well as the music between program segments and at the end of the show. An experienced director uses that music to help set the pace and establish the tone of the program, or to vary that pace and tone depending on the flow of material—a flow that often changes course while the program is on the air. Directing a live program forces you to think on your feet—literally, since a director often works standing up. It is a job that demands technical skills, artistic sensibility, and often nerves of steel. It's not for everybody; but the people who are successful studio directors play a key role in creating and maintaining the distinctive sound of public radio.

The Director as Number Cruncher

To be a successful studio director, you have to be able to add and subtract. And tell time.

Though the studio director on the news magazines plays a key creative role, one of his tasks every day is just making sure that all of the pieces of the program *fit*—that an eight-minute segment, for example, has no more than eight minutes of material. Pieces tend to "grow" as they are reported, edited, tracked, and mixed, and interviews slated to fill four minutes sometimes get edited down to 4:10. As a director, you have to check that everything is the length it's supposed to be.

That job can begin even before you get a script. Add up the times of the pieces as they're written on the program's roadmap—the list of stories included in each segment of the show—just to make sure the right amount of time has been budgeted. Once the script comes in, check everything again, making sure that the times on each page jibe with the times on the roadmap. It doesn't usually matter if one item is a few seconds too long, as long as another comes in shorter than planned. But it matters a great deal if *everything* is a little too long. This is especially true if your program is being fed to more than one station, since they count on segments beginning and ending at precise times so that they can provide local information (on traffic, weather, etc.), run recorded announcements and promos, or play their own pieces and interviews instead of those coming from the network. "Busting" a segment on a national show by even a few seconds can create problems all over the country.

If you do discover that a segment is too long, you have a number of options, says Van Williamson, who has directed *Morning Edition* for years. "Usually, I'll call the producer and say, 'We're going to bust here; there's too much time.' And what they can do is to get the second piece and just whack something off of it really fast." Sometimes an editor can rewrite an upcoming intro to save a few seconds.

But in an emergency—and the prospect of "busting" a segment *is* an emergency, though there are worse ones—you may need to ask a producer to cut out some of an upcoming news report or interview. The goal is to look for part of a voice track or actuality—or in an interview, part of an answer—that can be removed without causing too much editorial damage. This sort of "samurai editing" isn't ideal, of course, since it's usually done at breakneck speed and without a great deal of thought. That's why it's essential to pass along whatever you know about the timing of a segment as soon as you know it.

If you wait too long, or don't notice the problem until late in a segment, there won't be enough time for *any* internal editing. Then your only alternative may be to "dump out" of a piece or interview—that is, to cut it off before it's supposed to end. Williamson says doing that elegantly takes coordination among the director, host, and studio engineers. "You say to the host, 'You're going to have to backannounce this reporter because we're going to dump out before the soc [the "standard out cue" when the reporter signs-off].' Then I pull up the script on the computer, I read the script, and I say to the engineer, 'Here is the out cue, and it's going to come about twenty seconds before the end. And when he says these words of the script, we're going to dump out.'" Williamson says that thanks to skilled engineers and digital audio software, he can usually get away with this sort of editorial sleight of hand without drawing too much attention. "Most of the time, I don't think the listeners even know."

Keeping track of a program's timing is harder than it might seem. There are calculators that can add and subtract minutes and seconds, but most directors say they don't use them. "You just have to know that forty-seven plus fifty-three equals a minute forty," says Kathryn Fox, who has directed *Day to Day*. If this sort of arithmetic doesn't come easy to you, you should practice it until it's as easy as counting dollars and cents.

Ironically, in these days of digital timers and stopwatches, a good director also needs to be able to read an *analog* clock with just a glance. "If the segment's going to be out at 58:19," Tracy Wahl says, "you need to be able to look at the clock, and know in a split second that it's 57:15 right now, and not 58:15. Where that second hand is, and where that minute hand is, completely determines whether you're going to bust, or whether you've got a minute of music—and everything else."

Timing all the elements of a program may be the least interesting part of the director's job, but it's one of the most crucial. "You're the last line of defense against the show exploding," says Van Williamson.

The Director as Conductor

As a studio director, your primary job once the program is on the air is to serve as a sort of conductor, bringing the show producer's "score"—the program—to life. "In this case, the director is not conducting musicians," says Bob Boilen, who directed *All Things Considered* for eighteen years, "but conducting the people he's around—he does it with his hands, a lot of times, as a conductor would—and those people are engineers, and on the other side of the glass, hosts." Among other things, it's the director's job to make sure everything is "performed" as smoothly as possible, and that the broadcast's rhythm suits its material. And that requires thinking a few minutes ahead of whatever's on the air at the moment. "It's like reading music," Van Williamson says. "When you're reading music, you're reading about a measure ahead of what you're playing. What's happening at the moment is already past your thought process; you're on to the next moment, and what's after that, and what's after that."

CHOOSING AND USING MUSIC

At NPR, the director is the person who chooses all the music that plays between pieces, at the end of segments, and during the station breaks—and that music can set the tone for a piece or an entire program.[1] "I enjoy the mechanics of directing the program," Ned Wharton says, "but it's the choice of music that's so important to leaving the listeners a mood—a moment—to help them reflect on what they've just heard."

Music can provide a buffer between stories that are on starkly different topics, or whose moods would otherwise clash harshly; you should check whether adjacent pieces would benefit from a bit of music between them, and then try to find something that will bridge the gap in moods. "If [the program is] heading somewhere lighthearted out of something heavy, the music has to have some sort of change in it," Bob Boilen says. "It's like a jigsaw puzzle piece—you're looking for the piece that goes from the building that's green to the river that's blue." Boilen says that it helps to be familiar with lots of instrumental music of all genres—but knowing a smaller number of pieces very well will do.

Of course, there are limits to what you can do in the control room to erase jarring combinations of subjects. "Sometimes people think musical buttons are magic, that they can smooth any transition," Kathryn

1. In fact, the torrent of listeners' requests for more information about the music played between stories led to NPR's putting the artist and record label for every musical button on the Web, along with the rest of each show's lineup. Eventually it even inspired an entire online music program, *All Songs Considered.*

Fox says, adding that it sometimes makes more sense to ask a program producer to rearrange the lineup to avoid disconcerting combinations of stories. You can't count on music to take listeners from a heart-rending story about Eastern European orphans to a frothy commentary about deciding to get your hair dyed.

Music doesn't only serve as a bridge between items; it can also provide a coda to a story—a few seconds for listeners to reflect on what they've just heard. Ned Wharton says that he always listens to the last forty-five seconds or so of each piece or interview, and he reads every intro, so he has a good sense of what the mood of the story really is. But avoid the production cliché of matching up, say, a story about Chinese laborers with some Chinese music, or a piece about West African crafts with West African music. "I've learned to stay away from the country's music or the region's music, unless you're sure it's correct," Bob Boilen says, "because you'll play a piece of Iranian music that you think is right, and it'll be from the wrong part of Iran—so you're just insulting someone!"

There are other ways to be creative, Boilen says. "I will play something that will have some other level to it—that might be funny—where someone who hears it first thinks, 'Oh, that's a nice piece of music.' And then gets the pun and says, 'Oh, wow!'" For instance, Boilen once followed a commentary from an American in New Delhi talking about trying to learn how to wear a sari with an Oscar Aleman instrumental of "Who's Sorry Now?" Van Williamson says that he considers it a "two-fer" if he can match the tone *and* make a pun. "I occasionally get a three-fer, which means the piece of music works on *three* different levels." He says that if he had a story about a diner that charged World War II veterans 1940s prices, he might follow it with a chorus of the old song "Java Jive" by the Ink Spots—giving him the mood, a pun, and a 1940s reference, all in a few bars. Here too, though, a director sometimes has to restrain himself; you don't want your choice of music either to be either obvious or inscrutable. "The conventional wisdom is that if it's a story about ducks, you have to choose a piece of music with 'duck' in the title," Ned Wharton says. "More often than not, especially if it's an obscure piece of music, it's not going to mean anything to listeners."

Occasionally, even when you're not *trying* to make this sort of pun, the title of a piece of music can rear up and bite you. Bob Boilen tells a story about searching for a button to get from a sad report about a prolonged drought in Ethiopia to a more neutral story. "I threw a CD into the player just to start somewhere, and the music was just right! I only listened to the first twelve seconds of it, but I knew I wasn't going to use more than ten or fifteen seconds of it anyway. So I gave it to the engineer." As soon as the music began to play, Boilen says, he could hear the program producer running thunderously down the hall into the control room. "And he asked, 'What are you *doing*?' And I said, 'What do you mean? I'm playing this button because we're going to go—' And he said, 'You're playing Handel's *Water* Music!'"

WORKING WELL WITH OTHERS

As a director, you also "conduct" the people who take part in each broadcast. The hosts and program producer depend on you, and you depend on the engineers. "It's like a dance that we all do in the studio," Carline Watson explains. "It's very much a team effort."

Almost all directors strive for consistency—making sure they give cues to hosts and engineers at predictable times. You'll need to tell everyone what's coming up next, so that engineers can make sure the audio is ready and hosts know what they're expected to read. "I give the hosts at least a minute's notice," Watson says. "I can see what the hosts are doing, and very often they're tweaking scripts, they're reading the wires, they're talking to an editor outside, and so you want to make sure they have enough time to look at the next thing they're reading." In the same way, you need to get the engineers' attention so they can focus on what's ahead. "There's a lot of stuff going on in the studio, and you have to make sure that they hear what you're saying and acknowledge it," Ned Wharton says. "Speak clearly and loudly: 'After this piece, the music's going to come up; we're going to take it under for a time check, and then the host is going to read us to the next piece.'" Wharton, and most studio directors, tend to give all their cues at regular intervals. "The more machinelike you are, the better. Specifically, you give a drill sixty seconds before the event, you give a command to the engineers ten seconds before—a stand-by cue. 'Ready' at two seconds. The more that becomes a machinelike process, the more you're ready for the things that are unexpected." The most important part of giving directions is *clarity*; as the director, you may have a perfect idea of how you want a segment to sound, but it won't turn out as you imagined unless you can give the engineers crisp, unambiguous commands. "They are your hands," says Bob Boilen. "All you have to do is describe the picture you want painted, and they'll paint it."

The host also needs a clear stand-by cue, but some of the most critical communication between the director and host is *nonverbal*. For instance, after you've told the host what's coming up, look for some sort of acknowledgment—even if it's just a thumbs-up gesture or a nod of the head. "When I worked with Noah Adams, I'd ask him a question, and he would ever so slightly lift his head," Bob Boilen recalls. "I know by looking at Michele Norris whether she's got the next piece of copy by the way she's poised, by the way she sits—by her body language. I know when Robert Siegel doesn't have his next bit of copy by the look on his face." You'll also need to tell the hosts when to speak—or not to speak—using hand signals.[2] As soon as a host's mic is open, hold up your hand like a traffic cop telling a driver to stand and wait; then think about whether it will sound better to have the host come in quickly—right after the piece ends or the music posts—or to leave a slight pause. Having a host talk immediately or after a short delay can make him sound impatient, or thoughtful, or energetic, or uninterested—all depending on the material he's coming out of. "And then, when you point to the host, you point right at them—especially when you have a two-host show," Boilen says. "That way they know they're the one who's going to talk."

You can also make it clear whom you're talking to just by using a different tone of voice. You might give instructions to the engineers loudly and forcefully, and speak much more softly to the hosts over the intercom. "I almost want to be a ventriloquist," Bob Boilen says. "Although the engineers hear you both times, they know you're not talking directly to them because you're using two different voices."

The other vital communications link during any program is between the studio director and the program producer—especially when the program lineup is changing or problems arise. The producer needs to update you if news is breaking, or if a planned interview is falling through, or if anything else on the show's roadmap needs to change. And you need to inform the producer if there are technical problems, or the hosts don't have their script, or a segment is too tight. Some studios have an intercom—a sort of "hot line"—between the director and producer just for this purpose. "Each director has to figure out how they can most efficiently

2. Decades ago, directors (or engineers, whenever shows were broadcast without directors) often had to communicate with hosts using *only* hand signals. People used hand signs to indicate when to take a station break, when to speed up or slow down, when the next item up was on open-reel tape or cartridge, and so on.

talk to the line producer," Tracy Wahl says. "The program producer has his own sort of crazy world out there, and you've got your world in the studio. Sometimes making a phone call is best."

WHEN THINGS GO WRONG

A two-hour news magazine might have sixteen or twenty elements—news reports, interviews, and commentaries. On some shows, it is common for the majority of those pieces and interviews to be completed—and for most of the host's script to be written and edited—while the program is on the air. Throw in the possibility of a computer problem, a source returning a phone call late, a guest arriving at the wrong time, or a cell phone battery dying, and you've got a prescription for daily chaos. One important function of the studio director is to make sure as little as possible of that chaos gets on the air. "What you are in that control room is almost like a shock absorber," Bob Boilen says. "You're a big buffer. The panic and the craziness is going on in the other room outside of you—with the producer and the copy and all that stuff, and stories are changing—and you understand that, and you hear all that, but your job is not to pass that craziness on to the hosts."

There are several ways experienced directors can absorb those shocks. One is simply not to overreact to the routine problems that crop up on almost every program. "Your panic affects people," Tracy Wahl says. Some people are not cut out to be studio directors because they can't deal unemotionally with the sorts of uncertainty that are always a part of the job. "Each director has his own comfort zone about not having a piece, not having an intro, that type of thing. If you don't have a piece that's coming up next, or you don't have copy, two minutes is a good time to say, 'I need the copy now' or 'I need the piece now.'"

You can also help keep the program on an even keel by heading off potential problems. If a live interview is coming up soon, check several times that the guest is on the line and listening to the program. And always have a backup plan in mind. "Just anticipate," Carline Watson says. "This is a live interview. Things can go wrong. You can lose a phone line. Just think about what will I do if this doesn't work out?" Ask yourself that same question if a producer says that a piece or interview may not be finished on time. "It's like when you're flying an airplane," Van Williamson says. "In addition to all the stuff in the airplane, you're also thinking about if the engine stops right now, where am I going to land? Where can I glide to? Is there a field? Water? You should always know how many scripts the host has ahead of time, what the engineers have pulled up in the computer, how much time those things will take up and whether the music is loaded." If you have a backup plan—or multiple backup plans—"you can get something on the air pretty fast," Williamson says.

That doesn't mean you should always let the host know what your plan is. As a director, you should give the hosts all the information they need to be prepared, but not so much that they start worrying about what *might* happen if something goes wrong. That's your job! If Master Control is having difficulty establishing a connection with a guest at a studio in Kentucky for an interview six minutes from now, you'll want to have a backup plan, but it's probably too soon to let the hosts know there might be a problem. However, if you know for sure that the interview isn't going to materialize, you'll want to let them know how it's changing and what they do next. "They don't need to hear you say, 'I don't know if we're going to have this piece! We might not have it! And if we don't have it, we might have to flip three pieces around!'" Bob Boilen says. "All you say to them is, 'We're going to [reporter Sylvia] Poggioli next. And Melissa, you're going to read the intro.' That's it! All they have to concentrate on is, 'I have the Poggioli copy in front of me. Are the facts correct?' They don't need to have their brains scrambled with all the other stuff." Once you're past the problem, you can explain what went wrong.

On the other hand, you *don't* want to keep hosts in the dark if there is any kind of technical glitch that they may need to disclose to the listeners. If an audio file dies, for instance, and it's taking time to get it started again or to bring up the next story, the hosts need to know what's happened so they can tell the audience that there is a problem with the report.

Finally, when you make a mistake yourself, you need to be able to recover from it quickly. All directors have horror stories of "unforced errors" they have committed—goofs they have made just because they were not paying attention, or weren't thinking about the consequences of a decision. Many people have mistimed segments so that a program ends a minute too early. Some have played the wrong reports, or called for audio while a host was still talking. One director played what she thought was an instrumental piece of music, only to hear a voice on the CD say, "Somewhere a wildebeest is giving birth"—followed by the *sound* of wildebeest giving birth. "As sure as the sun rises in the morning and sets in the evening, something will happen on your watch and it will be your fault," Van Williamson says. "So get prepared. Get used to it now." And don't let miscues—however serious they may be—derail the show. "Whether it's you at fault or the engineer," Tracy Wahl says, "it's important to get everybody back on track. You have to be able to say, 'Here's what's next. Focus on what's next. We'll deal with the mistake later.' You have to be a little of a psychologist."

Getting the news on the air is serious work, but you can't allow the atmosphere in the control room to become too serious or strained; people don't work well when they're tense or agitated. When he was directing *All Things Considered*, Bob Boilen would always try to keep things upbeat; he'd start each program by telling the hosts, "Have a great show." Many directors make a point of chatting with the engineers, or making occasional jokes. Keeping the mood relaxed *most* of the time allows everyone to be more focused and alert when they need to deal with unexpected events. "I suppose it's like an emergency room physician who listens to country music," Wahl says. "You can't be too tightly wound."

Directing Live Events

Directing coverage of a live event—a politician's news conference, a Senate confirmation hearing, or any other scheduled proceeding—requires special skills. For one thing, you may be working without much of a script. If the news conference is to begin at the top of the hour, the host may simply introduce himself and the special coverage, and then improvise until the politician comes out. For another, you won't know how long the various elements of the program will be; the person holding the news conference may talk for five minutes before taking questions, or fifteen. You may also need to keep track of a number of guests at the same time—one analyst in the studio, another on the phone, a reporter at the state house, and a lawmaker at her home. At the last minute, a show editor might write a summary of the event that includes several actualities, which you'll have to find, and cue, on short notice. The producer may suddenly add or remove guests from the lineup. You may feel like a choreographer trying to plan a ballet as someone keeps changing the dancers.

Bob Boilen, who has directed many live events, says he always likes to have some sort of roadmap from the program producer "even if it's just written up in pencil by a producer sitting next to me, just so I have a sense of where we *may* go." As a director, you'll want to know in what order the producer imagines the guests will appear. After a presidential address, for instance, you might go to the White House reporter first, then both analysts, then a senator on Capitol Hill; you'll need to relay that information to engineers and others so they can make sure the sources of audio are brought into the control room in plenty of time. "You have people in Master Control switching in audio, you have phone calls that need to be made, you have backfeeds that need to be set up. As much as you can tell people, 'We may go here,' they can at least start setting things up," Boilen says.

Once all the necessary connections have been made, label any switches and buttons accordingly, so you know which of the many remote or phoner inputs will allow you to talk to which guest. "When the show starts, most of your energies are spent trying to figure out who is where," Ned Wharton says. "You need little reminders taped to each switch—rem 1 [remote line #1] is Poggioli, phoner 1 is a senator, phoner 2 is this other senator. Then you take them off and are constantly updating who is where."

As with any live interview, you want to talk to the guests off-air to make sure they are listening to the program and have at least a rough idea of when they will be called on. Van Williamson describes how he gets guests ready. "I'll say to them, 'This is the director. You'll be talking to Steve [i.e., the host] in about five minutes. First we need to get a level on your voice. So could you just talk to us for a minute.' And then I'll just say, 'Okay, you're going to be talking to the host in about three minutes, I'll get back to you when you're thirty seconds away.'" In a fluid program, like a live event, those three minutes can stretch into six or nine or twelve; you need to stay in touch with the guest so that he doesn't conclude he's been cut from the show. "I always hated it, and it used to happen a lot, when the first thing the host would say was, 'Senator? Are you there? Can you hear me?' It was an insult to me as a director," Bob Boilen says. "But in order to make that stop, I had to be really sure that the person would be there."

Things can get even more complicated during "rolling coverage"—when a natural disaster, the crash of a spacecraft, or some other major news event leads to nonstop programming. Then you may be lucky to get a scribbled-out roadmap for even the next half hour. "I remember arriving at five o'clock to work on a Sunday morning, and the news had come out that Saddam Hussein had been found," Ned Wharton recalls. "It became clear that we were going to go on the air an hour early, and I ended up directing for seven hours straight, without a break." During rolling coverage, you need to stay extremely flexible, and exercise what Bob Boilen calls "peripheral hearing." The program producer may be lining up guests or dealing with the hosts, and sometimes forgets to tell the director what the plan is, so you may need to eavesdrop to know where the show may be heading. "If there's an explosion somewhere, you get a sense of where it is and if it's a place you might get audio from."

When big news breaks, many people descend on the control room, in addition to editors and program producers. You'll find everyone from interns to senior executives (sometimes with friends or relatives trailing behind) asking to get an inside look at how the coverage is going. Some of them will even offer programming suggestions. ("Aren't we hearing from too many people with similar views? Shouldn't we try to get an opposition person on the line?") The chorus of voices will only add to your confusion. "This is where the program producer's role is just vital," Ned Wharton says. "The most frustrating thing is when you don't know who's calling the shots." So make sure you determine as soon as possible who you should be listening to, and who you can ignore. Since you all will be working without scripts, you may find the best strategy is to take your cues from the host—and not vice versa—about whether an interview should continue or be cut short, whether it's time for a recap of events or to go live to a politician's speech or news conference.

The Joy of Directing

Directing calls on an unusual combination of skills. You have to be as fussy and organized as a CPA, so that you can make sure all the program elements add up to the right amount of time; you should have a sense of rhythm, so you can weave the music artfully into the program, knowing when to let it "breathe" or when to pick up the pace; you must be able to stay calm when things go wrong, and reassure the people around you that you're in control; you need to know when to joke around and when to get serious; and you have got to develop sharp journalistic reflexes, so you can shift gears quickly during live coverage as the news evolves. On many days, it looks to outsiders as if the director is walking on a tightrope without a net. "I've shaken the hands of many guests who have sat in on the show, and their hands are always sweaty," Bob Boilen says. "They're scared!"

And yet, most longtime directors love the job.

"It's exhilarating," Ned Wharton says. "It's exciting when the pieces all fit together, when the music sounds perfect and it's packaged in a pleasing way. Choosing the music and directing the show is so enjoyable—you still have a hand in music and you have a hand in the sound of the show." Carline Watson says that even a rocky show can have its high points—usually when it's all over. "On the days when you go to the studio and everything runs like clockwork, it's a beautiful thing. But the days when you go to the studio and things fell apart, but you were able to salvage the show, and the people listening outside didn't know that it wasn't running like clockwork—that's also a beautiful thing."

ut while the Internet has given previously ephemeral radio stories —and seemingly perpetual—life, it has also presented some unex- ed challenges. Public radio journalists now face editorial, production, ethical questions that were never relevant when they were concerned with audio. Among them: Does a host's byline on the Web suggest r she has written the story, when the radio version may actually have n "tape and copy" written by a staff producer and only presented by host? Does providing a Web link to an organization imply an endorse- t of that group? Since a podcast may be downloaded days after it is lable, and listened to days after that, how does an editor ensure that ven podcast remains accurate? Can a slideshow on the Web have a sic soundtrack, even if we would never consider putting music behind voice tracks of a reporter's on-air piece?

NPR and its member stations are only beginning to answer these sorts uestions. But already it's clear that digital media turn some of the axi- s of conventional radio production and transmission on their heads. ple now can listen to pieces and interviews completely divorced n the shows on which they were broadcast. They can go backwards ime—by hitting the replay button on their audio player. A program ck doesn't govern the length of a story when it appears only on the b or in a podcast. News reports often include pictures, and lots of er visual elements. We frequently have headlines on the Web, as well story summaries and extended text. And spelling does matter.

ie Web-Radio Relationship

an online search for an important news story, and you'll find variations it on hundreds of Web sites representing all sorts of organizations— m the *Rutland Herald* in Vermont to Al Jazeera (produced in Qatar) the *Washington Post*. News is everywhere on the Internet. So it's fair to k how a radio network or individual station can distinguish itself from the other online news sources.

For one thing, it can use the Web as a vault for storing audio. That as one of the first ways public radio took advantage of the Internet; the rliest NPR stories online date back to 1996. "It's still the single most portant thing that we do," says Jeffrey Katz, who was a print reporter r twenty years before he came to NPR, first as a radio editor and then work for NPR.org. Today, a listener who hears an *All Things Considered*

CHAPTER 18

Beyond Radio

As the twenty-first century was approaching, organizations were forced to evolve at a been inconceivable just ten years earlie when NPR was created). The tremendous expansio the sophistication of search engines and the rise of aggregators," allowed people to get the latest new from across the globe as easily as from their local rad per. The invention and proliferation of portable MI designed as devices for holding music—unexpect podcast, which let anyone with a computer and m his or her own programming. Web logs rapidly mu personal diaries (although there are still plenty of sider reports on everything from an intern's life at a l Baghdad citizen's view of the war in Iraq. The ubiqui wireless connections, of digital cameras and cell pho for people with no previous journalism experience contributing to or even breaking a big story.

Today, a radio journalist often carries a digital cam camera, as well as a MiniDisc or flash recorder and ers may be asked not only to take photographs, but audio for the Web, to write a sidebar to be posted or in a podcast. As a result, a radio reporter who may be producer for the piece slated for *Morning Edition* or *D* laborates with an *online* producer for the Web versior journalists who have spent decades learning how to ears" are now thinking about the best graphs, docum video, or Web links to flesh out their reports. For m sponsibilities are also liberating: reporters working or need to worry about squeezing their reports into pre The Web is vast, and expandable; bits and bytes are ch story on a new treatment for a rare disease can find the audio of the piece online, even if he doesn't know the name of the reporter, the show that broadcast the story, or the date it aired. If he can't recall the name of the disease, he can try searching by the reporter's name or by the program title. "It's sort of like the old days of the library, when you could look for a book by the subject, or the author, or title," Katz says. "It's the same way [on the Web]: you can get it by topic, you can get it by the reporter, the show, the day—it's all there."

The Internet can also be a way to distribute radio-style reports that are not actually broadcast. Doug Mitchell has coordinated workshops at universities and conventions where students and young journalists experience what's involved in reporting and producing for public radio; a similar project for NPR interns is called "Intern Edition." In both cases, participants learn how to shape story ideas, find sources, record interviews, write a script, and read it aloud. The finished pieces then go on the Web. Originally, Mitchell says, the Web helped market the training programs. "It was sort of guerilla grass-roots marketing. Let's put it up on the Web, and the participants can tell their friends, and they can say, 'I want to do that, too!'" Today, radio producers and editors who listen to "Intern Edition" online frequently hear something that's airworthy—and invariably one or two of those stories end up on the network news magazines, or get broadcast by member stations.

Today, digital media can extend and support radio programming, and at the same time offer users material they won't find on the Web site of even a top-tier newspaper. For editors, producers, and reporters who are moving beyond radio, there are a few key ideas to keep in mind:

USE THE WEB AS A PLACE FOR AUDIO THAT DOESN'T GET ON THE AIR BECAUSE PROGRAM TIME IS SO LIMITED. Recorded interviews with newsmakers—which usually have to be cut to between three and eight minutes to fit in program segments—may be posted online at much greater length. A reporter may even decide to do a story exclusively for the Web, using material that was left over from the original reporting. For example, when Ellen Weiss was heading NPR's national desk, she edited a series of reports on solitary confinement in the U.S. "We ended up doing a radio profile of a prisoner who had been in solitary confinement for eighteen years," she recalls. "We originally thought we were going to broadcast it with a story about a guard who had also worked in solitary confinement. But his side of the story just didn't work that well on the air, and they didn't seem to work together." So the reporter, Laura Sullivan, wrote and voiced a piece that "aired" only online.

USE DIGITAL MEDIA TO EXTEND AND ENHANCE THE BROADCAST VERSION OF THE STORY—"to be a service arm of radio," as Katz puts it. When you conceive of a Web build-out of a story, remember that it can take many forms. On one typical day, the lead *All Things Considered* story on the U.S. response to a North Korean nuclear test was represented online by a few paragraphs of text and links to related stories that ran the same day on other shows. On that same program, there was a piece about special interest groups urging judicial candidates to answer written questions about where they stood on controversial issues; the Web page for that story included an adaptation of the radio script, plus downloadable versions of questionnaires from three different states. A broadcast story on "graduated licensing" of teenage drivers—imposing driving curfews, limiting the number of allowable passengers, or banning cell phone use until they have gotten some experience—appeared on the Web with text, a photo, a chart from the National Highway Transportation Board, and nine links to related stories about teen drivers dating back six years. And a host's interview with an author showed up on the Web with a long excerpt from the book, as well as a sidebar on one of the book's subjects. All of the Web pages included icons that took users to the audio for the radio stories, in case they missed the original items, wanted to hear them again, or came to the Web pages without tuning in first to *All Things Considered.*

ORGANIZE WEB PROJECTS THEMATICALLY, AS "SPECIAL REPORTS." "Rather than thinking of a one-to-one relationship between a story that airs and a Web page that goes with it, we're thinking more of the overall coverage," says Maria Godoy, who in her role as a Web producer has worked primarily with radio journalists covering national news. She ticks off some examples: A special site devoted to election coverage, featuring contested House and Senate seats around the country; one on the history and powers of the National Security Agency and the issue of warrantless wiretaps (which was big news in 2006); another on whether the Founding Fathers intended the United States to be a Christian nation, with excerpts from and links to various historical documents. Laura Sullivan's three radio pieces on solitary confinement spawned a special online report on the subject, Godoy says. "Sometimes when there's a special issue of a magazine, the first page will be a letter from the editor. Well, that was concept we had of how we could tie all these stories together. We gave people an overview about what prompted Laura to do this series, and what the conditions were in prisons—an overview of issues that only emerged slowly in the stories, that were told through prison visits." The text and audio of the Sullivan radio reports were only part of the Web coverage. "We also did a Q&A with the U.S. director of Human Rights Watch about concerns human rights advocates have about the practice. And we had something from a prisons expert at the Cato Institute about why the U.S. still needs to preserve [solitary confinement]—even if it's changed, why it's of value to law enforcement. And we had a history of solitary confinement—a timeline—because it might surprise people to know that it started as a Quaker practice in the 1820s."

START THE WEB-RADIO COLLABORATION AS EARLY AS POSSIBLE. A Web build-out like the one on solitary confinement takes time to develop—there are facts to be uncovered, guests to book, and interviews to edit, just as there would be on any radio program. Ensuring the Web story's success requires close and early cooperation between an online producer and the reporter and editor from the radio side—in the same way a good field producer works alongside a correspondent or host.

That similarity isn't accidental. Jeffrey Katz says he and other people who work in digital news media "think of ourselves pretty much as a *show*—that we're not there as an audio repository only, we're not just there only as a service arm." An online show might center on an issue, like the tension between national security and civil liberties, or it might focus on a breaking story. During the 2006 Israel-Lebanon War, for instance, Mary Louise Kelly briefly moved from her usual position as an intelligence reporter to the Web side of NPR. "Mary Louise did a particularly good job of pulling together what was in the news, what was on the wires, what was going to happen next, and most particularly, what was on our air," Katz says. "And because it was such a big story, we were able to connect four different audio pieces at the very top [from reporters in the Middle East and in Washington]." Kelly updated the site that same day—and again the following day. "For someone who values NPR's expertise on a big story, this was an incredible resource," Katz says.

AS A RADIO PRODUCER OR REPORTER, THINK ABOUT THE ASPECTS OF YOUR STORY THAT ARE BEST SUITED TO DIGITAL MEDIA. If you are covering a high-profile trial, you may be able to file a dispatch describing the mood in the courtroom, or the security procedure; if you're on the road with a congressman, you might want to write about all the different foods the politician had to eat in his weeklong campaign swing. The great advantage of digital media is that you can cover a story in whatever depth is appropriate; there's no such thing as a fixed segment length, as there is on the radio, and no "budget" for news stories, as there is at a newspaper. "The single most frustrating thing I faced as a print reporter was the news hole," Katz says. "I remember various times, when the cost of newsprint went up or advertising revenue went down, that suddenly instead of having twenty inches on a daily story and thirty on Sunday, now it was fifteen inches daily and twenty-five on Sunday. Well, it's never an issue on the Web. The news hole is what we make it."

MAKE SURE LISTENERS KNOW WHAT THEY'LL FIND ONLINE. Someone who's just heard a five-minute report on the air isn't likely to go to his computer if you only say, "For more on this story, go to our Web site." So let listeners know what they'll find: "To find out whether your hometown is one of the hot real estate markets . . ." or "We've put pictures of Marisa and the other orphans mentioned in this report . . ." or "To see whether the congressional race in *your* district is considered a toss-up . . ." And if you're posting an extended version of an interview, don't just say that it's "longer," but give people an idea of what they might hear online: "You can find out what the Defense Secretary thinks about the military budget, and hear his position on reinstituting the draft . . ."

The Manifold Media of the Web

It is hard to imagine any radio network or station mounting a Web site that didn't make good use of its audio. Public radio's long experience at recording and mixing sound is what differentiates it from other news organizations. But for a number of reasons, simply linking to the audio version of a report or interview doesn't serve most Web users very well. "People who are looking at the Internet on the job often can't listen to audio at their workplace," says Maria Thomas, who joined NPR in 2001 to head its digital media operation. They may fear that the sound will disturb the person working in the next cubicle or the corporate IT department may not allow them to download any audio players. Even if they *can* listen at their computer, Web users may not have the patience to do so. On the Internet, people tend to "graze," Jeffrey Katz says. "You're often looking for something particular, and you want to get right to it.

Users don't necessarily want to commit to hear a seven-minute or twenty-minute segment."

Fortunately, there are ways to adapt radio news reports for various digital media to make them more appealing to online users.

FIND NEW WAYS TO USE RECORDED AUDIO. Look for what Katz describes as "easier entry points" into the sound. For instance, the NPR Web page designed after Coretta Scott King died in 2006 not only featured a twelve-minute interview recorded several years earlier; it also incorporated a sidebar, "In Her Words." Users could sample a number of short audio clips, captioned, "King speaking at Memphis rally in April 1968, after her husband's death," "On why she married Martin Luther King Jr.," and "Civil rights struggle was a 'commitment,' not a 'sacrifice.'" These sorts of short sound bites are often just what users are looking for online. "To me that's a much more rewarding way to spend time on the Web than making a big commitment to listen to a long piece that may or may not give you what you want to hear," Katz says.

PROVIDE TEXT VERSIONS OF THE RADIO SCRIPTS—AND MAKE SURE THEY ARE ADAPTED FOR THE WEB. Because some users come to NPR's Web site through search engines or via a news aggregator, they're not expecting to *listen* to a story—they want to *read* it. At NPR.org, some of the online adaptations are only a few paragraphs long; others run 700 or 800 words. In general, more text is better. "When I go to the Web, I love being able to read the text of the story," says Ellen Weiss, who has spent her whole career on the radio side of the business. "There's a huge difference between being able to read a story in two minutes and listen to it for eight."

When you put the text of a radio story on the Web, you have to do a lot more than just transcribe the actualities, according to NPR Web editor Todd Holzman. "The best radio scripts are tailored to the voice and delivery of the individual correspondent," he points out. "Key points may be carried by inflection. Phrases are conversational and not always as precise as we might want for text." The Web text can be more precise and less redundant than the radio script. "Since a reader can stop and return to an earlier paragraph, or pause and reflect, there is less need to reemphasize key points, such as people's full names." Holzman says writing for the Web isn't even the same as writing for a newspaper, because people scan the Web more than they do printed text. "We shoot for shorter paragraphs and more opportunities to break away cleanly with a sense of the story. Long introductory paragraphs do not work, for the most part." That sort of concision can be a particular challenge for public radio, which often deals with serious issues in great depth. "We're dealing with some tough concepts at NPR," Jeffrey Katz says, "whether it's things happening in the Middle East, or immigration, or whatever. But I think we still have to give people information in digestible chunks."

IDENTIFY SIDEBARS CLEARLY. "When we have one big story and little sidebars, I want those sidebars, those inserts to tell you right from the get-go how this piece, this sidebar, differs from what else is on that page," Katz says. "In a couple of well-chosen paragraphs, you give them that summary, and then 'Click here to read more.'"

In September 2007, when General Motors reached a tentative contract with the UAW after the union staged a one-day strike, the reporter's *Morning Edition* report was only the starting point of the NPR Web version of the story. The online account included one sidebar on the details of the agreement ("At a Glance: GM's Deal with the UAW"), another on its possible ramifications for other automakers ("GM Labor Deal May Change U.S. Auto Industry"), and a third, titled "In Depth: The Man Leading the UAW," about union leader Ron Gettelfinger. That item included links to a Gettelfinger profile and a Q&A about why the negotiations were considered among the most important in the history of General Motors.

The ability to write strong, accurate, and catchy headlines—a skill few radio reporters and editors needed to acquire in the past—plays a crucial role in directing Web users to particular stories or sidebars. Here again, the Web reveals itself as a unique medium that calls for its own style of writing. "One of the things Web technology both permits and encourages is multiple use of the same information in multiple displays," Todd Holzman observes. A headline written for an item on new dental treatments, for instance, may not only show up on the *Morning Edition* Web page for the day on which the piece was broadcast, but also in a special online report on health care technology, as part of another Web special on medical spin-offs from the space program, on yet another page about new applications for plastics, and so on. "A newspaper headline that is just perfect for a given page—in context with photos and text, perhaps a sub-headline—might not permit multiple uses elsewhere on a Web site or in syndication on partner sites." In other words, Holzman says, "a good

Web headline is a multipurpose headline that tells the story in a functional way. And it's probably not too clever by half, although it can still have personality."

ADOPT A CONSISTENT STYLE. Web text has to be stylistically consistent—unlike a radio script, whose appearance doesn't matter as long as it makes sense to the reporter, editor, and studio engineer. A script may have idiosyncratic punctuation, pronouncers, and production instructions, such as "***Hit ambi cold (donkey braying) and bring sheep bleats under next trak***." The reporter may put all of her actualities in uppercase to distinguish them from her voice tracks. There may be misspellings, or abbreviations that only the reader and her editor understand (e.g., "wh" for "the White House"). And one person's style may differ completely from another's. None of this is acceptable on the Web, where text needs to have a consistent look and style.

To make that possible, NPR Digital Media hired its first copyeditor in 2006, and has developed an online style guide for its Web site to augment the *Associated Press Stylebook*, and to reflect decisions made by editors and management. It deals with many technical issues, but also with spelling ("Airstrike—one word"); capitalization ("Cabinet/cabinet—Capitalize references to a specific body of advisers heading executive departments for a president, king, governor, etc."); titles for NPR journalists ("Write around formalized NPR titles. Example: David Kestenbaum, a science correspondent"); and many other issues that never cropped up before about 2000.

MAKE SURE PHOTOGRAPHS HELP TELL THE STORY. Most public radio reporters now *expect* to be taking photographs when they're out on a story. Of course, not every radio journalist is a good photographer, and a successful Web site doesn't post photographs that are dull, poorly composed, or out of focus. "If it's a bad picture, we're not going to run it," Maria Godoy says. "If it were a bad piece of audio, you wouldn't put it on the air. So you don't want to use just any snapshots." Radio reporters should get at least basic training in photojournalism if their pictures are going to be used online.

USE GRAPHS, Q&AS, AND PRIMERS. A photograph may not be the appropriate illustration for a given story. Godoy gives the example of a reporter's profile of David Swensen, a Yale money manager who reaped an average 16 percent annual return over twenty-one years—better than any portfolio manager at any other university. When the reporter asked her what she might want to accompany the Web version of his story,

Godoy immediately asked for a Q&A listing some tips for average investors. "That's what listeners are going to want," Godoy says. "If he hadn't asked me, he might just have taken some pictures. But you don't care what the guy *looks* like—you want to know how he's going to help you make money!"

Q&As and primers are hallmarks of public radio Web sites; they fill in the spaces in a news story by adding details on related subjects, and by answering questions that might occur to users. The main NPR Web page on bird flu, for instance, not only lists more than fifteen broadcast stories—the number grows as more pieces air—and includes links to outside sites, like the Centers for Disease Control and Prevention and the World Health Organization; it also features a Q&A by two science editors, written specifically for the Web, on the potential for a flu pandemic and the status of influenza treatments and vaccines. Other Q&As on the same page were transcribed from broadcast interviews, in which experts fielded questions sent in by listeners. That "Ask the expert" format also works well on the Web; after *Morning Edition* broadcast a report on treatments for back pain, NPR.org took in questions from listeners and produced three online Q&As—on pain management and the back, physical therapy and back pain, and back surgery. These and other items were all part of a Web special called "Ah, My Aching Back!"

TAKE ADVANTAGE OF "HYPERLINKS"—links in a document to information that's either within that document or in another document. One part "Ah, My Aching Back!," for example, is an interactive feature called "Where the Back Fails," where a user can examine a detailed picture of the lower spine. Three sections are labeled "Spinal Stenosis," "Degenerating Spinal Joints," and "Ruptured Disc." Click on a label, and it brings up another screen with an illustration and additional text. Try that on the radio!

COMBINE AUDIO WITH OTHER MEDIA. Some of the most effective Web presentations combine pictures or video and audio—taking advantage of public radio's experience in radio recording and production, and the Web staff's knowledge of what makes a forceful visual presentation. In September 2006, for instance, NPR's Lourdes Garcia-Navarro filed a *Morning Edition* report on Michoacan, the Mexican state where tens of thousands of people live off the drug trade, and where hundreds died in a single year in drug-related violence. For a narrated slideshow that accompanied her story on the Web, NPR hired a professional photographer, and Garcia-Navarro provided the voice tracks and ambient sound. "Can the *New York Times* and the *Washington Post* do narrated slide shows? Absolutely," says Jeffrey Katz. "But what do we have? We have better audio than anyone else. The same sort of radio skills that we value so much on the air are just as valuable on the Web."

Even better: combine audio with pictures *and* text. The special Web site "Katrina: One Year Later," launched in the fall of 2006, illustrates how a broadcaster can exploit all of the different strengths of the Web. The site included photos and links to NPR stories broadcast around the anniversary of the hurricane. There were entries from the blog Noah Adams kept when he was in New Orleans. ("My final visit today was to the Fair Grounds Race Course in the Gentilly neighborhood. The 'sad Fair Grounds,' is the way I've come to think about the place, flooded, with plywood replacing big sheets of grandstand glass. . . .") *Morning Edition* host Steve Inskeep, who did a lot of reporting from New Orleans in the previous year, narrated an audio slide show—again mixing ambience with the voice tracks, and relying on a professional's photographs. There were other features designed specifically for the Web, among them an interactive map of the city's levees; a feature "Tracking the Diaspora," looking at where the evacuees were living after a year; and a sidebar called "Envisioning the Future," with essays by five historians, writers, and area residents imagining what New Orleans will be like fifty years from now.

"REIMAGINE" THE RADIO STORY. For Web producers, Maria Godoy says, the goal is not just adapting what's on the air, but "reimagining how to tell a radio story on the Web." For the fifth anniversary of the September 11 attacks, NPR produced a series of radio profiles of people whose lives were somehow transformed by 9/11. Each feature was available online on its own page—but NPR.org found a way to knit all of the stories together. "What we did for the Web," Godoy explains, "was to send out a professional photographer to take pictures of these people—mostly in New York but in other parts of the country, too—and then I worked with each of the reporters to get just one twenty-second sound bite that captured their story." She then created a multimedia slideshow by combining the actualities culled from the radio pieces with the pictures taken by the photographer. She added music and captions ("Mohommed Razvi ran a small family-business empire in Brooklyn. . . He gave it all up to become a community activist") to create something that borrowed from the radio stories, but was really born on the Web.

Podcasting

One of the digital innovations that coincided with the start of the twenty-first century was on-demand audio, or podcasting.[1] Public radio already had the infrastructure for recording and digitizing its programming; it was a relatively small step to make it possible for users to "subscribe" to individual shows, so that programs could be downloaded automatically at regular intervals.

It took only a few months for NPR to become either directly or peripherally involved in the production and distribution of all sorts of syndicated audio. Users can now sign up for weekly feeds of shows broadcast by NPR or its member stations. (They can also subscribe to *All Songs Considered*, the first NPR program *designed* as an Internet, as opposed to radio, broadcast.) Other podcasts consist of stories that have been broadcast on the air, but never in the same sequence in which they are heard on the podcast. NPR introduced a thirty-minute podcast consisting of the most emailed stories of the day, culled from various programs; it also created podcasts on movies, books, technology, business, and many other subjects, each one a compilation of previously broadcast pieces or interviews. Several NPR staff members put together original podcasts. In 2005, reporter Mike Pesca began a five-minute podcast on gambling; in the run-up to the 2006 elections, Washington editor Ron Elving and political correspondent Ken Rudin collaborated on the weekly podcast *It's All Politics*. NPR also started distributing podcasts—like *Benjamin Walker's Theory of Everything* and *Love and Radio*—created by independent producers.

Today the podcasting landscape is varied, vast, and growing; it may be unrecognizable in a decade. But people involved in podcasting say there

1. Who knows whether this term—a portmanteau word combining "broadcasting" with Apple's iPod—will survive? Digital Media Vice President and General Manager Maria Thomas recalls that NPR considered alternative descriptions for its on-demand audio—"pubcast," for instance, or "audiocast." But she says, "We decided that the word 'pod' had already come into some level of market acceptance to mean 'MP3 player,'" even though it was etymologically tied to a particular device.

are a few things that any public broadcaster should keep in mind as it expands into this and other digital media.

STICK TO YOUR CORE VALUES. Eric Nuzum—who helped to launch and expand NPR's podcasts—says the diverse programs and features offered as podcasts share the same values that are at the core of the public radio's news magazines and entertainment programming. Among them: "the authenticity of what we do; that we're trying to find the 'why' of things; that the personalities of those involved are secondary to the personalities of those we're trying to illuminate—it's not about us, it's about them; and that people's ideas have value—that people can disagree, but they do so in a way that's not dismissive."

DON'T COMPROMISE ON AUDIO QUALITY. The *sound* of these podcasts also tells listeners that there's a connection with public radio. They are professionally recorded; the hosts seem comfortable behind a mic; often the podcasts include interviews or actualities. That production quality should never neglected, says Stacey Bond, a twenty-year radio veteran who now teaches people to produce "content" on air and online for public radio stations. "Ever since we had the first recording—'Mary Had a Little Lamb,' recorded on a piece of tinfoil—we've been evolving the craft of audio. Why abandon all of that now just because the delivery system is different?" Bond says some people may be willing to put up with garage-quality recording if they're particularly interested in the topic, "but I think as more and more people start listening to audio content via podcast, the tolerance for poor-quality programming is going to erode."

DON'T FORGET WHAT RADIO HAS TAUGHT US ABOUT KEEPING LISTENERS' INTEREST. Bond also teaches would-be podcasters not to assume that they have the full attention of their listeners. "Just as with radio, they might be getting their children ready for school, or getting in their car, or making some bacon, or walking in and out of their room getting dressed," she says. "They may be listening on earbuds or on their computer, but they're doing other things while they're listening." So she trains podcasters to write for the ear, to start with a billboard, and to make sure their stories have a beginning, middle, and end—to "take the listener on a little journey."

FIND YOUR OWN VOICE. If all this sounds like the prescription for producing any radio story, it is—with a few differences. One is tone. When NPR editor Beth Donovan pitched the idea for the podcast *It's*

All Politics, she knew she didn't want it to sound "like something you would hear on NPR." Her favorite podcasts always sounded casual and spontaneous—"two guys sitting around shooting the breeze about, say fly-fishing, or whatever they liked"—and she wanted to capture that same informality in a feature about politics. Her goal was to make it sound like the listener had just dropped by the office and was hearing these two political junkies chat together.

The result was a feature whose subject matter could be heard on many NPR broadcasts, but whose style and pacing were decidedly zippier. In a typical podcast, Ken Rudin and Ron Elving interrupt one another, they digress to mention political trivia, they sometimes make puns or self-deprecating comments—even as they examine and discuss the biggest political stories of the week. In this podcast, for example, they riff on a sex scandal involving a member of Congress:

ELVING: Welcome to this week's episode of *It's All Politics* from NPR News. I'm Ron Elving.

RUDIN: And I'm Ken Rudin.

ELVING: And this has been perhaps the most remarkable four or five days of political turnaround and turmoil that I have seen in Washington in 22 years!

RUDIN: You know, the Republican Party—the President of the United States, has been trying to sell this election, this midterm election, as one on which party can better conduct the "war on terror."

ELVING: National security.

RUDIN: Democrats say, "No, the argument is really about the conduct of the war in Iraq—and throw in the economy, and [lobbyist] Jack Abramoff, and Katrina." But there's one little incident that happened with a member of Congress, a former member of Congress, from the 16th District of Florida, that has just reshaped this entire election.

ELVING: Well, we should say he's only been a former member of Congress since this Friday, when he was confronted by the instant messages he had sent to a former page...

RUDIN: And his name is—

ELVING: Mark Foley.

CATER TO YOUR AUDIENCE. From the start, *It's All Politics* differed from most of the political discussions on the radio news magazines because it presumed that its audience was *already* interested in the subject. That allowed the hosts to bring up fine points of a story that might be edited out of a two-way on *Morning Edition* or *Day to Day*:

RUDIN: There is an interesting sidebar—what happens to his seat in Florida—

ELVING: West Palm Beach.

RUDIN: The fact is, Mark Foley's name remains on the ballot, and the Republican—the new Republican-designated nominee—has to basically urge voters—has to say, "Please vote for Mark Foley"—

ELVING: "Vote for me, but you'll have to mark the name 'Mark Foley'"—

RUDIN: "—if you want the seat to remain Republican." And that's a—

ELVING: —a tough one.

RUDIN: An ignominious task.

ELVING: No one would want to be in that situation.

It's All Politics—or any other podcast on a specialized topic—is really a "narrowcast." The hosts can discuss things that a more general radio audience might not be interested in—arcana on a century-old special election, for instance. "Part of the idea for the podcast was to give political junkies that nugget, that one fresh fact that's the smartest thing they hear that day," says Beth Donovan. If you're producing a podcast for that sort of select audience, you don't need to provide the same degree of background information and context that would usually be added to a radio piece. "We do try to identify people and not be too oblique," Donovan says, "but we assume people are following the story and they do want the inside baseball." Stacey Bond, the podcast instructor, describes it as "niche marketing"—"It's a much narrower audience, but it's a more concentrated kind of audience."

REMEMBER THAT YOU, NOT A PROGRAM CLOCK, DETERMINE THE PROPER LENGTH OF A PODCAST. If you're a radio journalist who's just getting into podcasting, you'll be relieved to find the time slot is flexible; you're not confined by a segment, as you are on the air. "In radio, it's a crisis if you're five seconds long! And sometimes finding those five seconds is horrible," Beth Donovan says. "We don't have to do that on the podcast." But don't allow the length to fluctuate so much that users don't know what they're subscribing to, Stacey Bond warns. "The length can be whatever the content requires, but I don't think there should be a lot of variation. If you create an expectation in your listeners that a show is going to be approximately twenty minutes long, it would be weird to have a huge format breaker of a seventy-five-minute-long show."

BE SENSITIVE TO MATTERS OF TASTE. Podcasts can take on subjects and use language that would be taboo on the air. The Federal Communications Commission regulates radio broadcasts, and restricts those deemed obscene or profane. But in the online world, the syndicator, not a government agency, is responsible for policing what goes out in a podcast. "Our unwritten policy for podcasting is if it couldn't appear on the radio—the language, the subject matter—it should get an 'explicit' tag on it when it's listed online," Eric Nuzum says. "There's an implied understanding by the listener of what it means to be NPR, even though it isn't written down anywhere." Nuzum doesn't mind podcasters using the sort of language you might hear on the street—the occasional four-letter word or crude expression—as long as it's there as an integral part of the story and not merely for its shock value. "The decision about whether you would have profanity or vulgar subjects included is justified by whether I learn something about the world, or I understand things at a deeper level," he says.

Public radio producers involved in podcasting say they provide an inexpensive way to supplement the conventional broadcasts, and to establish connections with listeners who are passionate about certain subjects. There may be other pay-offs as well. "We're developing talent," says Nuzum, describing one podcast hosted by two recent college graduates. "I give them notes every time they do an episode, we have a lot of conversations—so they're kind of getting an education." In that respect, experimenting with podcasts now can be an investment in the future. "These guys could be great public radio producers some day—and the kind of producers we're going to need at that time, who see the world a little differently, but have an incredible amount of skill and have built up an editorial sense and know how to tell a story," Nuzum says. "We're teaching them how to be good at radio, according to the values we all share."

Editorial Issues Online

For the most part, journalists working in digital media confront and resolve the same sorts of editorial questions journalists have always grappled with—verifying allegations, fact checking, avoiding bias. But there are a few issues that arise on the Web that don't usually surface in radio broadcasts.

Take quotes, for instance. On the one hand, a radio journalist literally *records* what people say; his obligation is to make sure that an interview or actuality is never edited in such a way that it distorts the speaker's meaning. On the other hand, newspaper reporters often just take notes as they interview sources, and then reconstruct the quotes as accurately as they can (or limit the quotes to words or phrases that they know were uttered); the journalists are responsible for ensuring that an interviewee feels he or she has been quoted accurately, even if some of the words differ from what was actually said. But many stories appear on NPR.org both as the original audio and as text adaptations of radio scripts. That means listeners can easily determine whether any written quotes are actually paraphrases, rather than transcriptions of what was broadcast. (In fact, listeners can also obtain transcripts for a fee.) For that reason, online quotes excerpted from the radio stories have to be absolutely accurate, says Jeffrey Katz. "We shouldn't have to explain why a quote was different in one place than in another."

Similarly, visitors to the Web can often compare an interview they heard on the air with an extended online version. They can decide for themselves whether a producer took unwarranted liberties when he or she was cutting the interview for broadcast—by grafting together the beginning of one sentence with the end of another, for instance, or by moving the answer to one question up so it appears to be the response to another. As someone who has worked on both the radio and digital sides of NPR, Katz thinks having a longer online version of an interview may help to keep radio producers honest. "As a radio editor, I got pretty creative about ways I could get producers to edit interviews. And I have to tell you, in those cases where I knew we were going to put an extended version on the Web, I thought about it more," he says. He believes letting listeners see how the editorial process works—"opening the curtain a little bit"—is a positive development. "I think allowing some sunshine on how we do things is good. But we better make sure we have our act together when we show people that."

The use of stock photographs on the Web also can raise issues that are unfamiliar to radio journalists. There's generally no problem illustrating, say, a profile of a Maine lobsterman with a picture of the man in his boat. But how do you add a photo to a story on obesity? Or Alzheimer's? Or alcoholism? NPR.org's online style guide cautions Web producers not to use photos of an identifiable person—someone you would recognize by the picture if he or she were already known to you—and associate it with a situation "where it may not apply factually." It gives this example: "A middle-aged woman is not necessarily a middle-aged woman with a drinking problem just because she's holding a glass of wine. If it's a silhouette of a person who is not identifiable and she's holding a glass of wine, that's acceptable."

Web links can also be problematic. Users may hold Web producers responsible not only for the content they originate, but also for the sites they link to. So "link responsibly"; make sure you know something about the site you're sending users to. If it's blatantly commercial, obscene, or filled with unsubstantiated rumors, it reflects badly on you and your radio service.

Podcasts have their own set of editorial problems, especially since NPR syndicates some podcasts that it does not originate. Independent podcast producers may not feel the same editorial and ethical pressures that public radio has been confronting for three decades. But you should apply the same standards you would use on the air. For example, Eric Nuzum says he faced an issue that has also come up in radio programs—distinguishing a re-creation from an actuality. "*Benjamin Walker's Theory of Everything* occasionally uses fiction," he says. "So the question was how do we identify that as fiction—that this is something that's made up?" Nuzum says he decided they couldn't deceive podcast listeners any more than they would a radio audience, but they were careful not to be so ham-handed about labeling the fiction that it spoiled the story. A slightly different situation arose with the podcast *Brini Maxwell's Hints for Gracious Living*, which Nuzum describes as "a guy playing a character who's a woman from the 1950s, who gives household advice." The issue in that case was whether to identify the actor behind the character. "When we first started, we just said it was *Brini Maxwell's Hints for Gracious Living*. Then we realized this doesn't really follow people's expectations of us. We needed to identify it as 'Ben Sander as Brini Maxwell.'"

It can be harder to remove factual errors from a podcast than it is to correct mistakes on the air. On the radio, a misstatement can be corrected—immediately, when it's caught in time, later in the program or in a subsequent program if it's not. The first feed of a news magazine is updated as events change, and updated a second or third time, if necessary. (Only the most up-to-date feed goes into the online archives.) Podcasts also need to be accurate, even if that accuracy has to be achieved with the digital equivalent of a razor blade. "We never let a mistake go out in the podcast—ever," says editor Beth Donovan. "No matter what I have to cut out, we just don't do that." It may be even more difficult to spot and correct mistakes in podcasts from independent producers, since you may not have any way to determine if they are accurate. "If there is a mistake that gets by us, we can do a 'reburn,' and replace the podcast with a new one," Eric Nuzum says. "If someone has already downloaded it, they'll see it twice. If someone has not downloaded it yet, they only get the new one and they never knew the old one was even there."

Remember: online errors persist in a way that is unknown in live broadcasting. On the radio, Jeffrey Katz says, "if you make a big mistake, you might hear from listeners for a couple days; if it's small, someone would have to listen carefully even to hear that somebody misspoke. And usually, it's a little problem that day, or the next day." But he says editorial lapses on the Web seem to live on forever. "They're like little time-release mini-bombs that explode without fail every day!" Someone listening on the Web to a year-old interview or feature may complain about a grammatical mistake, or a host's failure to follow up a question, or a guest's referring to "Senator Jones" as "Congressman Jones." As Katz puts it, "Radio may be ephemeral; our audio isn't." And a Web page that was complete and accurate when it was originally put up may be out-of-date by the time a user finds his way to it, Katz says. "On the radio, no one would think of saying 'You need to update that story you did three years ago, because so-and-so has since died.' But we get these all the time!" Katz's staff fixes grammatical errors whenever they learn of them, and always removes outdated links, but he draws the line at updating every page in perpetuity. Just make sure every Web page has the proper dateline so users know when it first appeared.

The Future

Some of digital media's best-known creations—from the World Wide Web itself, to podcasts, to blogs—have come on the scene with little warning, grown rapidly, and evolved in unexpected ways. Anticipating online trends appears to be a risky and self-defeating undertaking. But there may be no choice: with public radio listenership starting to plateau, future audience growth may have to come via digital media. For both commercial and public news broadcasters, the pace of change is only likely to quicken. Digital Media chief Maria Thomas says managing that change successfully will require people to change their view of broadcasters generally and of NPR specifically—from a company that creates one product (radio programs) and has one distribution channel (radio stations) to a "multi-product, multi-channel company." A digital "product" might be a Web site, a podcast, or a blog; a "channel" could be anything from a computer screen to a cell phone.

Thomas foresees that news stories will increasingly be presented to online users a little at a time. "Today on the radio side, there's a lot of thinking around shaping the story, and molding the story, and choosing the bits and pieces of the story that are thought to be relevant to the listener," she says. "It's a very different mind-set to operate on a platform that's highly dynamic, where the expectation of the audience is that the story will change." She says people who get much of their news from the Web "understand that when an event happens, there might be just one line on a Web site, and then in an hour there might be five paragraphs, and in two hours two pages."

Posting pieces of a story as they are reported and verified requires more than simply a commitment of energy, especially if people have spent their careers in radio alone. People's jobs need to change; the table of organization has to be rearranged, Thomas says. For example, when a broadcaster puts out more stories as text, it needs a copy desk; as more people get their daily news via search engines, news aggregators, and news syndicators, radio programmers have to hire people with the knowledge of how to "optimize content" for those channels—to make their stories more prominent when people are looking for news on a specific topic.

As people get a growing share of their news by computer, wireless email devices, and cell phones, Thomas predicts that news stories will have to be "disaggregated"—taken out of their original context. This happened first in some podcasts, where items plucked from different programs were rearranged according to a common subject. In the future, the radio connection may be far more tenuous. A story filed for a particular radio newscast or news magazine may simultaneously show up—in text form—on a user's computer or cell phone, bundled together with items from other news organizations. That separation from any specific program—and even from the radio altogether—forces changes in the way stories are written, Thomas says. "Every story has to be complete unto itself." And each story needs to be "packaged" in a number of different ways, for the different channels through which it's distributed. Just as a radio pieces need to fit into the program "clock"—with its well-defined segments, cut-aways, promos, and station breaks—digital news has to be packaged appropriately, Thomas says. "Packaging in the digital media world is more dynamic, because there are more products and there are more possible packages. You can have the best content in the world, and if it's not searchable, not easily discoverable, not packaged in a way that's easily consumed, I don't think we're going to keep the audience—at least, not the new audience." What might a digital news package look like? If it is a story destined for a cell phone, it may be just 160 characters long—about the length of the first sentence in this paragraph. If it's a daily podcast designed for high school or college teachers to use in a classroom, it might consist of a half-hour of topical news reports.

Maria Thomas also foresees a growing role for users as *sources* of news, as well as consumers of it. In recent years, many radio programs have encouraged listener involvement by soliciting their questions for experts to answer, by getting them to recount first-person experiences in commentaries or on talk shows, and even by appealing to them to send in audio they had recorded on their own. NPR.org likewise interacts with people through program blogs. Minnesota Public Radio has gone much further with its Public Insight Journalism project, and has enlisted thousands of people as potential sources. MPR's Web site emphasizes, "We want to get to know you better so we can ask you about issues and events that you have experienced directly," and includes an online form where people can note their specific areas of interest and expertise. Members of MPR's Public Insight Network contributed to stories on how Minnesotans were faring financially, whether they were making adjustments to their lifestyles to try to reduce global warming, and many others. MPR also welcomes story ideas from the general population, with the qualification that "in some cases, your idea may be a good one, but the newsroom may have other priorities."

Maria Thomas thinks that listeners and users may be setting those priorities more and more in the future, and that media companies should be "thinking about how to engage the users *first* to inform the content." She gives the example of the BBC's coverage of the terrorist attack on the London subway system on July 7, 2005. "They got twenty thousand emails in a few hours, and many photos from people who were in the Tube, taking photos on their cell phones or their digital cameras." New technology makes it possible for any news organization to have thousands of potential citizen reporters in the field at any moment. If an earthquake were to strike a U.S. city, a hundred accounts from around the metropolitan area—with text messages and photos sent directly to NPR or its local member station—could present a detailed picture of the event before the usual wire services had even had time to run a story. "This is one of the really big ideas that's going to transform media significantly in the next five or ten years," Thomas predicts.

Such scenarios may send chills down the spines of seasoned radio reporters and editors. Will they have to spend much of their time in the future sifting through and trying to confirm unsubstantiated reports from thousands of individual citizens? Could a determined group of people foist a fake news report—replete with bogus photos and text messages "confirming" one another's "facts"—onto an established newspaper or broadcaster? Will reporters in the future have to file three or four different versions of each story—and if so, when will they have any time to report? If a story is going to be put on the Web a little at a time as it is reported, who will decide when it is finished? If all news stories are "disaggregated"—if users can order up only the stories they *want*—who will ensure people get the news they *need*?

Perhaps these worries will prove to be unfounded. After all, broadcasting is hardly obsolete. "I think the different media have a great opportunity to play together," Maria Thomas says. Indeed, Web editor Todd Holzman envisions the different media converging in a way that may breathe new life into radio, even as "broadcast" news moves into new channels. "The computer Web screen, per se, is not the future of digital media," Holzman says. "The portable device is the future of digital media—at least in the short term—and there is nothing more portable than good audio content. It's really the only content that is useful to people who also want to be doing something else with their time." The public radio enthusiast of the future may be able to read or listen to the latest news on a cellphone/MP3 player/email reader; the journalistic skills needed to produce accurate, engaging, informative, and inspiring radio programs today will stand people in good stead in the rest of the twenty-first century. "It's not so much about leaving the world of radio to work in the Web. It's finding a way for the world of digital media to extend radio to a larger, younger audience," Holzman says. "There are many ways to tell a story."

APPENDIX I

Glossary

Every field has its own specialized vocabulary. While many of the terms below would be familiar to journalists at most news organizations, others are unique to public radio or NPR. A few, such as those that mention tape, are reminders of a previous era, though you may still come across them.

acts and tracks: A simple **mixed** piece that does not include **ambience**. A reporter might say, "I'm sending you two pieces—but the first is just acts and tracks."

acts, tracks, and ambi: See **elements**.

actuality: An excerpt from interviews, news conferences, speeches, etc.—in other words, the recorded sound of someone speaking. Also known as *cuts* and, for the plural, *acts* or *ax*.

airworthy: Having sound quality that meets public radio's standards. An **actuality** might be deemed unairworthy because the sound was poorly recorded, because a phone line was distorted, or because the guest's accent made his speech unintelligible. A reporter may have to improve his delivery before his editor considers him airworthy.

ambience, ambi: Sustained background sound that is captured at a remote location. Ambience gathered at the site of an interview—whether it's a relatively quiet office or a noisy police station—permits an audio engineer to fade in and out of an **actuality** as it starts and finishes. For that reason, reporters are encouraged to provide a one-second "ambi **tail**" and "ambi **head**" to their **actualities**, and to provide an "ambi bed" of least a minute of ambience recorded wherever they conduct interviews. Much more ambience is needed whenever the sound is a key part of the report. Good ambience can be used to advance a story and to create a **scene**—we can *hear* the people on the lobster boat hauling in their catch, or the protesters trying to block the motorcade, or the confusion backstage before the performer starts her act. Also known as *natural sound* or *nat sound.* See also **sound effects**.

analog: Open-reel **tape**. The electromagnet inside a conventional tape recorder is an analog of the complex sound waves it captures—it responds continuously as the sound waves change—so audio tape is an analog medium. Compare **digital audio**.

backannounce: The **copy** that follows a report, commentary, or interview. It's usually used to identify the person we have just heard, or to add a bit more information to the report (e.g., "You can see a timeline of the Supreme Court's decisions on school integration at **NPR.org**.") Also known as *outro*, by analogy with **intro**.

backfeed: Involves sending audio from one **studio** to a remote location. For example, during an interview between a host or reporter in Washington and a guest in Denver, the guest's voice is **fed** to Washington—probably by **ISDN**—and the host's or reporter's voice is fed *back* to Colorado so the guest can hear the questions. Since the fidelity of the sound being recorded is what's important—not what the guest hears in his headset—the backfeed may be done over the phone. See also **mix minus**.

backtime: A calculation to determine when a specific part of a recording will be heard. It's a way of ensuring that a chorus to a song will come just when a reporter mentions it, the theme music for a program will conclude precisely at the end of the hour, and so on. The word may not be commonly used outside of broadcasting, but the concept is part of daily life. For example, a cook makes sure that all of the parts of a meal are ready at the same time by backtiming the preparations: he starts cooking the meat at 5, the dessert at 5:15, the vegetables at 6:00, so that dinner is ready at 6:30. In a radio **piece**, we might backtime the sound of a demonstration so that the chant, "Keep hope alive!" begins just after the **SOC**. See also **post**.

billboard: The first minute of each hour of most public radio programs, during which a host mentions some of the items to be heard in that hour, after the **newscast**. Also known as *opens.* Some programs include **teases** in their billboards. See also **lines**.

blade: To cut something out. The term is a vestige of the **analog** era, when razor blades were used, but still common at NPR. An editor might say to a producer, "See if you can blade the stutter out of that last question."

bleeble: NPR slang for the twenty-nine seconds of music following the **newscast**, provided as a **music bed** for local stations to voice over. Also known as *trixie*.

board: At NPR, the dry-erase board or Web page where the show producers map out their programs. They list host and producer assignments, interview times and **studios**, and the names of reporters and their editors.

The *board* can also refer to the audio console in a **control room.**

bridge: See **button.**

bust: To accidentally continue broadcasting a **segment** through a scheduled time post. For example, if a segment is supposed to end with two seconds of silence at 19 minutes and 58 seconds after the top of the hour, and the host is still talking or music is still playing at 20 minutes past the hour, the program has busted the segment by two seconds.

button: A bit of music played between **pieces** in a program. Also called *bridge* or *jingle.*

card: A four-ply index card on which NPR news programs used to keep track of who was being interviewed, at what time, and in which studio. Cards became obsolete when computers and laser printers came along, and all that info is now maintained electronically. But many producers still ask "Where's the *card*?" when they want the lowdown on an interviewee.

clip: To edit audio badly so that a sound at the end has been cut off abruptly. See also **ambience, upcut.**

Audio engineers also use the terms *clipped* or *clipping* to denote distortion in a recording.

clock: A graphic representation of a program's format—a pie chart in which each station break, **segment**, or **newscast** is a slice of the pie.

close: To turn off a microphone.

Close can also refer to the conclusion of a program.

coda: The final passage that ends a work of art. In radio, a coda may be **copy** read by a host after a report has ended or an **actuality** related to a story that comes after the reporter signs off—i.e., after the **SOC**:

REPORTER: Until five-year-old Matt Finish enters the Sunshine Academy for the Exceptionally Intelligent next September, his parents intend to keep teaching him every day at home—even at mealtimes. Russell Papers, NPR News, Boise.

AUDIO: (MATT AND HIS PARENTS) :45—**Backtime** so that the words "Time now for algebra, Matt" come just after the **SOC**; fade after the phrase, "Don't get peas in the keyboard."

codec: A box that converts an **analog** audio signal into digits that can be transmitted down an **ISDN** line. A similar box is needed on the

remote end of the **feed** to convert the digits back into an **analog** audio signal.

control room: Where the engineer, the director, and sometimes a producer or editor work during a broadcast, located on the other side of the glass from the **studio**. The protocol for both the studio and the control room calls for extraneous noise to be kept to a minimum.

copy: Any written material the hosts or reporters read on the air.

crash: To work on deadline—often an imminent deadline. If you're crashing, you don't have time for socializing; phone calls to reporters or editors often begin with the question "Are you crashing?"—and if the answer is yes, the caller apologizes and hangs up.

crossfade: In **mixing**, the process of simultaneously decreasing the audio level of one source and increasing the audio level of another—in other words, one sound goes away while another comes in. Sometimes a crossfade takes place under a voice track or **actuality**.

cross promo: A twenty-eight-second announcement at the bottom of the hour for another news program. See also **forward promo, jump promo**, and **promo**.

cue: The signal that an **actuality** is about to begin or a person should speak. When a reporter has to go live from a remote location, the producer preparing the **actualities** for his report has to make sure the director has cues to the **tape**—that he or she knows when to play the **actualities**. (Now that we're in the era of **digital audio**, the sound usually comes out of a computer, but many people still talk about "tape.") The cues are provided on a page that spells out the words the reporter will say just before the **actualities** are called for. For example:

CUE: ". . . spoke to reporters in the Rose Garden:"
AUDIO: 29
  i.c. "The United States . . ."
  o.c. ". . . of all others."

More generally, a *cue to tape*—sometimes abbreviated "Q"—is the last part of any **copy** that leads directly into an **actuality**.

curtain-raiser: A report that previews an upcoming event. An editor might say that a reporter is "putting together a curtain-raiser for next week's Democratic presidential nominating convention."

cut: See **actuality**.

cutaway: A part of a radio program designed as a module that can be removed by a station and replaced with local announcements, news, or other material.

cycle: See **news cycle.**

DACS [pronounced "daks"]: Direct Access Communications System—the computerized system that connects public radio stations. Stations often use the information on the **rundowns** for on-air promotion during the day. For that reason, information on the DACS should be accurate, grammatical, and ready for air. See also **lines.**

DAT: An acronym for **digital audio** tape. A DAT looks like a little videocassette; among other things, it makes it possible to record audio for up to two hours at a time without the tape hiss of old-fashioned cassettes.

dead air: Unwanted silence during a broadcast, usually a result of a technical, production, or talent error. Some stations have "silence sensors" that will set off an alarm if dead air persists for more than a few seconds.

deadroll: Audio that is started at a precise time, but not made audible until it is needed on the air. A deadroll is often begun so that it will conclude at a predetermined time. A good example is the theme at the end of most programs. If the theme is 2 minutes and 20 seconds long, and the program ends at 58:58 (58 minutes and 58 seconds after the hour), the music will be started as a deadroll at 56:38 (56 minutes and 38 seconds after the hour). The theme can then be faded up whenever the host has finished, and it will always end at just the right moment. See also **backtime, post.**

digital audio: Sound that is stored as a series of numbers. Conventional tape recorders are **analog** devices—the electric currents inside them vary continuously in response to changes in the frequency and amplitude of the sound waves captured by a microphone. Computers, on the other hand, are digital devices: they represent sounds as a series of discrete bits of information. Devices like **DAT** and **MiniDisc** recorders and digital audio workstations turn sounds into numbers by rapidly sampling the sound waves at specific intervals, and then reconstructing the sound waves from those digits.[1]

Direct Access Communications System: See **DACS.**

double-ender: See **sync.**

1. The greater the sampling rate and the more binary digits (bits) used to record those samples, the better the digital device can represent the actual sound. A typical sampling rate is 48,000 times per second. A computer that reserves 16 bits per sample can store any of 65,536 different numbers—more than enough to accurately represent the original sound.

downlink: An earth-based terminal capable of receiving data from a satellite. See also **uplink.**

drop-in: NPR jargon for a **piece**—often a commentary—that is broadcast without an introduction, in which the host introduces the speaker after a sentence or two. It's also sometimes called a *Who Dat?* If the beginning of a piece is a complete thought, a drop-in can be an effective way of changing the **texture** of a program. Here's an example:

AUDIO: :10

No holiday has suffered more from the combination of political correctness and bureaucratic tampering than the ill-named and ill-celebrated Presidents Day.

HOST: News analyst George Washington.

AUDIO: 2:21

i.c. "For decades, Americans happily . . ."

o.c. ". . . change it back. And that's no lie."

HOST: George Washington is an analyst at the Heritage Foundation in Washington.

D-sheet: NPR slang for a discrepancy report—a record of a **train wreck** or some other problem on the air.

dub: A copy of an audio recording, usually in real time.

dump out: To end a report or broadcast prematurely. For example, if a **studio** director realizes that a **piece** is longer than it's supposed to be and is therefore going to **bust** a **segment**, he may suggesting dumping out of the piece before the **SOC** and having the host provide a **backannounce**—e.g., "That report was from Oscar Meyer in Madison, Wisconsin."

echo: An inadvertent repetition of information—in an **intro** and report, an intro and interview, or a voice track and **actuality.** You can have an echo without repeating the exact same words. If an intro says, "Congressman Jones is the first Democrat ever to be elected in northern New Hampshire" and the reporter says "Northern New Hampshire has always been represented in Congress by Republicans," it's an echo.

Echo is also an audio term referring to an out-of-sync duplication of the primary audio signal. An echo of this sort can occur, for example, when the ghost image of a host or reporter's voice returns on an incoming **ISDN** line. This is most often caused by the remote guest's headphone volume being turned up too high and leaking back into his or her microphone.

elements: The component parts of a news report—the **actualities**, voice tracks, and **ambience** that will be **mixed** to make the final **piece**. Colloquially called *acts, tracks, and ambi.*

ender: A logical and elegant ending for an edited interview—which may or may not occur at the end of the interview when it is originally conducted. Program hosts and producers look for an ender by trying to think of a question whose answer will provide a satisfying conclusion, so that the edited interview does not seem to be cut off arbitrarily.

EQ, equalize, equalization: To adjust certain frequencies in an audio source in order to improve its quality. An engineer might EQ a recording by reducing some of the highest frequencies and enhancing those in the mid-range. See also **fix**.)

fade to black: To take a sound down and out during **mixing**. This can be done under a **track** or **actuality**, or used as a production device to indicate a change of **scene**. In the latter case, the fade to black is usually followed by a pause before a new sound or track is introduced—the shopping center **ambience** slowly fades out, and the reporter begins the next track by saying something like, "Across town, the police are checking Jim Smith's fingerprints against the ones they have on file . . ."

feed: To send **actualities**, voice **tracks**, and **ambience** to a station or network from a remote location. A producer or engineer then takes these elements of the report, arranges them in order, and produces the final **mixed piece**. Also said of entire programs—e.g., the second feed of *All Things Considered* begins at 6 PM Eastern time.

field producing: Producing with a host, reporter, or commentator at a remote site.

fish job: NPR slang for a report that is already **mixed** when it is **fed**, and that doesn't have any **pickups** in it. The producer's job in preparing it for broadcast is just to cut off the "**head**" and the "**tail**"—as you'd prepare a fish.

fix: The application of **EQ** or the adjustment of volume levels to improve the quality of a phone line, **actuality**, or other audio source. Audio **fed** on the **WAND**, for example, requires a fix before it goes to air.

flash recorder: A portable device for recording audio on flash memory cards. One of the advantages of using flash recorders is that **digital audio** files can be transferred directly and quickly from the recorder to a computer for editing. A twenty-minute interview might be copied from a flash recorder to a laptop computer in a minute or less.

foldback: See **talkback**.

format-breaker: At NPR, a report that is too long to fit within even the longest **segment** of a program. Format-breakers usually have to be scheduled in advance; the date of their broadcast is sent via the **DACS** to NPR member stations.

forward promo: A small block of **copy**, usually read at the end of a **segment** (before a station break, for example), promoting the stories in the next segment. See also **cross promo, jump promo,** and **promo.**

FTP: File transfer protocol, by means of which radio producers and reporters can send text or **digital audio** over a telephone, a high-speed line, or the Internet.

funder: An on-air underwriting announcement. At NPR, hosts or other journalists do not read funders.

head: The part of a quarter-inch, open-reel **tape** where the audio starts (archaic). A tape is *heads-out* if it is ready to play when it's placed on a tape machine's left-hand reel. See also **tail.**

Heads can also refer to the parts of a tape recorder used to record and playback audio. To edit audio tape successfully, it is essential to know which head is the playback head.

The beginning of a **digital audio** file is also called the *head.* A producer may tack on a little bit of **room tone** to the head of a file so that it can be faded in gradually.

hot: Loud; usually used to describe a **piece** of audio that is louder than other audio in the same piece or interview. A reporter might tell an engineer, "I've got four **acts** in this piece, but be careful because the second one is hot."

Hot can also mean "at full volume." A producer might want one piece of music to be "hit hot"—played at full volume from the beginning—and another to be hit **warm** or as a **sneak.**

i.c.: See **incue.**

IFB: See **talkback.**

incue, in cue: The first few words of an audio file. In a reporter's script, the incue indicates where an **actuality** should start. On a program script, the incue tells the hosts and the show editor how a report begins. It's often abbreviated *i.c.*—as in:

i.c. "For more than thirty years, NPR has . . ."

index mark: See **track mark.**

interruptible foldback: See **talkback.**

intro: The **copy** that a host will read on the air; it usually includes information for the program director about **tape time** and other peculiarities of the audio. See also **lede.**

ISDN (Integrated Services Digital Network): A service provided by local telecom companies that modifies regular phone lines so they can transmit data at high speed. ISDN allows a broadcaster to get high-quality audio from member stations, universities, and sometimes from hotel rooms or reporters' homes. ISDN capability can be found in places where you might not expect it—e.g., at big-city newspapers and nonprofit organizations. Because the fidelity of ISDN is so much higher than that of a conventional telephone, reporters and producers should always determine if an interviewee has access to an ISDN line.

jingle: See **button.**

jump promo: A small block of **copy**, usually read at the end of a **segment** (before a station break, for example), promoting the stories in the next segment. See also **cross promo**, **forward promo**, and **promo.**

kill fee: A payment made to a radio reporter, commentator, or producer whose work has been authorized and produced but for some reason ultimately rejected—or "killed."

lead: The beginning of a newspaper story. In a radio **piece**, the lead is usually the **intro**, since that's where a report actually starts. A reporter may also refer to his lead when he's talking about the first paragraph of his piece. Confusion about the meaning of the term can sometimes result in an **echo**. Also spelled *lede.*[2]

leader: White paper or plastic tape attached to magnetic recording tape that, among other things, allows it to be wound around the reels without being damaged (archaic). The difference in color between leader and magnetic tape indicates where the sound begins. Leader can precede the audio portion or follow it.

lede: See **lead.**

line: At NPR, a short description of a **piece** that editors provide so that program producers know what's in the works on any given day. Lines

2. *Lede* was originally a newspaper term, and is said to be misspelled to eliminate confusion with "lead," used in the days of hot metal typesetting. Other odd spellings or abbreviations, like *graf* (paragraph), or the printer's marks *dele* (delete) and *stet* (reinstate), have unusual etymologies; *stet* for instance, is Latin for "Let it stand." These curious spellings have been adopted by many radio journalists.

are used to create the **DACS**, and may also be relied on for writing **billboards**. For that reason, it's essential that the facts and pronunciations in the **lines** be accurate and up-to-date. If a story changes over the course of the day, the lines have to change as well.

live mix: The **mixing** of a **piece** at the same time that it is being broadcast. Mixing live obviously doesn't allow a producer to fix **pickups** that occur during the course of the mix.

live read: A live reading of the script by a reporter as the program is being broadcast, while the **actualities** are played off a computer. A live read is usually called for when a reporter has not had time to **track** his **piece** in advance. Tracking live means there's no opportunity to fix **pickups**; a reporter who stumbles or fluffs when he is reading live generally just has to plow ahead.

live roll: The insertion of a **feed** of a **mixed piece** directly into a program that is on the air. For example, if a reporter is at NPR's New York bureau and is not able to **feed** his or her piece before it is broadcast, the report may be played out of New York directly into a Washington **studio** and onto the air.

log: To transcribe audio. Some reporters routinely log all or parts of their interviews; others say they never log their **tape**, and instead rely on their memories (or the track marks on their **MiniDiscs** or flash memory cards) to decide which **actualities** to use.

lose: To fade a sound out. A **studio** director may tell an engineer, "Let's lose this convention noise when the host begins the next intro."

man on the street: See **vox**.

Master Control: The place at NPR or at an individual station that makes many of the connections with remote **studios** and other sources of audio.

MD: See **MiniDisc**.

MiniDisc: A digital medium that uses a laser to record and play back audio. Like a computer floppy disc or a hard drive, the MiniDisc uses a digital table of contents to keep track of data that may have been recorded in many different places on the disc. This random-access feature makes it easy for the MiniDisc recorder to erase and re-record quickly—the machine never has to rewind. Like **DAT** machines, MiniDisc recorders sample data at a high rate; unlike DATs, however, MiniDiscs store that data in a condensed form. This data compression can cause problems when audio is copied from one MiniDisc recorder to another. Often abbreviated *MD*.

mix: To combine two or more audio sources—either **analog** tapes or digital sound files—to make one finished **piece.** In an **analog** mix, an audio engineer would physically control the starting and stopping of tape machines and adjust sound levels according to a producer's directions. In a **digital** mix, the producer will arrange the **acts, tracks,** and **ambi** on a computer screen before he or she goes into the **studio.** The audio engineer in that case may still be responsible for improving the quality of the audio—putting in a **fix**—and adjusting the levels as the producer directs the mix. Many reporters mix their own pieces without producers or engineers.

mix minus: A **mix** that is minus one **element.** Every remote radio interview—whether it's on the phone, **ISDN**, or fiber optic line—includes a mix minus. The guest needs to hear everything that the host or reporter hears *except* himself coming back up the line.

Mix minus is also a way of setting up the audio **board** in the **control room** so that a producer, host, and interviewee can hear all of the parts of a **piece** or interview, even though only some of them are being recorded. For example, during an interview with a musician, a producer may ask that a CD be played in mix minus, so that the host and interviewee can hear the excerpts of songs that will be mixed in after the interview is edited.

MOS: See **vox.**

mult, mult box: A device that takes a signal from an audio source and feeds it to multiple jacks so it can be easily distributed. The presence of a mult box at a public hearing or news conference allows several reporters to record the event at the same time without each one having to put a microphone up on the lectern.

music bed: Music held for an extended period under a **track** or **actuality.**

natural sound, nat sound: See **ambience.**

news cycle: All of the programs in a given twenty-four hour period, which is not necessarily a calendar day. For instance, editors planning coverage at 2 p.m. Tuesday for the "next cycle" would be thinking about *All Things Considered* on Tuesday, but other news programs on Wednesday. Also known as *cycle.*

newscast: An NPR news program between four and eight and a half minutes long. Two newscasts are broadcast every hour when *Morning Edition* and *All Things Considered* are on the air; at other times, they are broadcast hourly. Listeners hear newscasts as part of the program, though they are produced, written, and read by a different group of people. For that reason, information in **billboards** and **pieces**—death tolls, pronunciations, etc.—has to correspond exactly to what's being presented in newscasts.

Newsheimer's: Slang for the common newsroom condition of not being able to remember the story you covered two days ago. Longtime editors and hosts are particularly susceptible to this affliction, which also manifests itself in the belief that they did a story or interview on a subject "recently," when in fact they did it five or ten years ago.

Nipper: NPR slang for the one-line network identification, "This is NPR—National Public Radio."

"Nipper" was also the name of the mixed-breed terrier that was mesmerized by the Victrola in the old RCA logo, subtitled "His Master's Voice."

NPR.org: The NPR Web site, featuring current news and feature stories, an audio archive of NPR programming going back to 1996, links to the music played between items on the news magazines, a corporate area that includes staff biographies, NPR's podcasting directory, the NPR Shop, etc.

NPR West: The network's west coast production facility in Culver City, California, which opened in 2002 and houses a news bureau and the staffs of several NPR programs.

o.c.: See **outcue**.

on spec: Work that is provided without the promise, implied or explicit, that it will be paid for. Commentaries, for example, are often sent to NPR or its member stations on spec.

open: To turn a microphone on. Hence the phrase "open mic."

Open is also a synonym for **billboard**.

Ops Desk: The central desk for coordinating live **feeds**, special coverage, and the booking of **studios** at NPR. ("Ops" is shorthand for "news operations"—a term also used in other organizations.) The Ops Desk routinely records all audio feeds from the White House, including the daily press briefing. Other regularly recorded feeds include briefings from the State Department, the Pentagon, and the Attorney General. See also **WAND**.

outcue, out cue: The last few words of a recording. In a reporter's script, the outcue indicates the end of an **actuality**. On a program script, the outcue tells the hosts and the show editor how a report ends. Often abbreviated *o.c.*—as in:

o.c. ". . . only time will tell."

If a **piece**, such as a commentary, ends with a repeated phrase, a short outcue may be misleading. In such a case, it's a good idea to provide the engineer or **studio** director a much longer-than-usual quote from the piece, and to draw attention to the potential misunderstanding:

AUDIO: 3:12 (Note double out-cue!)
i.c. "Whose woods these are . . ."
o.c. ". . . promises to keep, And miles to go before I sleep, And miles to go before I sleep."
HOST: Robert Frost is a dead New England poet.

outro: See **backannounce**.

out-take: A part of an interview or report that doesn't get on the air.

package: At NPR, two or more related elements in a program—an "immigration package" might consist of a reporter **piece** on pending legislation followed by an interview about a new poll on Mexicans' views toward working in the U.S.

PAND [pronounced "pond"]: Presidential Audio News Distribution. When the president travels outside Washington, his events are **fed** via PAND over the **WAND** system. See also **white line**.

pass-off: At NPR, any information passed along from one editor or producer to another, or to a program host.

phoner: An interview conducted over the telephone.

Audio engineers also use the term *phoner* to describe the box used to conduct a phone interview. It's also called a "phoner hybrid," or a "phoner unit."

pickup: A place in a sound file where a reporter or host has made a mistake or for some other reason had to restart—pick up—his or her track. One of the producer's jobs is to edit out any pickups before a **piece** airs. There may also be pickups in a **mixed** piece, where a section may have been recorded more than once in order to get things just right.

piece: At NPR, a news or feature report broadcast in the body of a program. Pieces can be of almost any length, from about a minute and a half to twenty minutes—or even longer. Compare **spot**, **superspot**.

pitch: A description of a proposed story. Reporters pitch their stories to editors, producers pitch their stories to hosts, and so on. Making an effective pitch involves examining a possible story from many different angles—anticipating how it might be reported, what the listener would learn, even how you might write an **intro** for the story.

post: Any of several things, all of them related to the timing of a **piece** or broadcast. **Newscasts** and programs have designated time posts—points at which the **segments** have to end (usually so that member stations know when they can begin adding their own material). A post is also the result of a **backtime**—it is the point in a piece of music or other sound that we want to be heard by the listener. The producer of a music piece, for instance, may say "There's a good post at thirty-two," meaning that thirty-two seconds into a particular selection there's a place where the music should be heard "in the clear," without anyone talking over it. "Post" is also a verb—a reporter might say, "I want the sound to post right when you hear the car horns beginning."

postcard: The radio version of a picture postcard. It's short—usually three minutes or less—and the "picture" is some remarkable sound, or series of sounds. For example, a reporter may produce a postcard whose tracks consist of little more than, "Dear *All Things Considered*. I came to central New Hampshire to do a story about state funding of public schools. But what I found instead—everywhere I turned—was the sound of the January thaw." The postcard might then have the sounds of ice breaking up in the river, or snow falling off roofs, or whatever the reporter found.

promo: A special promotional announcement for upcoming **pieces** or programs. Promos are usually **fed** to member stations at least a day or two ahead of time so they can air more than once before the item they're promoting is broadcast. See also **cross promo, forward promo,** and **jump promo.**

pronouncer: An informal way of making it clear how a word, or name, should be read out loud—e.g., the pronouncer for "Ivan Basso" might be "ee-VAHN BAH-soh."

pull tape: To isolate **actualities** or sound bites. There was a time when producers and reporters actually pulled the sections of **tape** they wanted off of a reel; the phrase survives even though the process today usually involves a mouse and a computer.

Q: See **cue.**

QC: Quality control.

reader: A story told entirely through **copy**, as opposed to a **spot**, or **tape and copy.**

return: At NPR, the twenty-eight-second block of **copy** read by program hosts to begin the second half of each hour. A return is sort of a mini-**billboard.**

roadmap: At NPR, a list of **pieces** in a program; it's usually a paper version of all the information on the **board.** It also contains information about underwriting announcements.

room tone: **Ambience** recorded at the location of an interview. Even a relatively quiet office has some ambient sound—from the hum of fluorescent lights to distant traffic. Recording a minute or so of that sound immediately after an interview ensures that the background noise in **actualities** does not suddenly disappear when the **piece** is finally **mixed.**

roll: To play or record audio. A reporter may tell a guest "We're not rolling yet" to indicate that anything she says won't be recorded. A **studio** director may instruct an engineer to "roll **tape**" when he wants him to play a report or **actuality**. Many radio people still use the term "roll" in this digital era, when often nothing is "rolling"—or moving at all—on either the recorder that captures sound or the computer that plays it back.

rollover: A delayed version of a broadcast. Public radio news programs, like *Morning Edition* or *Weekend Edition Saturday*, are generally recorded as they are broadcast on the east coast so that they can easily be rolled over for broadcast in other parts of the country. At NPR, rollovers are always kept up-to-date; if the news changes, **segments** may be reworked, interviews may be redone, and **intros** may be read **live** in an otherwise recorded program. In addition, any mistakes in the first broadcast are fixed for the rollovers.

rundown: The list of items in a program, reflecting the information on the **board**. The rundown is sent out to NPR member stations on the **DACS**. Compare **roadmap**.

scene: The location where some action takes place; it has much the same meaning in radio as in a movie or play. The best radio **pieces** are usually built around scenes; they can make even complex, issues-oriented pieces intelligible. For example, a report on how welfare reform is affecting poor, non-English-speaking immigrants might include **actualities** of a university professor describing how he conducted his research on welfare and what it implies. But the report will be effective and memorable only if it includes some scenes—a worker trying to apply for a job, for example, or a welfare official trying to explain to a Spanish-speaking woman how to take public transportation to her new workplace. Creating a scene usually involves both good sound and careful, descriptive writing. See **ambience, sound effects**.

segment: A portion of a program of fixed length. Segment lengths differ both within and among programs.

segue [pronounced "SEG-way"]: Going from one sound to another without an interruption—e.g., a producer may decide to segue from one

piece of music to another, or from one report to another without a host introduction. (See **crossfade.**)

slug: A word or phrase used to identify a news story. At NPR, program producers assign slugs to **two-ways, pieces,** and commentaries. A slug may appear on the **board**, on the **DACS**—and on the Internet. For this reason, producers should choose their slugs carefully: a phrase that's intended to be amusing or seen only by one's colleagues—something like, "WACKO GEEZER PILOTS"—can end up being part of a public document.

sneak: To fade audio in gradually, usually under a voice track. Sound that begins as a sneak should eventually be brought up full; experienced producers avoid sneaking sound or music in and then fading it out without ever letting listeners hear it in the clear at full volume.

SOC [pronounced "sock"]: Standard Out Cue—the preferred way for a reporter to end his **piece**. A typical SOC for an NPR news reporter would be "Edmund Fitzgerald, NPR News, Superior, Wisconsin." For a non-staff reporter, it would be, "For NPR News, I'm Edmund Fitzgerald in Superior, Wisconsin."

sound effect (SFX): Sound used to *create* an effect—to establish a **scene** or make a point. It can and should be used, but you should never use artificial sound effects in news reports; if a **piece** concerns the birds of Brazil, the birds singing in the background must have been recorded in the course of reporting the story. On rare occasions, producers will create sound effects in the **studio**, or draw on recorded sounds compiled on CDs, but they should never pass off those sounds as if they were recorded in the field. Often used as a synonym for **ambience.**

spec: See **on spec.**

split track: A **tape** on which one source of audio (e.g., a host) is recorded on one half and a second source (e.g., a guest) on the other, or a **digital audio** file where the sources are recorded separately.

spot: At NPR, a short news report (less than a minute) for inclusion in a **newscast.** Compare **piece, superspot.**

stand-up: A **track** recorded on location, often without a script; as a production device, it often works best where the reporter's perspective and observations play a critical role in the story: "I'm standing at the corner of Tenth Street and Jefferson Boulevard—which police describe as the most dangerous spot in this city. From here, you can see drug dealers, prostitutes, and panhandlers. All of the stores on this corner are closed and boarded up."

submix: A partial or intermediate **mix**—one stage in the production of a complex **piece**. For example, it may be necessary or advisable to mix the **acts** and **tracks** first—that would be the submix—and then add the **ambience**, music, or other sound in a subsequent mix.

superspot: At NPR, a **piece** that is little more than an overgrown **spot**—something around a minute and a half long. Show producers will sometimes ask a reporter to carve a piece down to a superspot length so that it can fit a small hole in the program. Compare **piece**, **spot**.

sync: A way of improving the quality of an interview that would otherwise have to be a **phoner**. It involves sending a producer or engineer to a guest's home or office to record his answers, while a reporter's or host's questions are recorded in a **studio**. If the conversation is going to be broadcast as an interview, the two parts are synchronized and re-recorded, so that both ends are high-quality. If the guest's answers are going to be used as **actualities** in a new report, the sync is not necessary. Some reporters call this sort of remote recording of an interview a *double-ender*.

tail: The end of a **tape** (archaic). A tape that is *tails-out* has not been rewound after being recorded or played. See also **head**.

Tail is also used to describe the end of a **digital audio** file. A producer may complain that the reporter didn't supply an "ambi tail"—i.e., sound at the end of an **actuality** to be faded out.

take out: To keep something from being heard, often by **closing** a mic. A **studio** director may instruct an engineer to "take out the host" after she has finished her **intro** to a report. The engineer might also take out a guest who has been waiting on the phone because the director knows he won't be returning to him.

talkback: The intercom between the **control room** and **studio**, or between the **control room** and a remote location. The technical name for talkback is *IFB* (*interruptible foldback*). *Foldback* is the audio that goes from the main **mixing** console into the host's or reporter's headphones. When a director pushes the intercom switch to talk to someone in the studio, it interrupts the foldback, and allows his or her voice to be heard.

tape: Magnetic tape (archaic). Even today, at least at NPR, it often means any audio. Though sound is recorded and edited on computers, many people in public broadcasting continue to talk about "tape editors," "tape-and-copy blocks," "tape **syncs**," "great tape," etc.

tape and copy: At NPR, a story, often written by a producer for a host to read on the air, that includes both text and audio.

tape time: The length of a **piece** from the first words of a reporter or interviewee or host to the **SOC** or conclusion. Many people still refer to "tape times" even when they are talking about **digital audio**. The tape time is represented in minutes and seconds:

AUDIO: 4:06
i.c. "The Secretary of Defense..."
o.c. "... Jones, NPR News, the Pentagon."

Computer software calculates the time automatically.

The tape time usually does not include the length of the live **intro**. For instance, a producer may allot five minutes for the piece **slugged** "PENTAGON APOLOGY," but because she's leaving forty seconds for the intro, the tape time can only be 4:20.

tease: A short **actuality** that precedes any other information in the story. A good tease serves as a "hook" that makes the listener eager to hear more. Some shows include teases in their **billboards**. Billboard teases should be catchy—their whole point is to make the listener stay tuned—and they should reflect the basic theme of the report.

texture: The variation of program elements—by length, type, sound, etc. It may be best defined by a negative example: a show made up exclusively of five-minute reports lacks texture. That's why show producers work hard to make sure their program includes a mix of interviews, news reports, extended **actualities**, music, and other elements.

top: The beginning of a recording or script. A **studio** director may ask a reporter to "take it again from the top"—to reread his story from the beginning—or he may caution the engineer to watch the levels of a cut because "the sound is a little **hot** at the top." See **head, tail.**

track: A "voice track"—a part of a report read by the reporter or host. A typical public radio news **piece** may include **tracks**, **acts**, and **ambi**. Also spelled *trax* in plural.

track mark: A pointer in a digital file (usually on a **MiniDisc** or flash memory card) that allows a user to **cue** audio quickly at a specific place. With some recorders, adding a track mark does not create a new audio file. Even though a single file may have dozens of track marks, labeled sequentially—e.g., from 1 to 40—finding a specific one usually only takes a few seconds. Also known as an *index mark*.

train wreck: At NPR, an error in the **studio** or **control room** that somehow starts a sequence of problems. A train wreck will be documented in a discrepancy report, or **D-sheet**:

Coming out of last C **Segment piece**, Director calls for end of show; after sequence has begun, Director realizes that show is 1:00 short. Host gives **Nipper**, and Director calls for **Funder** (which is very early). Director then asks to go to Emergency Music after the Funder. During Emergency Music, Funder was not called for at proper time. Director did not direct to go to Break #3, so that **post** was also missed. Break #4 hit on time. Will be fixed for **rollovers**.

trax: See **track**.

trixie: See **bleeble**.

two-way: Generally, a synonym for *interview*, though a few people reserve the term to refer to an interview between a host and a reporter, or to a host in a **studio** and a guest at a remote location.

upcut: To edit audio poorly so that there is a loss of a syllable or sound at the beginning—e.g., "'om NPR News, this is Morning Edition." Compare **clip**.

uplink: An earth station capable of transmitting full-fidelity audio signals to a satellite.

voiceover, v.o.: At NPR, an English translation that is **mixed** over an **actuality** in a foreign language. The best voiceovers suit the sex and age of the person whose **actuality** has been translated. Ideally, the person doing the voiceover comes from the country or the region where the story takes place. (In the commercial audio and film industries, the term "voiceover," or "v.o.," means any recorded narration or voice track.)

volume unit: See **VU meter**.

vox: Person-on-the-street **actuality**, usually gathered to present a variety of opinions on an issue. Some reporters also refer to vox as "MOS," short for "man on the street."

VU meter [pronounced "vee-yoo"]: An instrument that measures audio volume. VU meters use "volume units" as a standard of measurement.

WAND: Washington Area News Distribution, a service of ABC Radio News in Washington. Subscribers pay an annual fee for a selection of audio **feeds** of presidential, congressional, agency, and other public affairs events. WAND's mandate is to provide audio from "inside the Beltway." But the WAND can often provide audio feeds of major news events from around the nation. See also **PAND**.

warm: Audio that is played louder than a **deadroll**, but not at full volume. A producer **mixing** a music **piece** might call for a "warm hit" of a song, so that we hear the music begin as the host starts to talk about it.

white line: A grease pencil mark made on a piece of magnetic tape to help an audio engineer see where a particular sound begins (archaic). A white line may be used during a **mix**, for example, so the engineer knows when to start fading a **tape** out under an **actuality**.

White line is also the term used for the presidential audio pool when the President is on the road. See **PAND**.

Who Dat?: See **drop-in**.

APPENDIX II

Pronouncers

The best way to accurately describe the *sound* of a word in print is to use a specialized system of letters and symbols. Linguists rely on the IPA—the International Phonetic Alphabet—which has twenty-eight different vowel sounds alone. It's not a system well-suited to radio scripts, especially when a host may have only a few seconds to make sense of a pronouncer.

For that reason, most news broadcasters use the regular English alphabet to *approximate* the sounds of words—especially foreign words, or unusual names of people and places. The list below includes the most common phonetic spellings used to represent different sounds:

VOWEL SOUNDS:

a	apple, bat
ah	father, hot
ai	air, pair
aw	law, long
ay	ace, fate
e	bed
ehr	merry
ee	see, tea
ih	pin, middle
oh	go, oval
oo	food, two
ow	cow
oy	boy
u	foot, put
uh	puff
ur	burden, curl
y, eye	ice, time

CONSONANTS:

g	got, beg
j	gem, job
k	cap, keep
ch	chair
dh	there (i.e., voiced *th*)
kh	guttural "k"
s	see
sh	shut
th	thin (i.e., voiceless *th*)
y	yes
z	zoom
zh	mirage

You can indicate where you want the emphasis to fall by putting the stressed syllable in CAPITALS. Some people (and notably the Associated Press) also place an apostrophe after the emphasized syllable. So, for instance, a pronouncer for "Jonathan Kern" might be "JAH′-nuh-thin KURN."

Where there is any doubt about a sound, it's a good idea to add a note to clarify things: "Islamabad (ihs-LAH′-muh-bahd′—slight stress on final syllable)."

You might notice that the same vowel sound was represented by "i" in "Jonathan" and by "ih" in "Islamabad." That's because it can be hard for a reader to decipher a letter combination like "thihn." So use your judgment when you're trying to let a host or reporter know how to pronounce something. Clarity is as important as accuracy.

After all, these pronouncers are not perfect. For example, the "kh" we use for a "guttural k" actually comprises a range of sounds. In German, there's a difference between the "ch" sound in "ich" (the singular first person pronoun) and the one in "ach" (an expression of dismay), but we'd use "kh" for both.[1] Because our alphabet can be a crude tool for describing a foreign word, the best way to make sure an unfamiliar name or place is pronounced the way you intend is just to say it directly to the person who is going to read it on the air—even if you also provide a written pronouncer.

1. The difference has to do with the place in the mouth where the sound is made. Technically, one ("ich") is a voiceless palatal fricative, and the other ("ach") is a voiceless velar fricative.

Also, use your judgment about how much to Anglicize foreign names and places. Generally, reporters and hosts try to pronounce people's names on the air the way they do in their daily lives; they would call the former Polish president "LEKH vuh-WEN-suh," though his name is written "Lech Walesa." But don't go overboard trying get umlauted German vowels or Arabic consonants just right, or try to imitate Chinese tonal inflections. And use the common Anglicized names for Munich, Warsaw, Shanghai, and other foreign cities. Trying to sound like a native—especially if you don't speak the language—can make you sound pretentious.

The Associated Press issues daily updates of names in the news; you can find them by searching the wires for the word "pronunciation." If you're in doubt about how a name or place should be pronounced, you can also consult a librarian. If you can't reach a librarian and a name doesn't appear on the wires, call the embassy of the country in question and see if someone there can help you. Some Web sites include both IPA pronunciations and audio that you can listen to. The Voice of America—the U.S Government broadcast service—maintains an online pronunciation guide with pronouncers for many foreign leaders' names at names.voa.gov.

INDEX

C

N

S